Advances in Intelligent and Soft Computing 70

Editor-in-Chief: J. Kacprzyk

Advances in Intelligent and Soft Computing

Further volumes of this series can be found on our homepage: springer.com

Vol. 55. Y. Demazeau, J. Pavón, J.M. Corchado, J. Bajo (Eds.)
7th International Conference on Practical Applications of Agents and Multi-Agent Systems (PAAMS 2009), 2009
ISBN 978-3-642-00486-5

Vol. 56. H. Wang, Y. Shen, T. Huang, Z. Zeng (Eds.)
The Sixth International Symposium on Neural Networks (ISNN 2009), 2009
ISBN 978-3-642-01215-0

Vol. 57. M. Kurzynski, M. Wozniak (Eds.)
Computer Recognition Systems 3, 2009
ISBN 978-3-540-93904-7

Vol. 58. J. Mehnen, A. Tiwari, M. Köppen, A. Saad (Eds.)
Applications of Soft Computing, 2009
ISBN 978-3-540-89618-0

Vol. 59. K.A. Cyran, S. Kozielski, J.F. Peters, U. Stańczyk, A. Wakulicz-Deja (Eds.)
Man-Machine Interactions, 2009
ISBN 978-3-642-00562-6

Vol. 60. Z.S. Hippe, J.L. Kulikowski (Eds.)
Human-Computer Systems Interaction, 2009
ISBN 978-3-642-03201-1

Vol. 61. W. Yu, E.N. Sanchez (Eds.)
Advances in Computational Intelligence, 2009
ISBN 978-3-642-03155-7

Vol. 62. B. Cao, T.-F. Li, C.-Y. Zhang (Eds.)
Fuzzy Information and Engineering Volume 2, 2009
ISBN 978-3-642-03663-7

Vol. 63. Á. Herrero, P. Gastaldo, R. Zunino, E. Corchado (Eds.)
Computational Intelligence in Security for Information Systems, 2009
ISBN 978-3-642-04090-0

Vol. 64. E. Tkacz, A. Kapczynski (Eds.)
Internet – Technical Development and Applications, 2009
ISBN 978-3-642-05018-3

Vol. 65. E. Kącki, M. Rudnicki, J. Stempczyńska (Eds.)
Computers in Medical Activity, 2009
ISBN 978-3-642-04461-8

Vol. 66. G.Q. Huang, K.L. Mak, P.G. Maropoulos (Eds.)
Proceedings of the 6th CIRP-Sponsored International Conference on Digital Enterprise Technology, 2009
ISBN 978-3-642-10429-9

Vol. 67. V. Snášel, P.S. Szczepaniak, A. Abraham, J. Kacprzyk (Eds.)
Advances in Intelligent Web Mastering - 2, 2010
ISBN 978-3-642-10686-6

Vol. 68. V.-N. Huynh, Y. Nakamori, J. Lawry, M. Inuiguchi (Eds.)
Integrated Uncertainty Management and Applications, 2010
ISBN 978-3-642-11959-0

Vol. 69. XXX

Vol. 70. Y. Demazeau, F. Dignum, J.M. Corchado, J. Bajo Pérez (Eds.)
Advances in Practical Applications of Agents and Multiagent Systems, 2010
ISBN 978-3-642-12383-2

Yves Demazeau, Frank Dignum,
Juan M. Corchado, Javier Bajo Pérez (Eds.)

Advances in Practical Applications of Agents and Multiagent Systems

8th International Conference on
Practical Applications of Agents
and Multiagent Systems (PAAMS 2010)

Springer

Editors

Yves Demazeau
Laboratoire d'Informatique de Grenoble
Centre National de la
Recherche Scientifique,
Maison Jean Kuntzmann
110 av. de la Chimie
F-38041 Grenoble,
France
E-mail: Yves.Demazeau@imag.fr

Juan M. Corchado
Departamento de Informática
y Automática
Facultad de Ciencias
Universidad de Salamanca
Plaza de la Merced S/N
37008 Salamanca
Spain
E-mail: corchado@usal.es

Frank Dignum
Department of Information and
Computing Sciences
Universiteit Utrecht
Centrumgebouw Noord, office A117,
Padualaan 14, De Uitho
3584CH Utrecht,
The Netherlands
E-mail: dignum@cs.uu.nl

Javier Bajo Pérez
Escuela Universitaria de Informática
Universidad Pontificia
de Salamanca,
Compañía 5,
37002 Salamanca,
Spain
E-mail: jbajope@upsa.es

ISBN 978-3-642-12383-2 e-ISBN 978-3-642-12384-9

DOI 10.1007/978-3-642-12384-9

Advances in Intelligent and Soft Computing ISSN 1867-5662

Library of Congress Control Number: 2010924117

Typeset & Cover Design: Scientific Publishing Services Pvt. Ltd., Chennai, India.

Printed on acid-free paper

5 4 3 2 1 0

springer.com

Preface

Research on Agents and Multi-Agent Systems has matured during the last decade and many effective applications of this technology are now deployed. The advances in practical applications of Agents and Multi-Agent systems is continuous and it is necessary to provide an international forum to present and discuss the latest scientific developments and their effective applications, to assess the impact of the approach, and to facilitate technology transfer, in this field.

PAAMS, the International Conference on Practical Applications of Agents and Multi-Agent Systems is an international yearly stage to present, to discuss, and to disseminate the latest advances and the most important outcomes related to real-world applications. It provides a unique opportunity to bring multi-disciplinary experts, academics and practitioners together to exchange their experience in the development of Agents and Multi-Agent Systems.

This volume presents the papers that have been accepted for the 2010 edition. These articles capture the most innovative results and this year's advances. Each paper has been reviewed by three different reviewers, from an international committee composed of 82 members from 26 different countries. From the 66 submissions received, 19 were selected for full presentation at the conference, and 14 were accepted as short papers. Moreover, PAAMS'10 incorporated special sessions and workshops to complement the regular program, which included 85 accepted papers.

We would like to thank all the contributing authors, as well as the members of the Program Committee, Auxiliary Reviewers and the Organizing Committee for their hard and highly valuable work. Their work has helped to contribute to the success of the PAAMS 2010 event. Thanks for your help; PAAMS'10 wouldn't exist without your contribution.

Yves Demazeau
Frank Dignum
PAAMS 2010 Program Co-chairs

Juan Manuel Corchado
Javier Bajo Pérez
PAAMS 2010 Organizing Co-chairs

Organization

General Co-chairs

Yves Demazeau	Centre National de la Recherche Scientifique (France)
Frank Dignum	Utrecht University (The Netherlands)
Juan M. Corchado	University of Salamanca (Spain)
Javier Bajo Pérez	Pontifical University of Salamanca (Spain)

Program Committee

Yves Demazeau (Co-chairman)	Centre National de la Recherche Scientifique (France)
Frank Dignum (Co-chairman)	Utrecht University (The Netherlands)
Francesco Amigoni	Politecnico di Milano (Italy)
Luis Antunes	University of Lisbon (Portugal)
Olivier Boissier	Ecole Nationale Superieure des Mines de Saint Etienne (France)
Magnus Boman	Royal Institute of Technology (Sweden)
Juan A. Botía	University of Murcia (Spain)
Vicente Botti	Polytechnic University of Valencia (Spain)
Jeffrey Bradshaw	Florida Institute for Human and Machine Cognition (USA)
Bernard Burg	Panasonic Ltd (USA)
Valerie Camps	University Paul Sabatier (France)
Longbing Cao	University of Technology Sydney (Australia)
Pierre Chevaillier	University of Brest (France)
Owen Cliffe	University of Bath (United Kingdom)
Juan M. Corchado	University of Salamanca (Spain)
Rafael Corchuelo	University of Sevilla (Spain)

Keith Decker	University of Delaware (USA)
Jurriaan van Diggelen	TNO (The Netherlands)
Virginia Dignum	Utrecht University (The Netherlands)
Alexis Drogoul	IRD (Institut de Recherche pour le Developpement) (Vietnam)
Julie Dugdale	University Pierre Mendes France (France)
Edmund Durfee	University of Michigan (USA)
Amal El Fallah	University of Paris 6 (France)
Torsten Eymann	University of Bayreuth (Germany)
Takayuki Ito	MIT (USA)
Klaus Fischer	DFKI (Germany)
Rubén Fuentes	Complutense University of Madrid (Spain)
Francisco Garijo	Telefónica I+D (Spain)
Khaled Ghedira	National School of Computer Sciences (Tunisia)
Sylvain Giroux	Unversity of Sherbrooke (Canada)
Pierre Glize	University Paul Sabatier (France)
Vladimir Gorodetski	University of Saint Petersburg (Russia)
Dominic Greenwood	Whitestein Technologies (Switzerland)
Kasper Hallenborg	University of Southern Denmark (Denmark)
Koen Hindriks	University of Delft (The Netherlands)
Shinichi Honiden	National Institute of Informatics Tokyo (Japan)
Tom Holvoet	Catholic University of Leuven (Belgium)
Toru Ishida	University of Kyoto (Japan)
Vicente Julián	Polytechnic University of Valencia (Spain)
Achilles Kameas	Delft University of Patras (Greece)
Franziska Kluegl	University of Örebro (Sweden)
Matthias Klusch	DFKI (Germany)
Martin Kollingbaum	University of Aberdeen (United Kingdom)
Jaroslaw Kozlak	University of Science and Technology in Krakow (Poland)
Beatriz López	University of Gerona (Spain)
Adolfo López Paredes	University of Valladolid (Spain)
Zakaria Maamar	Zayed University (United Arab Emirates)
Rene Mandiau	University of Valenciennes (France)
Philippe Mathieu	University of Lille (France)
Eric Matson	Purdue University (USA)
Fabien Michel	University of Reims (France)
José M. Molina	University Carlos III of Madrid (Spain)
Mirko Morandini	University of Trento (Italy)

Bernard Moulin	University Laval (Canada)
Jörg Müller	Clausthal University of Technology (Germany)
Jean-Pierre Muller	CIRAD-TERA-REV-GREEN (France)
James Odell	James Odell Consultancy (USA)
Eugenio Oliveira	University of Porto (Portugal)
Andrea Omicini	University of Bologna (Italy)
Sascha Ossowski	University of Rey Juan Carlos (Spain)
Julian Padget	University of Bath (United Kingdom)
Van Parunak	New Vectors (USA)
Juan Pavón	Universidad Complutense de Madrid (Spain)
Michal Pechoucek	Czech Technical University in Prague (Czech Republic)
Paolo Petta	University of Vienna (Austria)
Jeremy Pitt	Imperial College of London (United Kingdom)
Antonio Rocha Costa	Universidade Católica de Pelotas (Brazil)
Juan A. Rodríguez Aguilar	IIIA-CSIC (Spain)
Fabrice Saffre	British Telecom (United Kingdom)
Jaime Sichman	University of Sao Paulo (Brazil)
Kostas Stathis	Royal Holloway University of London (UK)
John Thangarajah	RMIT (Australia)
Paolo Torroni	University of Bologna (Italy)
Rainer Unland	University of Duisburg (Germany)
Domenico Ursino	University of Reggio Calabria (Italy)
Birna van Riemsdijk	University of Delft (The Netherlands)
Javier Vázquez Salceda	Polytechnic University of Catalogna (Spain)
Jacques Verriet	Embedded Systems Institute (The Netherlands)
Danny Weyns	Catholic University of Leuven (Belgium)
Niek Wijngaards	Thales, D-CIS lab (The Netherlands)
Michael Winikoff	University of Otago (New Zealand)
Franco Zambonelli	University of Modena (Italy)

Organizing Committee

Juan M. Corchado (Co-chairman)	University of Salamanca (Spain)
Javier Bajo Pérez (Co-chairman)	Pontifical University of Salamanca (Spain)
Juan F. De Paz	University of Salamanca (Spain)
Sara Rodríguez	University of Salamanca (Spain)
Dante I. Tapia	University of Salamanca (Spain)
M. Dolores Muñoz Vicente	University of Salamanca (Spain)

Auxiliary Reviewers

Maxime Morge	LIFL/SMAC, Université Lille1 (France)
Giovanni Quattrone	University of Reggio Calabria (Italy)
Pasquale De Meo	University of Reggio Calabria (Italy)
Yasuyuki Tahara	National Institute of Informatics (Japan)
Rubén Ortiz	University Rey Juan Carlos (Spain)
Matteo Vasirani	University Rey Juan Carlos (Spain)
Moser Fagundes	University Rey Juan Carlos (Spain)
Hakan Duman	BT Innovate & Design (United Kingdom)
Aistis Simaitis	University of Leeds (United Kingdom)
Norman Salazar	IIIA-CSIC (Spain)
Jose L. Fernandez-Marquez	IIIA-CSIC (Spain)
Christoph Niemann	University of Bayreuth (Germany)
Stefan Koenig	University of Bayreuth (Germany)
Luis Gustavo Nardin	University of Sao Paulo (Brazil)
Sara Jane Casare	University of Sao Paulo (Brazil)

Contents

Keynote

Real-Time and Personalisation

Modelling and Computation

Methodology and Engineering

CRISIS Management and Robots

Development and Evaluation

Search and Problem Solving

Multiagent Modelling and Simulation as a Means to Wider Industrial Deployment of Agent Based Computing in Air-Traffic Control

Michal Pěchouček

Keynote Abstract

Agent based computing is an innovative sub-field of artificial intelligence and computer science, which provides theories for understanding behaviour of distributed, multi-actor systems and provides methods and algorithms for control of individual components of such system and protocols ruling their interactions. Despite agent-based computing being a highly successful basic research community, it still struggles for wider industrial deployment. The major limitation of massive exploitation of research results is the fact that they have not been successfully tested on large scale complex systems and that such testing would be expensive and in many cases risky. Sophisticated multiagent modelling and simulation can provide high fidelity, high-performance and scalable computational models of the targeted applications. The models can be used for empirical validation of the agent algorithms, methods and theories and can be used as a vehicle for transfer of the technology.

Multiagent modelling and simulation can support deployment of agents solutions in monitoring and management of the urban utility infrastructure, modeling and administration of public transport networks and private transport alternatives (such as smart carpooling), planning future energy billing models for electric cars, design and daily operation f intelligent buildings and smart sensors, but can also assist in testing collective robotics scenarios supported by unmanned aerial vehicles, various security and rescue applications (surveillance, tracking or search). At the same time multiagent simulation can also support modelling of social networks, on-line trading, or malicious behaviour on the Internet (such as network intrusion).

The great challenge of multiagent simulation as a gateway to reliable and industry robust application depends on how well has the simulation system been

Michal Pěchouček
Czech Technical University in Prague
Technická 2, 16627 Prague 6, Czech Republic
e-mail: pechoucek@fel.cvut.cz

Y. Demazeau et al. (Eds.): Advances in PAAMS, AISC 70, pp. 1–3.
springerlink.com

designed. Often there the following aspects one that need to be considered for a successful simulation system:

- **Fidelity:** it is important to size well fidelity of the simulation so that it reflects the properties of the multiagent system is a very good approximation of the targeted system. On the other hand, the programming resources devoted to development of simulation need to be well balanced so that only relevant properties of the system are modelled.
- **Scalability:** often, agent based solutions work well for limited number of members of the multiagent community. It is important to design scalable multiagent simulation of the targeted system so that the tested multiagent solution can work well on the real application of the required size and scale.
- **Generality:** the simulation application have got different requirements for reusability that navigates the design between open, more general application and single-purpose multiagent simulations. Also existing multiagent system development environment or frameworks for developing multiagent simulations can be used.
- **Transferability:** it is important that the tested agent-based solutions can be easily transferred form the mulitagent simulation to the targeted system, with minimal solution changes and so that the properties of the simulation are maintained on the target system.

These aspects of multiagent simulation will be discussed in the context of the AgentFly multiagent system for planning the collision-free free-flight campaign of the collectives of unmanned aerial vehicles. AgentFly—a scalable, agent-based technology for free-flight simulation, planning, and collision avoidance has been developed in the Agent Technology Center, Czech Technical University in cooperation with the US AirForce Laboratory. In AgentFly, each flying asset represents a specific software container hosting multiple intelligent software agents. Each agent either models a specific hardware functionality such as sensory capability, dynamic flight control, or communication, or encapsulates an intelligent decisionmaking technology that supports planning or collision avoidance. Such an architecture supports three principal AgentFly use cases:

- multiagent modelling and simulation of free-flight,
- control of free-flight unmanned aerial platforms, and
- alternative approaches to planning, which supports civilian air traffic control.

Multiagent simulation of free-flight operations has been used in order to empirically analyze various planning and collision avoidance algorithms before physically deploying them on hardware platforms. Tn empirical analysis of realistic simulation of the findings provides valuable information about the properties of free flight in various circumstances (such as surveillance tracking, worsened weather conditions, dense civilian traffic, or emergency situations). AgentFly is designed so that no centralized component is needed; all the planning and collision avoidance algorithms are based on the flying assets' sensory capability and distributed (peertopeer) decisionmaking capability. The AgentFly system is based on Aglobe multiagent technology, which supports seamless migration from computational simulation to hardware deployment. Previously, researchers successfully migrated an Aglobe-based model of a ground based robotic scenario to the RoboCup soccer environment.

The most promising direction for applying AgentFly is in the area of air traffic planning. The US Federal Aviation Authority (FAA) is interested in testing AgentFly's planning capacity for heavily overloaded civilian air traffic across the entire national air space. The idea is to relax the planning problem and perform multiagent flight simulations. Instead of planning a collision-free operation for numerous aircraft, there is a possibility to construct a flight plan for each individual craft without considering possible collisions. Subsequently, such an operation is simulated in the AgentFly environment, where possible collisions are detected, and solved through either individual replanning or peer-to-peer negotiation.

The fidelity of the AgentFly system is very good as it works with GPS coordinates, integrates BADA flight models of the physical aircrafts, integrates the changing weather patterns as well as no-flight zones in the US. While the AgentFly has been originally designed for coordination of tens of the UAVs, the cooperation with the FAA made the scalability to rise to the whole national airspace in the United States that represents about 50.000 aircrafts a day. Generality and reusability of AgentFly is reasonably good ad it provides an open system for the agent interaction (AGLOBE) and a separate reusable component for the environment simulation (Xsimulation). Transferability of the algorithms onto the hardware platforms is current research challenge as AGENTFLY is being transferred onto the PROCERUS unmanned aerial asset.

Real Time Learning of Behaviour Features for Personalised Interest Assessment

Sylvain Lemouzy, Valérie Camps, and Pierre Glize

Abstract. This paper deals with an adaptive and personalized algorithm to dynamically determine the interest of a user in a document from the observation of his behaviour during its consultation. Several existing works propose an implicit feedback of user's interest from observations of his behaviours when he is reading a document. Nevertheless, the used algorithms are *a priori* defined and then cannot be really adapted to each user who has personal habits when he is working with the Web. This paper focuses on the generalisation of this problem and proposes a multi-agent algorithm that, dynamically and according to each user's specific behaviours, computes the personal interest assessment of each of them. Several experimentations show the efficiency of this algorithm, able to self-adapt its functionality in real time when the user habits evolve, compared to existing methods.

1 Introduction

The incessant growth of the World Wide Web increases the number of responses provided by the search engines to a request. The user is then faced to an information overload from which it is difficult to distinguish information that is directly relevant from information that is secondary or even irrelevant. Furthermore, the answers to requests are determined regardless of the user that sent them or the context in which they are issued; in particular, the answers to a same request expressed by two different users are identical, even if these users do not have the same expectations, the same preferences and the same interests.

The use of personalisation is a way to reduce this informational overload. It enables to implicitly or explicitly limit the answers according to the needs and specificities of the user. Personalisation is based on the notion of user profile ([8], [1] which represents his interests, his preferences and his needs. User's preferences

Sylvain Lemouzy · Valérie Camps · Pierre Glize
Université Paul Sabatier Toulouse 3 – IRIT, Toulouse, France
e-mail: `{lemouzy,camps,glize}@irit.fr`

Y. Demazeau et al. (Eds.): Advances in PAAMS, AISC 70, pp. 5–14.
springerlink.com

usually relate to various aspects such as the access to the search system, the display of the returned answers, the criteria for limiting the desired information, the data related to his geolocalisation, etc. The interests of the user are related to fields which can be expressed through preferences. Finally the user needs can either be explicitly expressed by him or be discovered by a system that observes his actions, learns about his behaviour and makes a representation of his habits from which it can deduct his need(s).

This paper focuses on this last point: the implicit determination of the interest of a user facing document provided by a search engine in response to his request. Having highlighted the importance of the adaptive needs of the implicit user preferences assessment, a generalization and a formalization of this problem are proposed. The section 3 is devoted to the proposed algorithm which is based on the AMAS approach. Various experiments and results to evaluate its relevance are then presented before concluding and giving some perspectives to this work.

2 Personalized Assessment of User's Interests

2.1 Towards a Personalized Implicit Feedback

Whatever the interest of a user may be, it changes over time. The user profile has to be updated accordingly, by taking into account and by incorporating these changes. Several methods exist to account for this evolution depending on whether the system appeals to the user explicitly (explicit feedback) or not (implicit feedback).

In an explicit feedback, the user has to indicate his assessment of the results returned by the system (a document in the context of information retrieval). Three main approaches exist to express this assessment: a binary approach (like / not like, interesting / not interesting or relevant / irrelevant) [3] a more incremental approach using discrete values expressed by a note [9] and a textual approach. Such an explicit evaluation is admittedly easy to integrate into the system; but if one takes the user's point of view, it is not quite easy to express an opinion using numerical scales and particularly, the user does not necessarily have the time or the inclination to devote too much energy to this task.

Implicit feedback tries to remedy these problems by attempting to automatically infer the preferences of the user, without appealing to him but by observing him interacting with the system. This observation relies on some *implicit interest indicators* such as the study of links followed by users when searching [7], the study of the history of purchases (Amazon), etc. Others considers the indicators commonly used [5], [4] such as the time spent on a document, the number of mouse clicks, the page bookmarking, the printing or saving a document, etc., these indicators needing to be later interpreted to extract an interest.

However, to be faithful to the user and contrary to what is usually done in such cases in the literature [12], interpretation of these indicators should not be identical for all users but should truly adapt to specificities of each of them. In particular, a user might prefer to print a document of interest while another might prefer to bookmark it. The particularity of each user has to be taken into account to extract as closely as possible his interests and to provide a truly personalized system because tailored for one specific individual. It is therefore necessary to learn and to find relevant indicators but also their involvement in the function enabling to determine the most interesting documents for a particular user, while bearing in mind that they can evolve for a same user but also from one user to another.

Nevertheless, the implementation of such implicit feedback can not be fully disconnected from the user and requires some minimal interaction with him. In particular it requires to construct and to take into account two highly interconnected aspects: (*i*) the first one, based on a user's feedback and on a range of indicators obtained from observations of the user's behaviour, consists in extracting discriminating indicators that contextually reflect his interests in the consulted documents; (*ii*) the second consists in defining and in implementing mechanisms which, as few as possible, appeal to the user for a feedback (for example only when mechanisms which implement the implicit feedback cannot deduct anything).

This paper focuses on the first point and supposes that the second issue is solved; we assume that the decision function of the user is known at every moment, even if it is evolutive.

2.2 Learning User Behaviour Features to Asses Its Centres of Interest

We aim at designing a system able to assess the interest of a user for a document from some observations on his behaviour when he is consulting it. Each observation is a potential criterion enabling the determination of a piece of the interest of a user for this document. Thus, the adaptation of the assessment function of the user's interest consists in dynamically defining the importance of each criterion. That is why we propose a real time learning algorithm based on behaviour features which aims at deducing the importance of each criterion for a personalized interest assessment (i.e. contextual and adapted to each user).

Two main classes of learning algorithms can be distinguished: the supervised one where a learning "teacher" giving input and output samples exists and the unsupervised one where a "teacher" does not exist, where the desired output is not used [6]. As our objective is to be the most independent as possible of the user, a supervised learning algorithm is not suited. That is why neuronal networks (NN) cannot be a solution to solve our problem. Such systems strongly depend on the intended solution that is, in our case, on the user's feedback. The reinforcement learning (RL), first introduced by [15] as $\mathcal{Q}$-learning and improved and implemented in many other works [13], is an intermediate class; "after choosing an action the agent is told the immediate reward and the subsequent state, but is not told which action would have been in its best long-term interests" [6]. Even if they have been extended to MAS [14], $\mathcal{Q}$-learning inspired algorithms are only efficient in Markovian environments but in our problem, the environment, *i.e.* the user, is known as not being Markovian at all. Furthermore, we also believe that faced with the diversity of observed behaviours to take into account and faced with the dynamic of these behaviours, the intended assessment function cannot be checked and managed by an external supervision. The system implementing this function has to be autonomous and to adapt itself locally to environmental changes, according to what it perceives and its internal state. Thus, the learning phase is a never-ending process.

Because of its intrinsic nature, human decision making cannot be reduced to a precise formal model. Moreover, this decision making function evolves during time. Thus, according to [11] the environment of the system to build – *i.e.* the user – is dynamic, inaccessible (its state cannot be totally known) and non deterministic (it is not possible to known the effect of a given action upon its behaviour). The AMAS (Adaptive Multi-Agent Systems) approach [2] is particularly adapted to solve such kind of problem; therefore, we used it to implement our real time learning algorithm.

In this approach, a system is said *functionally adequate* if it produces the function for which it was conceived, according to the viewpoint of an external observer who knows its finality. To reach this functional adequacy, it had been proven that each autonomous agent that composes an AMAS and pursues a cycle composed of three steps (perception/decision/action) must keep relations as cooperative as possible with its social (other agents) and physical environment. The definition of cooperation we use is not conventional (simple sharing of resources or common work). Our definition is based on three local meta-rules the designer has to instantiate according to the problem to be solved: **Meta-rule 1** (c_{per}): Every signal perceived by an agent has to be understood without ambiguity; **Meta-rule 2** (c_{dec}): Information coming from its perceptions has to lead the agent to produce a new decision; **Meta-rule 3** (c_{act}): This reasoning must lead the agent to make the actions which have to be useful to other agents and the environment. If one of this meta-rule is not checked, the agent is faced to a "Non Cooperative Situations" (NCS) that can be assimilated to an "exception" in traditional programming. It occurs when at least one of the three previous meta-rules is not locally verified by an agent. Different generic NCS have been highlighted: *incomprehension* and *ambiguity* if (c_{per}) is not checked, *incompetence* and *unproductiveness* if (c_{dec}) is not verified and finally *uselessness*, *competition* and *conflict* when (c_{act}) is not checked. This approach is proscriptive because each agent must first of all, anticipate, avoid and repair a NCS. This has strong methodological implications: designing an AMAS consists in defining and assigning cooperation rules to agents. In particular, the designer, according to the current problem to solve, (*i*) has to determine what an agent is, then (*ii*) he has to define the nominal behaviour of an agent then (*iii*) he has to deduce the NCSs the agent can be confronted with and (*iv*) finally he has to define the actions the agent has to perform to come back to a cooperative state. The designer has to keep in mind that agents do not have a view of the global system and do not base their reasoning on the expected collective function realized by the system.

2.3 Problem Generalisation

In order to formalise the problem, we set some hypothesis. First of all, we suppose that two types of criterion exist : *(i)* boolean criterion (*i.e.* document printing, saving, *etc.*) ; *(ii)* continuous criterion (*i.e.* document consultation time, *etc.*). Let $\mathscr{C}_b$ be the set of boolean criteria and $\mathscr{C}_c$ be the set of continuous criteria. The consultation of a document leads to a set of observations, each of them matching with a criterion. Let $c_{i,j}$ be the normalised value of the criterion c_i for the jth document consultation.

If $c_i \in \mathscr{C}_b$ then $c_{i,j} = 1$ when the value of c_i is "`true`", otherwise $c_{i,j} = 0$.

If $c_i \in \mathscr{C}_c$ then $c_{i,j} \in [0;1]$. This standard range avoids taking into account the meaning of an observation. For example, if we assume that the consultation time doesn't relay any interest value before 5 seconds and reaches the highest interest after 2 minutes, then before 5 seconds $c_{i,j} = 0$, after 2 minutes $c_{i,j} = 1$ and inside this range $c_{i,j}$ evolves linearly.

Each criterion c_i is associated with an influence degree $\omega_i \in [0;1]$ that stands for the importance of the criterion during the user's interest calculus. When $\omega_i = 1$, c_i is very important for it, but when $\omega_i = 0$ c_i has no influence. Thus the local interest of c_i is equal to $c_{i,j} \cdot \omega_i$.

We suppose that higher the local interests are, higher the global interest is. Thus, the interest value of the user u for an observed document j, can be approximated by a weighted sum function $\mathscr{F}$ and the adaptation of the user interest evaluation function can be generalised to the search of the multi-criterion decision function $\mathscr{F}$ considering:

$$\mathscr{I}_u(j) = \mathscr{F}(j), \forall j \text{ with } \mathscr{F}(j) = \sum_{i=1}^{n} (\omega_i \cdot c_{i,j})$$

Where $\mathscr{I}_u$ is the multi-criterion decision function of the user u to approximate; n is the number of criteria; $c_{i,j}$ is the normalised value of c_i for the decision j; ω_i is the influence value of c_i.

Thus, solving this problem consists in searching the set of ω_i that verifies the previous equation. As already said in the §2.2, the decision function $\mathscr{I}_u$ can evolve. Therefore, in order to keep an adequate decision function, this equation has to be verified for only the few last decisions.

3 Real Time Learning of User's Interests

3.1 System Analysis

The AMAS that we have to conceive is a multi-criterion decision making function $\mathscr{F}$ defined by the weighted sum $\mathscr{F}(j)$ given in section 2.3. The goal of the system is to learn the set of ω_i in order to approximate the target decision making function $\mathscr{I}_u$. This function has to self-adapt in real time to its environment: the user. Therefore, we propose to incrementally improve it at each decision j, with the following life cycle: *(i)* the system perceives the values of the criteria; *(ii)* the decision value $\mathscr{F}(j)$ is calculated with these values and is sent to the environment; *(iii)* depending on this value, the environment returns a feedback $fb(j) \in \{\uparrow, \downarrow, \sim\}$ in order to inform the system if the searched $\mathscr{I}_u(j)$ is higher ($\uparrow$), lower ($\downarrow$), or equal ($\sim$) to the provided one; *(iv)* the system adjusts the values of ω_i in order to correct the possible errors.

Thus, at each new decision, the system adjusts its function in order to converge towards a more and more adequate decision making value. Because the decision making function is not *a priori* known, it is possible that this function can only be approximated. Thus, ε expresses a tolerance value such as: if $|\mathscr{F}(j) - \mathscr{I}_u(j)| \leqslant \varepsilon$ then the environment returns a "$\sim$" feedback.

3.2 Agents Identification

Here the only entities that have a local goal inside the system are the ones that are in charge of the local decision value of a criterion: so we can identify the *criterion agents*. In the rest of this paper, "agent" refers to "criterion agent". The goal of a criteria agent c_i is then to adjust its influence value ω_i so that its partial function – its local decision – enables the system to figure out an adequate global decision. Now, we have to define the local behaviour of our agents.

3.3 Identification of Non Cooperative Situations

During the perception phase, the perceived value of an agent is the current value of its criterion. This value is well defined and cannot generate an ambiguity, thus, agents cannot be faced to a NCS at this step. During the decision phase, each agent has to multiply two values, so no NCS can occur during this step. During the action phase that can generate three NCS (uselessness, concurrency and conflict), only the conflict situation is relevant for an agent. When an agent decides locally a value that contradicts the decision of the user, it is in conflict with the environment. This situation occurs when the agent provides a local decision value that is involved in the generation of a wrong global decision, that is to say, when the normalised value of the criteria is considered not null and when the environment returns a feedback different to "$\sim$". More formally, this cooperation failure is detected by the agent c_i for the decision j when $c_{i,j} > \varepsilon \wedge fb(j) \neq \sim$.

3.4 Cooperative Agent Behaviour

When a conflict is detected by an agent, he has to act in order to suppress this situation. Thus, when the feedback is "$\uparrow$" the agent c_i has to increase ω_i; when the feedback is "$\downarrow$" the agent has to decrease ω_i. Let be Δ_{ω_i} the increment or decrement step value of ω_i. In order to be really cooperative, each agent has to tune its ω_i according to its possible involvement in the conflict situation. If the value of $c_{i,j}$ is relatively small, an agent can assume that it was not really involved in the error, so it decides to tune its ω_i very slightly. In the opposite case, c_i must strongly tune ω_i. So, from a local point of view, ω_i values are tuned proportionally to the possible involvement of the corresponding agent c_i. More formally, the agent c_i increments or decrements ω_i with $\Delta_{\omega_i} \cdot c_{i,j}$.

An autonomous and cooperative agent must be able to decide by itself the modification strength of its ω_i. Therefore, each agent must be able to define the uncertainty of the current value of ω_i. Higher the uncertainty is, higher ω_i is modified. On the other hand, lower the uncertainty is, slighter ω_i is modified.

Thanks to the local observation of successive feedbacks, agents can consider the uncertainty of ω_i and then define its tuning step Δ_{ω_i} as follows. Let *(i)* $\Delta_{\omega_i}^{max}$ be the max value of the tuning step of ω_i; *(ii)* λ_{ω_i} be the maximum number of ω_i modifications that leads to the modification of Δ_{ω_i}. λ_{ω_i} stands for a kind of "tuning sensibility" for ω_i; *(iii)* $\Delta_{\omega_i}^{min}$ be the maximal value of the modification step of ω_i. It is defined as ❶. When the agent tunes successively ω_i towards opposite directions, the uncertainty decreases, thus Δ_{ω_i} is decreased as ❷. When the agent tunes successively ω_i towards the same direction, the uncertainty increases, thus Δ_{ω_i} is increased as ❸.

$$❶\ \Delta_{\omega_i}^{min} = \frac{\Delta_{\omega_i}^{max}}{2^{\lambda_{\omega_i}}} \ ;\ ❷\ \Delta_{\omega_i} \leftarrow \max\left(\frac{\Delta_{\omega_i}}{2}, \Delta_{\omega_i}^{min}\right) \ ;\ ❸\ \Delta_{\omega_i} \leftarrow \min\left(\Delta_{\omega_i} \cdot 2, \Delta_{\omega_i}^{max}\right)$$

Thus, without any other interaction with other agents, this behaviour enables each agent to tune in real time its ω_i in order to contribute as well as it is possible to the global assessment function. Although agents are fine-grained, they have a real autonomous behaviour. This behaviour is led by the local estimation of the goodness or badness of their contribution to the collective activity. Each agent is able to estimate the uncertainty (Δ_{ω_i}) of its ω_i. The observation of each criterion does not explain on its own the result of the collective activity which is effectively an emergent one. Indeed, the use of the AMAS cooperation ensures the functional adequacy at the higher level. This adequacy is verified in the following section.

4 Experiments and Analysis

In order to analyse the multi-agent system introduced in section 3, this algorithm has been implemented in Java. The system, which represents a $\mathscr{F}$ function, is initialised with n criteria c_i, each associated to a random ω_i. Its environment is the target function $\mathscr{I}_u$ which is initialised with other random ω_i values. At each decision phase, $\mathscr{F}$ receives feedback from $\mathscr{I}_u$ and then tries to adapt its function.

4.1 Checking of Convergence

The purpose of the first experiments was to check if the system is able to find a solution whatever its internal state is, with a decision function $\mathscr{F}$ containing 5 boolean criteria and 5 continuous criteria. During the adaptation process, the system is regularly disturbed by random modifications of agents' states (*i.e.* each 1500 cycles) each ω_i and Δ_{ω_i} are randomly modified. The figure 1 shows the evolution of the average distance of ω_i between the adaptive function and the searched one. After each disturbance, the system converges towards the solution and gets set after approximately 600 cycles. Thus, we can conclude that whatever the state of the system, it is able to converge towards the solution.

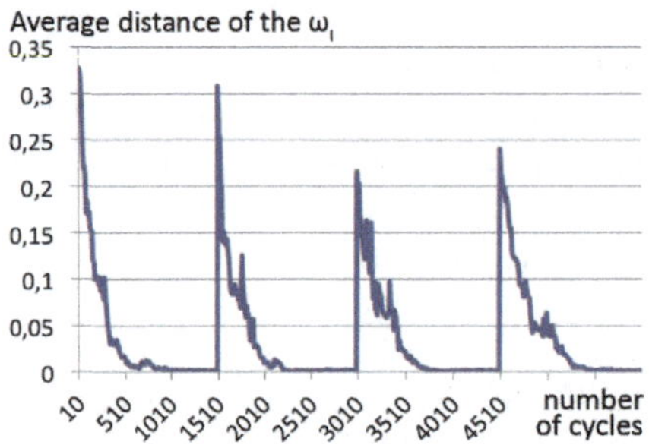

Fig. 1 Convergence of the system

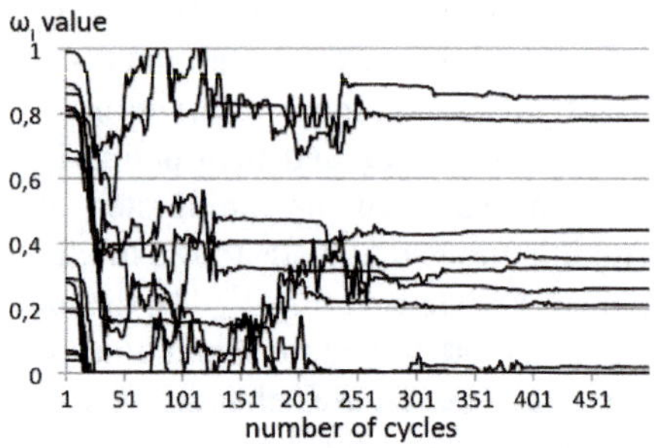

Fig. 2 The pruning of useless criteria

4.2 *The Pruning of Useless Criteria*

Because the relevant criteria are not *a priori* known, the algorithm has to perceive as many criteria as possible even if some of them can be useless. Indeed, the adaptation process has to find by itself useless and useful criteria. The figure 2 illustrates an example of this criteria pruning where only 8 out of 16 criteria are relevant ($\omega_i > 0$). We notice that the system filters out theses criteria by setting their influence to zero. The influence of these useless criteria converges quickly towards zero; this does not really disturb the influence of relevant criteria. Thus, for practical purposes, it will be worthwhile to add more criteria than those that are expected to be relevant.

4.3 *Convergence Speed*

We have studied the convergence speed of the algorithm when *(i)* the system is composed of boolean criteria and when *(ii)* it is composed of continuous criteria. After 100 searches, the average number of cycles required to find a solution is calculated for 4, 8, 16, 32, 64, 128 and 256 criteria. We consider that the system finds a solution if the environment returns a "$\sim$" feedback during 100 consecutive cycles.

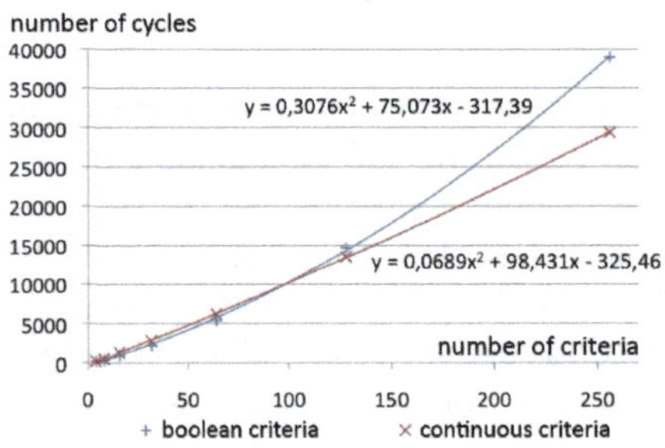

Fig. 3 Average convergence speed

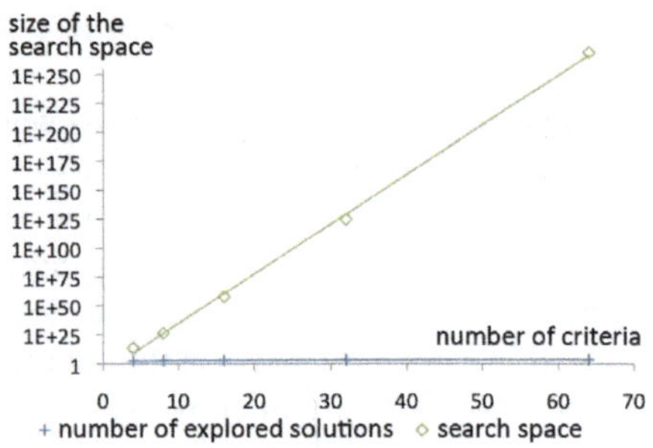

Fig. 4 Search space

The figure 3 shows the average number of cycles required to find a solution depending on the number of criteria belonging to the searched function. When the function is composed of boolean criteria the complexity is polynomial whereas when the function is composed of continuous criteria, the complexity, even though being actually polynomial, is very close to the linear complexity. This difference of complexity is due to the fact that boolean criteria have a value that is either 0 or 1 and then their ω_i tuning is more crude than the continuous criteria one.

In order to estimate more precisely the efficiency of the algorithm, we compared the number of explored states (before finding a solution) to the size of the solution space. Let $|C|$ be the number of criteria and $|\omega_i|$ be the cardinal of possible values of ω_i. This cardinal is obtained by dividing the range of ω_i by $\Delta_{\omega_i}^{min}$. The search space of $\mathscr{F}$ is then $S = \prod_{i=1}^{|C|} |\omega_i|$. As shown in figure 4, whereas the search space grows exponentially (the ordinate axis range is logarithmic), the number of explored solutions evolves quasi-linerally. We can then conclude that the algorithm converges quasi-linerally depending on the number of criteria and does not depend on the size of the search space. Thereby, this algorithm is extremely efficient, all the more so the processing time of each criterion (not introduced here) is linear.

5 Conclusion and Perspectives

This article proposes, implements and evaluates an adaptive model for the assessment of the interest of one particular user. This problem is generalised to the adaptive learning of a multi-criteria decision function which is a weighted sum of observed criteria values. This decision function is *a priori* unknown and can evolve. Solving this problem consists in searching the relevant criteria that are actually used by the user during the decision making process.

The proposed system is an Adaptive Multi-Agent System composed of criteria agents. Their local goal is to determine their relative influence in the decision function. We have defined and implemented a local behaviour that enables agents to correct the system errors. Thanks to a set of experiments, we have verified the convergence and the adaptiveness properties of the system. These experiments have also shown the efficiency of the proposed solution: it finds a solution with a polynomial complexity, even if the search space grows exponentially with the number of criteria.

Although this system works well with simulated users, we are now experimenting and implementing it in a real world problem. Therefore we are currently working to its integration in the iSAC project [10], an intelligent Citizens Information Service, in collaboration with the university of Girona.

References

1. Brusilovsky, P., Kobsa, A., Nejdl, W. (eds.): Adaptive Web 2007. LNCS, vol. 4321. Springer, Heidelberg (2007)
2. Camps, V., Gleizes, M.P., Glize, P.: A self-organization process based on cooperation theory for adaptive artificial systems. In: 1st Int. Conference on Philosophy and Computer Science Processes of evolution in real and Virtual Systems, Krakow, Poland (1998)

3. Chen, L., Sycara, K.: Webmate: A personal agent for browsing and searching (1998)
4. Claypool, M., Le, P., Waseda, M., Brown, D.: Implicit interest indicators. In: Intelligent User Interfaces, pp. 33–40. ACM Press, New York (2000)
5. Jude, J.G., Shavlik, J., Dept, C.S.: Learning users' interests by unobtrusively observing their normal behavior. In: Proceedings of the 2000 International Conference on Intelligent User Interfaces, pp. 129–132. ACM Press, New York (2000)
6. Kaelbling, L.P., Littman, M.L., Moore, A.W.: Reinforcement learning: A survey. Journal of Artificial Intelligence Research 4, 237–285 (1996)
7. Lieberman, H.: Letizia: An Agent That Assists Web Browsing. In: Mellish, C.S. (ed.) Proceedings of the 14th International Joint Conference on Artificial Intelligence (IJCAI-1995), pp. 924–929. M. Kaufmann Inc., Montreal (1995)
8. Montaner, M., López, B., Lluís De La Rosa, J.: A Taxonomy of Recommender Agents on the Internet. Artif. Intell. Rev. 19(4), 285–330 (2003)
9. Moukas, A.: User Modeling in a MultiAgent Evolving System. In: Workshop on Machine Learning for User Modeling, 6th International Conference on User Modeling, Chia Laguna, Sardinia (1997)
10. de la Rosa, J., Rovira, M., Beer, M., Montaner, M.: Reducing Administrative Burden by Online Information and Referral Services. In: Citizens and E-Government: Evaluating Policy and Management, Reddick, C.G., Austin, Texas (to appear in 2010)
11. Russell, S., Norvig, P.: Artificial Intelligence: A Modern Approach, 2nd edn. Prentice-Hall, Englewood Cliffs (2003)
12. Seo, Y.W., Zhang, B.T.: A reinforcement learning agent for personalized information filtering (2000)
13. Sutton, R., Barto, A.: Reinforcement Learning: An Introduction. MIT Press, Cambridge (1998)
14. Thomas, V., Bourjot, C., Chevrier, V.: Interac-DEC-MDP: Towards the use of interactions in DEC-MDP. In: Third Int. Joint Conference on Autonomous Agents and Multi-Agent Systems - AAMAS 2004, New York, pp. 1450–1451 (2004)
15. Watkins, C.: Learning from delayed rewards. Ph.D. thesis, University of Cambridge, England (1989)

A GPU-Based Multi-agent System for Real-Time Simulations

Guillermo Vigueras, Juan M. Orduña, and Miguel Lozano

Abstract. The huge number of cores existing in current Graphics Processor Units (GPUs) provides these devices with computing capabilities that can be exploited by distributed applications. In particular, these capabilites have been used in crowd simulations for enhancing the crowd rendering, and even for simulating continuum crowds. However, GPUs have not been used for simulating large crowds of complex agents, since these simulations require distributed architectures that can support huge amounts of agents. In this paper, we propose a GPU-based multi-agent system for crowd simulation. Concretely, we propose the use of an on-board GPU to implement one of the main tasks that a distributed server for crowd simulations should perform. The huge number of cores in the GPU is used to simultaneously validate movement requests from different agents, greatly reducing the server response time. Since this task represents the critical data path, the use of this hardware significantly increases the parallelism achieved with respect to the implementation of the same distributed server on a CPU. An application example shows that the system can support agents with complex navigational behaviors.

1 Introduction

The huge number of cores existing in current Graphics Processor Units (GPUs) provides these devices with computing capabilities that can be exploited by distributed applications. Since some years ago, GPU vendors introduced programmability to these devices in order to facilitate their use for scientific computation. One of the distributed applications that can benefit from the capabilities of the GPUs is crowd simulation.

Guillermo Vigueras · Juan M. Orduña · Miguel Lozano
Departamento de Informática, Universidad de Valencia
Avda. Vicente Andrés Estellés, s/n
Burjassot, Valencia, Spain
e-mail: {guillermo.vigueras,juan.orduna,miguel.lozano}@uv.es

Y. Demazeau et al. (Eds.): Advances in PAAMS, AISC 70, pp. 15–24.
springerlink.com

Regarding crowd simulations, some proposals have been made for exploiting the capabilities of multicore architectures. In this sense, a new approach has been presented for PLAYSTATION3 to distribute the load among the PS3-Cell elements[13]. Another work uses graphics hardware to simulate crowds of thousands individuals using models designed for gaseous phenomena [1]. Recently, some authors have started to use GPU in an animation context (particle engine) [11, 5], and there are also some proposals for running simple stochastic agent simulations on GPUs [7, 10]. However, these proposals are far from displaying complex behaviors at interactive rates. On the other hand, other proposals including complex agent systems have been made [9, 15], but they are not designed to provide the required scalability in number of agents.

In this paper, we show that large scale multiagent systems can benefit from the use of GPU computing. In order to achieve this goal, we have implemented a distributed server for crowd simulations [16] using an on-board GPU. The huge number of cores in the GPU is used to simultaneously validate movement requests from different agents, greatly reducing the server response time. Since this task represents the critical data path, the use of this hardware significantly increases the parallelism achieved with respect to the implementation of the same multiagent system using a distributed server implemented on a CPU. Thus, the performance evaluation results show that the system throughput is significantly increased, supporting a significantly higher number of agents while providing the same latency levels. These results can be used for simulating crowds with a larger size.

The rest of the paper is organized as follows: Section 2 describes in detail the use of GPUs for increasing parallelism in crowd simulations. Next, Section 3 shows the performance evaluation of the proposed architecture. Next, Section 4 describes an application example to show our system working in a real and complex scenario. Finally, Section 5 shows some concluding remarks and future work to be done.

2 A GPU-Based Action Server for Crowd Simulation

In a previous work, a distributed system architecture for crowd simulation was proposed in order to take advantage of the underlying distributed computer system [16]. That software architecture is mainly composed by two elements: the action server (AS) and the client processes (CP). The AS is devoted to execute the crowd actions, while a CP handles a subset of the existing agents. Agents are implemented as threads of a single process for reducing the communication cost (the communications exchanged are easily performed if they share the same address space). Each thread manages the perception of the environment and the reasoning about the next action. Each client process is hosted on a different computer, in such a way that the system can have a different number of client processes, depending on the number of agents in the system. The Action Server is divided into a set of processes so that each one can be executed in parallel in a different computer. Each of these processes is denoted as an Action Server, while the whole set of ASs is denoted as the Parallel Action Server (PAS). The rest of the section describes the version of the PAS for execution in CPU, and the modifications made to this version in order to take advantage of a GPU.

Each piece (process) of the Parallel Action Server can be viewed as a partial world manager, since it controls and properly modifies the information in a region of the whole simulation space. Thus, it can be considered as the system core. Each AS process contains three basic elements: the Interface module, the Crowd AS Control (CASC) module and the Semantic Data Base (SDB). Figure 1 illustrates a detailed scheme of an AS.

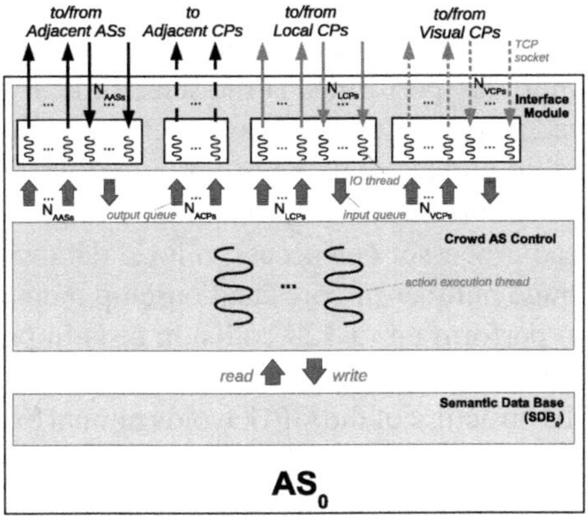

Fig. 1 Internal structure of an Action Server

The main module is the Crowd AS Control module, which is responsible for executing the crowd actions. This module contains a configurable number of threads for executing actions (action execution threads in Figure 1). For an action execution thread (AE thread), all messages sent to or received from other ASs and CPs are exchanged asynchronously (the details are hidden by the Interface module, see below). This means that the AE threads only may have to wait when accessing shared data structures such as the semantic database. Thus, experimental tests have shown that having more AE threads than cores allows each AS to take advantage of several cores.

Most action requests from agents are executed from start to end by the AE thread, that extracts the requests from the corresponding input queue and process them. These requests consist of collision tests, in order to check if the new position computed by each agent when it moves coincides with the position of other agent or object. The Interface module hides all the details of the message exchanges. This module provides the Crowd AS Control module with the abstraction of asynchronous messages. Two separate input queues exist, one for messages coming from local CPs (action requests) and the other one for messages coming from adjacent ASs (responses to requests issued because of local border actions or requests for remote border actions from adjacent ASs). Having two separate input queues is an efficient way of giving a higher priority to messages from adjacent ASs. The reason for improving the priority of these messages is that the border actions are the ones whose processing takes longer, and we should reduce as much as possible their response time to provide realistic interactive effects.

In order to process messages as soon as they arrive, the Interface module contains one IO thread dedicated to getting incoming messages from each TCP socket. There are no input threads associated with sockets connecting one AS to their adjacent CPs, because CPs only send messages to their local AS. In the same way, there is one IO thread and one output queue per TCP socket, so that messages are sent as soon as the corresponding TCP socket is ready for writing.

The GPU used for implementing a parallel action server has been a NVIDIA Tesla C870. This GPU has 128 thread processors, each one with its texture cache and low-latency shared memory. The communications of the different shared memories with global memory are performed at the same time, by means of Load/Store operations. Each thread processor executes a kernel in parallel with the rest of processors. Just before calling this kernel, the use of *cudaBindTexture()* instruction allows to copy into each texture cache the proper range of global memory addresses. In this way, each thread processor can access to local data with a very low latency. The integration of a huge number of cores and on-chip memories with low latency in this GPU allows to perform up to 128 collision tests in parallel, clearly outperforming the number of collision tests performed in parallel when using the CPU. Additionally, the SIMD structure of the GPU avoids mutual exclusions when accessing shared data structures, while the CPU-version AE Threads have to wait when accessing data structures shared among the existing cores, such as the semantic database.

A general overview of the collision checking process in the server is as following: agent requests received by the interface module are passed to the CASC module. However, the AE threads in the GPU-based server do not actually perform the collision tests (unlike the AE Threads in the CPU-based server). Instead, the AE threads collect the requests from the interface module and copy them to the SDB module as they arrive, until $N_{OPERATIONS}$ requests are collected. At this point, one of the AE threads signals the *GPU manager thread*. This thread controls the data path in the GPU. Concretely, this thread copies the requests into the GPU global memory and launches the collision tests. When the collision tests finish, the GPU manager thread copies the results into the CPU memory. When the AE threads finish their current task, they will start to process the GPU replies. That is, the AE threads in the GPU-based server collect client requests, but these requests are actually checked by the GPU manager thread. In this way, the parallelism achieved by having several AE Threads is decoupled from the parallelism provided by the GPU thread processors. This collision checking scheme can take advantage of the parallel computation capabilities of a GPU like the NVIDIA Tesla C870.

The internal structure of a parallel action server using a GPU is shown in Figure 2. In this implementation, the SDB module contains the *object positions array*, the *collision response array* and the *GPU manager thread*. The *object positions array* is the host-GPU input interface. This array, contains the agents positions, needed to check collisions. The structure *collision response array*, represents the host-GPU output interface, and it contains the result of the collisions tests performed by the

GPU. When the *GPU manager thread* is signaled by the AE Threads, it copies the object positions array into the GPU and launches the collision tests. When the collision tests finish, the *GPU manager thread* copies the results from the GPU memory into the SDB module. Then, it signals the AE threads to start replying to CPs or ASs.

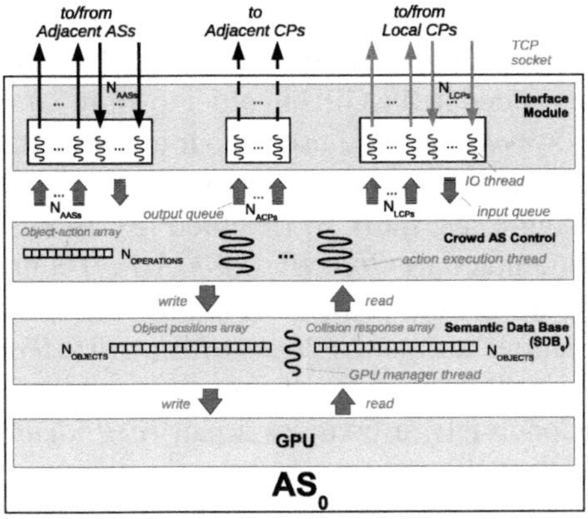

Fig. 2 Internal structure of an action server using a GPU

The *object positions array* has $N_{OBJECTS}$ elements, that is, as many elements as agents managed by a server. An agent identifier is associated to each request, in order to update the proper array position. Each element in the array has four floats for each object passed to the GPU. The first and second floats contain the x and y coordinates of the agent's position, respectively. The third float is not used. The fourth float is a flag indicating the GPU whether the element in the array has been updated by AE threads. If an object position has been updated (i.e. the flag is equal to a positive number representing the agent identifier number) then the collision test will be performed. Otherwise (i.e. the flag is equal to -1.0) the collision test is skipped. This flag is initialized by AE threads when the x and y coordinates are updated and is cleared when a collision test finishes.

The *collision response array* has, as the previous array, $N_{OBJECTS}$ elements. Each element of the array has two floats. The first one indicates whether a collision occurred (in this case is equals to 1.0) or not (in this case is equals to 0.0). The second float contains the agent identification number and indicates which agent is associated to the collision result. The agent identification number is obtained by the GPU from the fourth float contained in each element of the *object positions array*. The *collision response array* is accessed by the AE threads to send the collision responses corresponding to each collision request. Although this array has as many elements as agents, the AE threads will only collect $N_{OPERATIONS}$ instead of $N_{OBJECTS}$ collision results. In order to efficiently read the *collision response array*, random access is needed. The *object-action array* (that provides this random access) contains the elements that should be accessed for reading each collision result.

Collisions on the GPU are checked using a spatial hashing based method. For this reason, a two dimensional grid is created in the GPU memory when the simulation starts. The dimensions of the grid, grid cell size and grid origin coordinates are fixed, depending on the scene simulated. The easiest way to implement the collision grid into the GPU is defining an array in which each position represents a grid cell. The mapping of agents to grid cells is performed by the spatial hashing method, depending on the cell size and the position of agents. Because many agents can fall within the same cell, the GPU threads can simultaneously update the same array position. Atomic operations are needed in the GPU in order to allow simultaneous accesses to global memory [8]. However, the GPU used in our implementation does not support atomic operations. For that reason, a more complex approach (based on sorting) has been implemented, using a fast radix sort method [3]. In this approach, there is a global memory array (denoted as *ObjectPositionsArray*) containing the agents positions. Another array (denoted as *collisionResponse*) contains the collision results. Also, some other structures are used: *ObjectsHash*, *sortedPositions* and *cellStart*. The *ObjectsHash* array represents the collision grid, and it stores the cell to which each agent belongs. Concretely, it contains a pair *(cell identifier, agent identifier)* for each position. The *sortedPositions* array contains the same elements as *ObjectPositionsArray*, but sorted by cell identifier. In this way, neighboring agents can be efficiently obtained for collision checking. The *cellStart* array allows to determine the beginning of each cell in the *ObjectsHash* array. Thus, if position i in *cellStart* array contains the value j, it means that the first agent of grid cell i appears in position j in the *ObjectsHash* array. In this way, the *cellStart* array allows a quick access to the agents in neighboring cells.

3 Performance Evaluation

This section shows the performance evaluation of the GPU-based server described in the previous section. We have performed different measurements on a real system using the GPU-based server. For comparison purposes, we have also performed the same measurements on the same real system but using the CPU-based server. The most important performance measurements in distributed systems are latency and throughput [2]. The performance improvement that a GPU-based server can provide to the distributed crowd simulation [16] would depend on the number of distributed servers in the system. In order to evaluate the worst case, we have performed simulations with one server and with different number of agents. Concretely, we have measured the aggregated computing time for collision tests during the simulations, and the average response times provided to agents. In order to define an acceptable behavior for the system, we have considered 250 ms. as the maximum threshold value for the average response time. This value is considered as the limit for providing realistic effects to users in Distributed Virtual Environments (DVEs) [4, 14]. Since crowd simulation can be considered as DVEs where avatars are intelligent agents (instead of dummy entities controlled by users), we have used this value taken from the literature.

We have performed crowd simulations with wandering agents because all their actions should be verified by an AS. Each simulation consists of the crowd moving within the virtual world following k-length random paths. Nevertheless, in order to obtain reproducible and comparable results, we have saved the paths followed by the agents in the first execution of each configuration and we have used the same paths for all the executions tested with the same number of avatars. For all the populations tested, the average response time has become stable within the first thirty seconds of execution time. Therefore, we have used simulations lengths of thirty seconds for the configurations tested.

We have performed experiments using a computer platform with one server and six clients. The server was based on Intel Core Duo 2.0 GHz, with 4GB of RAM, executing Linux 2.6.18.2-34 operating system and had incorporated a NVIDIA C870 Tesla GPU. Each client computer was based on AMD Opteron (2 x 1.56 GHz processors) with 3.84GB of RAM, executing Linux 2.6.18-92 operating system. The interconnection network was a Gigabit Ethernet network. Using this platform, we have simulated up to nine thousand agents.

Figure 3 a) shows the aggregated computing time for operations involved in collision tests during the whole simulation. On the X-axis, this figure shows the number of agents in the system for different simulations. The Y-axis shows aggregated computing time (in seconds) devoted to compute the collision tests required by the simulations. This Figure shows that the plot for the CPU-based server has a parabolic shape, while the plot for the GPU-based server has a flat slope. These results show that the use of the replicated hardware in the GPU has a significant effect in the time required by the server to compute the collision tests. The population size (number of agents) considered in these simulations generates a number of collision tests that does not exceed the computation bandwidth available in the GPU. As a result, the computing time required for different population sizes is very similar. An additional benefit is derived from the fact that the GPU is exclusively devoted to compute collision tests, releasing the CPU from that task. This is the reason for the lower values shown by the GPU plot, even for the smallest population sizes.

Although Figure 3 a) shows a huge improvement in the GPU-based server performance, the effects of such improvement should be measured on the system. Thus, Figure 3 b) shows the average response time provided for the agents hosted by a given client. In order to show the results for the worst case, we show the values for the client with the highest average response time (it must be noticed that we have a single server and six clients).

Figure 3 b) shows that the average response times increases linearly with the number of agents for both plots until 6000 agents. From that point up, the same behavior is shown when no GPU is used. However, when using GPU the plot shows a flat slope. Beyond a given threshold, the average response time is not increased because all the collisions tests are performed in parallel in the GPU. Additionally, Figure 3 b) shows that the CPU-based crowd simulation can support up to 3500 agents while providing average response times below the 250 milliseconds threshold. When GPU is used, the number of agents supported grows up to 5300 agents, providing an improvement of a 50%. Taking into account that this improvement can

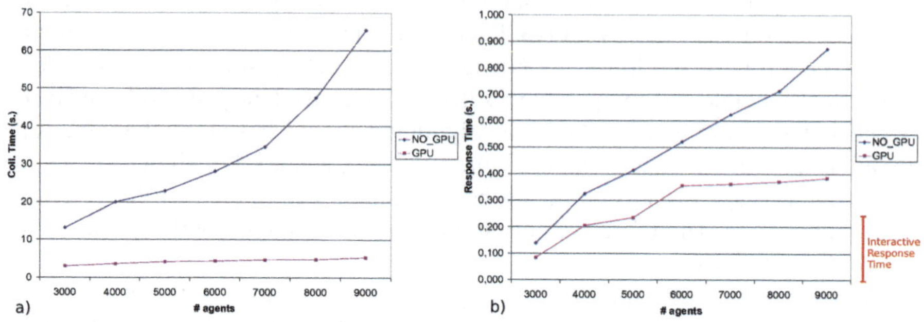

Fig. 3 a) Aggregated computing time for collision tests. b) Average Response times provided to agents

be achieved per each server in the system, these results show that the GPU-based server can actually have a significant impact in the performance of large-scale crowd simulations.

Additionally, it must be noticed that the performance achieved when using GPUs depends on the number of cores in the CPU, because the CPU threads are responsible for replying to the agents requests. Since we have used dual-core processors for evaluation purposes (in order to measure the worst-case performance), the improvements shown in this section can be increased when using platforms with a higher number of processor cores.

4 Application Example

In Section 3 we have shown the performance evaluation of our proposal using wandering agents, since this kind of agent generates the highest load in the server because all its actions should be verified by an AS. In this section, we show that the proposed approach can integrate more complex navigational behaviors in real and structured scenarios. Concretely, we have simulated the evacuation of an actually existing Faculty building.

We have captured different simulation data in order to visualize the movement followed by the crowd. These data are computed in the Client Processes, in such a way that the Action Server is not affected by these computations. The behavior of agents has been integrated in the client processes, as explained in Section 2. A hybrid navigation, composed of a two-modules model, has been used. On one hand, the high-level navigation module is in charge of pre-computing a set of paths from any cell to the exits. We have implemented this module as a Cellular Automata (CA). The number and length of the paths computed can be adjusted in order to reduce the memory used by the CA [6]. On other hand, a low-level navigation module determines how the paths computed by the CA should be followed. This module is implemented as a rule-based model [9, 12], and it allows to move agents in a continuum domain.

In this application example there are also static objects that agents should avoid. Since the rule-based navigation model implemented in the client process can provide agent positions colliding with static objects, some modifications are needed in the GPU collision checking algorithm. However, the performance of the GPU collision checking algorithm is not significantly affected by these changes, since the obstacles grid is computed and sorted before the simulation starts, and the grid data is kept in GPU memory during the whole simulation, avoiding memory transfers between CPU and GPU. Figure 4 shows a detailed 3D view of a congestion produced during the evacuation simulation. This example shows that the performance improvements shown in Section 3 can also be obtained when simulating agents with complex navigational behaviors. Moreover, these improvements are achieved without affecting the visual quality of the crowd.

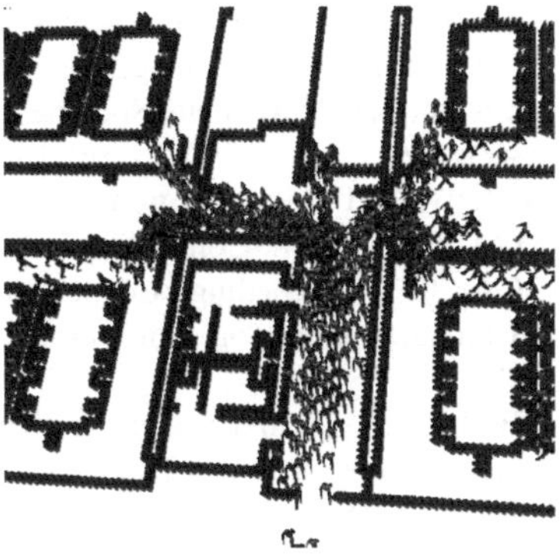

Fig. 4 3D snapshot of the evacuation scenario

5 Conclusions and Future Work

In this paper, we have proposed the implementation of a distributed server for crowd simulations using an on-board GPU. The huge number of cores in the GPU is used to simultaneously validate movement requests from different agents, greatly reducing the server response time. Since this represents the critical data path, the use of this hardware significantly increases the parallelism achieved with respect to the implementation of the same distributed server on a CPU. Thus, the system can support a significantly higher number of agents while providing the same latency levels.

As a future work to be done, we plan to study the performance improvements that the use of processors with a higher number of cores can provide.

Acknowledgements. This work has been jointly supported by the Spanish MEC and the European Commission FEDER funds under grants Consolider-Ingenio 2010 CSD2006-00046 and TIN2009-14475-C04-04.

References

1. Courty, N., Musse, S.R.: Simulation of large crowds in emergency situations including gaseous phenomena. In: CGI 2005: Proceedings of the Computer Graphics International 2005, pp. 206–212. IEEE Computer Society, Los Alamitos (2005)
2. Duato, J., Yalamanchili, S., Ni, L.: Interconnection Networks: An Engineering Approach. IEEE Computer Society Press, Los Alamitos (1997)
3. Harada, T., Tanaka, M., Koshizuka, S., Kawaguchi, Y.: Real-time rigid body simulation using gpus. IPSJ SIG Technical Reports 13, 79–84 (2007)
4. Henderson, T., Bhatti, S.: Networked games: a qos-sensitive application for qos-insensitive users? In: Proceedings of the ACM SIGCOMM 2003, pp. 141–147 (2003)
5. Latta, L.: Building a million particle system. In: Proc. of Game Developers Conference (GDC-2004) (2004)
6. Lozano, M., Morillo, P., Orduña, J.M., Cavero, V., Vigueras, G.: A new system architecture for crowd simulation. J. Netw. Comput. Appl. 32(2), 474–482 (2009), `http://dx.doi.org/10.1016/j.jnca.2008.02.011`
7. Lysenko, M., D'Souza, R.M.: A framework for megascale agent based model simulations on graphics processing units. Journal of Artificial Societies and Social Simulation 11(4), 10 (2008), `http://jasss.soc.surrey.ac.uk/11/4/10.html`
8. NVIDIA Corporation: NVIDIA CUDA Programming Guide, Ver. 1.1 (2007)
9. Pelechano, N., Allbeck, J.M., Badler, N.I.: Controlling individual agents in high-density crowd simulation. In: SCA 2007: Proceedings of the 2007 ACM SIGGRAPH/Eurographics symposium on Computer animation. Eurographics Association, Aire-la-Ville, Switzerland, pp. 99–108 (2007)
10. Perumalla, K.S., Aaby, B.G.: Data parallel execution challenges and runtime performance of agent simulations on gpus. In: SpringSim 2008: Proc. of Spring simulation multiconference, pp. 116–123. ACM, New York (2008)
11. Peter, K., Mark, S., Rudiger, W.: Uberflow: a gpu-based particle engine. In: HWWS 2004: Proc. of the SIGGRAPH/EUROGRAPHICS conference on Graphics hardware. ACM, New York (2004)
12. Reynolds, C.: Steering behaviors for autonomous characters. In: Game Developers Conference 1999, pp. 763–782 (1999)
13. Reynolds, C.: Big fast crowds on ps3. In: Proceedings of the ACM SIGGRAPH symposium on Videogames, pp. 113–121. ACM, New York (2006), `http://doi.acm.org/10.1145/1183316.1183333`
14. Rueda, S., Morillo, P., Orduña, J.M.: A comparative study of awareness methods for peer-to-peer distributed virtual environments. Comput. Animat. Virtual Worlds 19(5), 537–552 (2008), `http://dx.doi.org/10.1002/cav.v19:5`
15. Treuille, A., Cooper, S., Popovic, Z.: Continuum crowds. In: SIGGRAPH 2006: ACM SIGGRAPH 2006 Papers, pp. 1160–1168. ACM, New York (2006)
16. Vigueras, G., Lozano, M., Perez, C., Orduña, J.: A scalable architecture for crowd simulation: Implementing a parallel action server. In: Proceedings of the 37th International Conference on Parallel Processing (ICPP 2008), pp. 430–437 (2008) DOI 10.1109/ICPP.2008.20

CLIC: An Agent-Based Interactive and Autonomous Piece of Art

Laurent Lacomme, Yves Demazeau, and Julie Dugdale

Abstract. This work consists of integrating programming paradigms such as multi-agent systems and rule-based reasoning into a multimedia creation and display platform for interactive artistic creation. It has been developed in order to allow artists to build dynamic and interactive exhibitions based on pictures and sounds and featuring self-evolving and autonomous configurations.

1 Introduction

The project we present here is part of the biennial event called "Rencontres-i", organized by the Grenoble theatre Hexagone about art and science collaboration. Inspired by the behavior of social insects, the theme of this year's event is *gathering and swarming*. The event involves the participation of the MAGMA multi-agent systems team and a group of artists named Coincoin Production (www.collectif-coin.com) that has jointly coproduced the work. This collaborative project aims at using scientific work to produce a piece of art, matching the theme of the event, which represents the artists' views. This paper describes the scientific aspect of it.

The artistic idea for this work consists of presenting, reorganizing and developing the images and sounds that can be found on a place that is well-known to the visitors: the university campus. The result was a dynamic and interactive exhibition named CLIC – Conception d'un Logiciel Interactif Collaboratif (Conception of a collaborative and interactive software), which was shown to the public for three weeks. Thus, the computer science part of the work was to design and develop a software system, on top of the *Max/MSP/Jitter* media platform that is used by the artists, in order to design and build a multimedia exhibition. The goal was not only to build a system that can be used for this specific project, but to design a flexible and easy-to-use software that can be used by different artists to construct other multimedia exhibitions. This software has taken the form of a multi-agent system, where agents linked to media interact in several environments to produce commands for the multimedia platform.

Laurent Lacomme · Yves Demazeau · Julie Dugdale
Laboratoire d'Informatique de Grenoble – Grenoble INP, CNRS,
Université Pierre Mendès France
email: {Laurent.Lacomme,Yves.Demazeau,Julie.Dugdale}@imag.fr

Y. Demazeau et al. (Eds.): Advances in PAAMS, AISC 70, pp. 25–34.
springerlink.com

In the following section we firstly discuss the related works and the specificity of the project. The theoretical analysis of the artistic challenge and the design of a solution are presented in section three. Section four covers the practical problems that have appeared during the development as well as the solutions we have found. Section five describes the final software and the exhibition. Finally, we conclude with an evaluation of the work and draw some perspectives.

2 Related Works

Several works have already been achieved concerning the collaboration between art and artificial intelligence. From the beginning of AI, and the first expert systems, projects have been exploring the relation between artistic creation and the possibilities given by AI systems. One of the first projects was AARON [1], an expert system painter, which was able to create original drawings that represented scenes that were understandable and appreciated by humans. Although this project involved a creative artificial intelligence, the goal was to make it autonomously create finalized pieces of art. The project we are concerned with consists, on the contrary, in creating a dynamic – thus never completed – exhibition.

Projects that use AI paradigms such as cognitive reasoning and multi-agent systems for developing dynamic and interactive media production already exist. One of the most significant works is a series of dynamic self-evolving paintings named "Le Jardin des Hasards" [5], whose paintings are composed of several agent elements, which evolve with regard to the public activity and meteorological data, in a biological-inspired manner. Another significant work is a project named "M@trice @ctive" [4], which consisted of representing a painting from Kandinsky as a 3D environment in which shape-elements evolve and move according to initial conditions and interaction laws inspired from the painter's description of his own work; the visitor can navigate in this environment to discover new points of view on the painting.

Our project is different from these ones in the sense that it does not consist of modeling specific relationships between elements. Instead, it aims at providing a flexible, reusable platform for artists to create several dynamic and interactive exhibitions. It has also the goal of being easy to use by artists, without any help from computer scientists, whereas the above projects needed to be directly configured by computer scientists.

3 Theoretical Analysis

The artists wanted to build an exhibition that would be autonomous and dynamic. It means that the exhibition, controlled by a computer, was going to evolve by itself through time, without any direct action from the artists. Hence the main goal was to provide a *self-evolving*, *autonomous* and *dynamic* software component, whose behavior and evolution should be both scripted and unpredictable. Thus, the artists had to be able to set a global behavior for their exhibition, in order to express their creative views, but the precise evolution of the exhibition, such as the choice of one picture or another, or the display for a particular effect, should be determined by the software over time.

A good way to achieve this was to incorporate some *artificial intelligence* capabilities, so that the system can handle different situations and adapt its behavior, following goals or rules that reflect the artists' expectations. That would guarantee both variability – because of the non-algorithmic aspect in the software design – and stability – because of the ability of the software behavior to be matched to artists' wishes. That solution would also ensure another important property of the software: interactivity. Indeed, as the behavior of the exhibition is computed in real-time by the program, we aimed to provide the public with some feedback methods that could directly influence the exhibition's evolution. The main purpose of this is to catch the visitors' attention and to give them the feeling of being involved in the work's evolution.

Another difficulty has been to ensure that the program could also be easy to use for the artists, who are usually mostly unfamiliar with computer sciences. As the artists often have a precise idea of what their production should look like, they need to be able to understand a priori which global behavior would result from how they configure the system, at least to a certain extent. Hence, we had to link a media control system with a way for artists to express their preferences over the system behavior which should be easy to understand. The system should also be able to run in real-time, thus it should have a low computational complexity.

3.1 An Agent-Based Solution

The main problem was to link media control and artists' wishes. We needed a system which could handle a complex network of media elements linked with one another and work autonomously with it. These requisites and the participative design of the exhibition between artists and scientists have led to the choice of a solution that has already been used in artistic works [4, 5]: a *multi-agent system*. Multi-agent systems offer the possibility of maintaining a structured organization through time, referring to laws and rules for their behavior while giving varying results, depending on environmental events and internal configurations. Unlike other AI paradigms, multi-agent systems allow the user to configure locally the agents' behavior, which is simple and efficient, without having to specify the overall system's behavior. Moreover, the agents' autonomy and their evolving organization allow the system to have a coherent behavior through time – as agents' behavior is specified – while never returning to the exact same configuration.

The first question when designing a multi-agent system concerns identifying the agents. The solution we adopted was taken for its simplicity and its understandability for the artists more than for methodological reasons. It consists of associating each media – images, sounds or videos – with a single agent. This solution provides a way to distribute the control on the resulting work over local rules specific to sets of agents. Other solutions obviously exist, such as building a different agent for each simple behavior of the system, e.g. drawing, blinking, playing and looping. However, the chosen solution has the advantage of being easily understood by artists, who could manipulate sounds or images instead of unknown abstract entities.

3.2 VOWELS Methodology

As the system design is based on agents and their interactions, and as we have already defined what the agents should be, the VOWELS methodology [2] is a useful approach for designing the system. Following this methodology, we have analyzed the problem in terms of Agents, Environments, Interaction and Organizations. For each of these terms, we can specify the considered entities, methods, the expected behaviors and the imposed implementation restrictions.

Since *agents* are associated with media, their attributes must include their type (sound or image), a name or reference to their associated media and their duration when applicable. For their behavior to be adaptable, they require some memory. We have chosen a simple form of memory as a table of associated keys and values (*e.g.* "active" = "yes"). In addition to being simple to manage it is expressively powerful. Since the agents' behaviors must be understandable and computationally simple, *reactive agents* are used. They allow complex emergent behaviors with low computational load and less complex inputs than cognitive agents – as some of the inputs should come from the artists, they need to be kept simple.

As agents were associated to media, *environments* were logically associated to displays (screens and speakers). Each environment represents a set of displays of various types, in which the agents' interactions occur. When interacting, an agent can play its media on the displays included in its current environment that correspond to the type of its media (*i.e.* screens for an image, speakers for a sound).

An *interaction* between two or more agents can only occur within an environment if all these agents are situated in this environment. We limit the interactions to take place inside environments because of the conceptual idea of proximity: situated agents can only interact when they perceive each other, meaning here that they are located in the same environment. Interactions may change the agents' memory, make them move, stop or resume some action. Interactions also trigger outputs, which take the form of user-defined strings, for the system to command the media application. We define an *interaction pattern* as a model of interactions that consists of triggering conditions and effects of the interaction. Each interaction is then an instantiation of an interaction pattern.

Finally, the system's *organization* is defined by two elements: the agents' repartition in the environments and the interaction patterns that are defined and applied. The important point is that organization is the only element that the users – either the artists or the public through interaction – can modify in the system. To do this, users must then be able to move agents from one environment to another, to activate or to deactivate them, and also to control which interaction patterns each agent will use. This control is applicable both for the initial situation, and then in real-time, in order to provide a sort of control script to the system or to modify the system's evolution through interaction with the public.

3.3 AGR Interpretation

Another helpful approach for the system's design is the *Agent-Group-Role* paradigm [3], as it is a good complimentary approach to the VOWELS methodology. Following this method, we describe the system's organization along three dimensions:

agents that compose the system; groups, which are sets of interacting agents; roles, which are functions' templates agents can adopt in the system.

Firstly, agents are defined as media elements representatives. Since agents can only interact inside environments, we can associate the notion of groups to sets of agents that are in the same environment. Then, defining roles consists in linking agents to functions in the system, which should be semantically understandable for the artists. For more simplicity, we assume that agents have fixed roles at their creation and cannot change their roles during their existence. If we need a media to be linked successively to two or more incompatible roles, we create two agents for the same media and link each one with a different role. Hence, role definition can be done by associating a set of *tags* – simple user-defined strings such as "slideshow image", "brief sound", "blue image", etc. – with each agent. These tags are then used in interaction patterns' formulation.

4 Software Development: Problems and Solutions

Since the multi-agent system was to be part of a multimedia creation platform, Max/MSP/Jitter, there were many technical constraints. Firstly, the software had to be linked to an external API written in C, hence it had to be developed in C/C++ and then compiled as a dynamic library. Secondly, the software had to be usable on both Windows and Mac OS systems, because the software environment used by the artist was undefined at the time of development, and Max/MSP/Jitter exists on both operating systems. For that reason, the software had to be linked only with libraries that were available for both operating systems. Thirdly, only a few of all the possible agents would act at the same time, so we wanted to base their code on semaphores and monitors to minimize the computational load of the system; hence, we chose *asynchronous agents.*

Another problem has been the management of asynchronous inputs – commands given to the multi-agent system – and outputs – instructions passed from the system to Max/MSP/Jitter platform. Semaphores and monitors were sufficient to manage the distribution of inputs to appropriate agents or environments. Nevertheless, to give the system coherent orders – not passing a command to the platform and a contradictory one a split second later – we had to synchronize agents at two levels: environments and system. In each environment, agents must wait for every active one to propose an interaction before one is selected by highest priority. At the system level, commands to pass to the platform are added to a queue and sent out with a minimum delay in order not to saturate the platform capacity. This prevents the system from undertaking useless interactions (that occur but cannot produce a media output because the displays are already used by another interaction) and unnecessary computation. It is actually more efficient than a simple filter on the outputs because it requires less computation and because the delay it imposes on agents before they interact (waiting for others to compute their possible interactions) has no impact on the display, as it is only a few milliseconds long.

Another requirement is that the software configuration (entities and interactions' patterns) must be easy to define and to understand even for non-scientist digital artists, as it is the basis for all the exhibition's scripting. Hence, the system is initialized through two XML configuration files: one for agents, resources and environments and one for interactions' patterns. XML has been chosen because it is easily read and written by users. For scripting possibilities, users can move, activate and deactivate agents, and add or remove interactions' patterns to each agent in real-time, as defined earlier in the organizational description of the system.

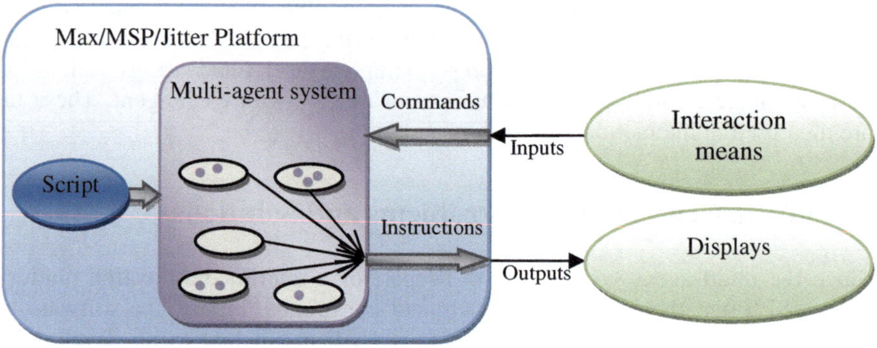

Fig. 1. System schema

Another major issue was to decide which behavior model should be chosen for the agents. The model has to be both simple to understand and efficient. It also needs to have the best expressive power possible, to allow more flexibility for the artists to choose the agents' behaviors. The agents have thus been given a *rule-based reasoning* capability [7]. This allows artists to easily understand the behaviors since the *interaction patterns* are coded by explicit rules. When agents interact, they take the needed resources for a given time, send commands to Max/MSP/Jitter platform and apply effects that are internal to the multi-agent system; then they release resources and resume checking for other applicable rules. A rule could be described by a sentence such as "WHEN (memory state "active" = "yes") and (a partner with tag "image" is present) THEN (set memory state "active" to "no") and (play sound for 5ms) and (display partner's image for 5ms)".

Rules are defined as a composition of: a set of conditions, a set of required resources, a set of effects to apply to agents in the system and a set of commands to pass to the media platform. Conditions are defined relative to the presence of a type (particular tag, meaning particular role) of agent or the value associated to a particular key in the agent's memory, which can be combined through Boolean operators. Effects (moving, activating, deactivating or changing memory values), and commands (user-defined character strings with wild-cards) can be delayed for a value between zero, meaning immediate, and the interaction's duration.

In order to compensate for the situations when rule-based reasoning could not provide the agents with the way to behave as expected (*e.g.* if an interaction through two random agents should occur only once in each environment, or only in one environment at a time), we have added the possibility of defining agents that are not linked to media. These agents are just considered as triggers for particular events, and can act as such when given the appropriate behavior rules.

5 Resulting Software and Exhibition

For the exhibition, the artists used three kinds of media: photographs of the campus, photographs of visitors to the exhibition, and sounds they had created. The hardware installation was composed of 3 computers, running the multi-agent system and controlling the displays, a buzzer and a keyboard for visitors' interaction, a camera to take photographs of the visitors when they entered the room, a mini-printer, 4 speakers and 6 LCD screens, one of them displaying a view of interactions taking place in the system and the others displaying the media.

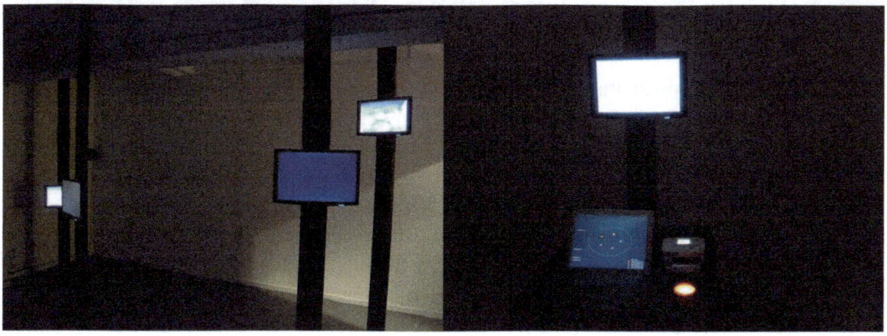

Fig. 2a and 2b The exhibition setup

Due to the number of screens available for media display, the software settings involved 5 environments, each corresponding to a screen and sharing all the speakers. The configuration also involved more than 500 agents, mostly linked to photographs, but with a few sounds and triggers. Only a maximum of 250 agents were activated simultaneously in the artists' script because of artistic needs.

About 30 different rules were defined in this configuration and used during the 4 steps of the script written by the artists. Approximately 30 different tags were used to describe the roles. During the first step, two agents (among about 40) representing groups of images interacted with a trigger to be selected. Then, image agents from the groups they represent interacted in pairs on each screen to be displayed simultaneously by applying some textural transformations on the images. During the second step, triggered by a timer, sounds interacted with images to display quickly the images on a screen while playing the sound, in order to give the feeling of an accelerating explosion of sounds and images. During the third step, also started by a timer, agents representing groups of images interact to trigger slideshows displays on each screen. Then, during the last step, triggered when anyone from the public presses the buzzer, an image agent, from a selection of about 200, interacts with a trigger to display an image. A member of the public has to enter a title for the image before it fades to its mean color, and a slideshow involving all titled images is displayed together with a representation of the set of all titles. Then the displayed single-colored picture and its associated keyword are printed as a summary of the visit (to the artists' point of view) and some rules are applied to the agents to reset some step-specific memory for another cycle.

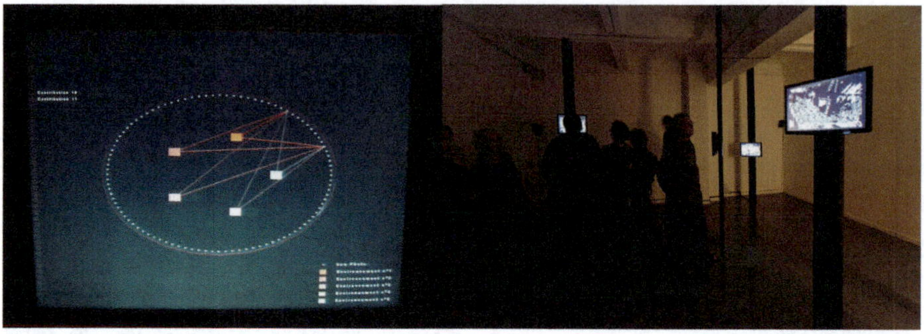

Fig. 3a and 3b Representation of the MAS's interactions and picture of the exhibition

6 Evaluation

As a result of the exhibition, we were able to evaluate if the software met its objectives or not. First of all, the system was able to run correctly in real-time on a dual core computer with a configuration of more than 200 agents acting at the same time. Whereas the system has not been tested with more than 500 agents, it seems that, on a reasonably powerful computer – a 3GHz quad-core – the program runs smoothly; so we can assert that the system is able to be used with enough agents for most artists' needs on a common computer.

Other evaluations resulted from questioning and interviewing the artists and the public (amongst were which other digital artists, able to evaluate more accurately the artistic work). Firstly, as far as we could see, the agent paradigm was easy to understand because of the association between agents and media, and the tagging system to define roles was also very simple for artists. However, devising the rules, and especially defining priorities and conditions was something theoretically hard to understand, and the loss of control over precise behavior of the system induced by its autonomy was something difficult to accept in the first place by artists. Manipulation of XML was not a major difficulty though, given a few sample files and explanation, even for non-specialist artists.

Additionally, we observed that the public found it difficult to understand the meaning of the interactions and the role of the multi-agent system despite the artists emphasizing the agent paradigm by displaying interactions on a separate screen and by giving the public a small written explanation of the project prior to the visit. Nevertheless, the exhibition generated good feedback from the public as well as the theatre representatives. Amongst the good points that were mentioned was interactivity. People saw themselves as part of the exhibition because it incorporated photographs of them and because they were invited to buzz at some time and enter words on the keyboard. Another mentioned good point was the non-repetitive displays due to the agents never having exactly the same configuration.

7 Conclusions and Perspectives

For this project, we have constructed a software component based on the multi-agent paradigm and integrated it into a media creation and control platform. This system allows artists, given a few explanations on how to construct rules, to easily build and script a dynamic exhibition based on sound and video display. Once constructed, this exhibition can be left on its own to evolve autonomously through time and react interactively to public presence or actions.

From the artists' and public's feedbacks, we can determine that the software component has been useful for artists. We can see that it allows them to create interesting possibilities of interaction and autonomy, which is something that the public would notice and appreciate. However, we can also see that, despite given explanations, the scientific part of the project is hard to understand for the public.

A first possible improvement concerns the positioning of agents. Currently there is no definition of a position for agents in the environments. That means that the only location we define for an agent is the actual environment in which it is situated. Adding some coordinates and maybe an environment topology could be a good way to improve the display abilities.

Another way to improve the system, regarding its interactivity, would be to change the way the public is considered. Currently, the public has the same abilities as the artists, i.e. they can add and remove agents and rules. However, they are limited by the interactions means, especially in this exhibition, where they could only interact by triggering portions of scripts that had been designed by the artists (i.e. they did not actually choose the modifications they wanted to apply to the system but only triggered those that have been coded by the artists). A powerful improvement would be to consider the public as one or more agents so they can interact with other agents by triggering rules, as agents interact with each other.

Finally, this work may be extended by integrating it into a larger piece of software concerned with assisting creativity [6]. The use of XML in the system makes this possibility easier, because it allows the software to smoothly interact with any potential creativity assistant (which could just manipulate the XML configuration files). This could allow artists to use intelligent assistants to help them defining their agents, the tags and the rules to obtain the configuration that fit their views the most.

Acknowledgments. We would like to acknowledge for their participation in the project the Grenoble theatre Hexagone, in particular Cécile Gauthier; the "Service Culturel de Grenoble Universités", specifically Bertrand Vignon; the members of Coincoin Production: Julien Castet, Mélanie Cornillac, Emilie Darroux and Maxime Houot; and of course Antoine Lefebvre, the engineering computer science student at ENSIMAG who programmed the multi-agent system.

References

1. Cohen, H.: How to draw three people in a botanical garden. AAAI 89, 846–855 (1988)
2. Demazeau, Y.: From Cognitive Interactions to Collective Behaviour in Agent-Based Systems. In: 1st European Conference on Cognitive Science, Saint-Malo, France, pp. 117–132 (1995)
3. Ferber, J., Gutknecht, O., Michel, F.: From Agents to Organizations: an Organizational View of Multi-Agent Systems. In: Giorgini, P., Müller, J.P., Odell, J.J. (eds.) AOSE 2003. LNCS, vol. 2935, pp. 214–230. Springer, Heidelberg (2004)
4. Gufflet, Y., Demazeau, Y.: Applying the PACO paradigm to a three-dimensional artistic creation. In: 5th International Workshop on Agent-Based Simulation, ABS 2004, SCS, Lisbon, Portugal, pp. 121–126 (2004)
5. Hutzler, G., Gortais, B., Drogoul, A.: Le Jardin Des Hasards: peinture abstraite et Intelligence Artificielle Distribuée réactive. In: Muller, J.-P., Quinqueton, J. (eds.) Journées Francophones d'Intelligence Artificielle Distribuée et Systèmes Multi-Agents, La Colle sur Loup, Hermès, Paris, pp. 295–306 (1997)
6. Lösch, U., Dugdale, J., Demazeau, Y.: Requirements for Supporting Individual Human Creativity in the Design Domain. In: Natkin, S., Dupire, J. (eds.) Entertainment Computing – ICEC 2009. LNCS, vol. 5709, pp. 210–215. Springer, Heidelberg (2009)
7. McDermott, J.: R1: A Rule-Based Configurer of Computer Systems. Artificial Intelligence 19(1), 39–88 (1982)

Collaborative Information Extraction for Adaptive Recommendations in a Multiagent Tourism Recommender System

Víctor Sánchez-Anguix, Sergio Esparcia, Estefanía Argente, Ana García-Fornes, and Vicente Julián

Abstract. In this paper we present an agent-based add-on for the Social-Net Tourism Recommender System that uses information extraction and natural language processing techniques in order to automatically extract and classify information from the Web. Its goal is to maintain the system updated and obtain information about third parties services that are not offered by service providers inside the system.

1 Introduction

Recommender systems are able to provide users with personalized information that covers their needs. There are many domains in which recommender systems have been applied, such as the tourism industry. Tourism recommender systems are able to provide tourists with custom recommendation based on their preferences[2, 6, 4]. One example of collaborative recommender system is the Social-Net Tourism Recommender System (STRS)[5], a tourism application that helps tourists to make their visits to a city more enjoyable and adapted to their preferences. It uses a mobile device, such a phone or a PDA, allowing tourists to make reservations in restaurants or cinemas, and providing a visit plan for a day. STRS integrates Multi-Agent technology and a recommender system based on social network analysis. It uses social networks to model communities of users, trying to identify the relations among them, to identify users similar to others so as to get a recommendation and recommend the items the users like the most.

Víctor Sánchez-Anguix · Sergio Esparcia · Estefanía Argente
Ana García-Fornes · Vicente Julián
Universidad Politécnica de Valencia
Departamento de Sistemas Informáticos y Computación
Grupo de Tecnología Informática - Inteligencia Artificial
Camí de Vera s/n 46022, Valencia, Spain
e-mail:{sanguix,sesparcia,eargente,
agarcia,vinglada}@dsic.upv.es

Y. Demazeau et al. (Eds.): Advances in PAAMS, AISC 70, pp. 35–40.
springerlink.com

However, some problems arise in most tourism recommender systems. First, each service provider is responsible for keeping up-to-date their business information in the recommender system. If it is not carried out by an automatic process, keeping information up-to-date is a costly task. Second, there may be third parties (tourist service providers that are not part of the system) which offer types of services that are not offered in system. The inclusion of information of third party services available on the Web may enhance recommendations given to tourists, and therefore improve the system. The main aim of this work is to create an add-on that can extract and keep up-to-date information from the websites of third parties and tourist service providers. Two are the advantages of the inclusion of the add-on. On the one hand, the tourism recommender system dynamically adapts its information. On the other hand, recommendations can be richer and thus more adaptable to the tourist profile. Wrapper agents and natural language processing based agents are used to extract and classify information respectively. The information classification process is performed by means of a voting process which is governed by a trusted mediator that can adjust the voting power of each classification agent.

The remainder of this paper is organized as follows. Section 2 gives a description of the add-on (the description for information extraction and information classification agents); section 3 presents the experiments used to test the system, showing the classification accuracy of the system and an experiment in which a simple update rule for the voting power is used; finally, section 4 presents some conclusions.

2 The Collaborative Information Extraction for Adaptive Recommendations Add-On

The proposed add-on for the STRS[5] is based on a collaborative strategy that extracts and classifies leisure services available on the Web. It is composed of two different types of agents: information extraction agents (IE agents) and information classification agents (IC agents). IE agents are needed to extract information that is available on the Web. We employ a set of collaborative agents that use natural language processing techniques (NLP) in order to classify information into a category of leisure service (concert, theater play, exhibition, etc...).

2.1 Information Extraction Agents

IE agents are designed following a wrapper architecture. Our wrapper agents transform the information available in a website into fields that are interesting for the application. In order to extract these specific fields, our IE agents examine the HTML structure of the website looking for specific patterns that point to the place where such fields are.

Even though some information has been extracted, some other fields may not appear explicitly. In our case, the event category needs to be inferred since many times it does not appear in the original website. Our IE agents send the event description they have extracted to an organization of IC agents, a team-based organization, and wait for their opinion. It must be noted that our wrapper agents were specifically created for some test websites. However, adapting wrappers to new websites does not pose a major problem since there are techniques that allow to generate wrappers automatically [3].

2.2 Information Classification Agents

Each IC agent is specialized in classifying into one specific event category. As a matter of fact, each IC agent gives a score that represents its confidence. The higher the score, the more confident the agent is in the classification of the service description into the agent's specializing category. In order to analyze how relevant a service description is with respect to a specific service category, each IC agent uses a rule based system based on NLP knowledge. Rules are applied over a preprocessed service description that only contains filtered words (using a stop-word list) and their corresponding lemma and lexical category. Two different types of rule can be applied over inferred lemmas:

1. Term Strength rules: Term Strength (TS) [7] is a measure of how relevant a word/lemma is with respect a specific category. More specifically, the TS of a word with respect a specific category can be calculated as follows:

$$TS(w_i) = \frac{\sum_{i=1}^{|D|} \sum_{j \neq i}^{|D|} occurs(w_i, d_i) * occurs(w_i, d_j)}{\sum_{i=1}^{|D|} \sum_{j \neq i}^{|D|} occurs(w_i, d_i)} \quad (1)$$

   where w_i is a word/lemma, $D = \{d_1, d_2, ..., d_{|D|}\}$ is a set of documents related to a specific category, and $occurs(w, d_i)$ is a function that returns 1 if the word/lemma w_i can be found in the document d_i and 0 otherwise. The TS of words with respect a specific category is precalculated using a corpus. We propose the following mechanism for TS rules: TS rules look for lemmas whose TS value has been precalculated during the training phase. If a match is found, the matched TS rule r_j produces a score $SC_{TS}(w_i)$ equal to the precalculated TS.
2. Hyperonym rules: Hyperonymy is the semantic relation between a more general word and a more specific word. In hyperonym trees, the root is the word/lemma that is analyzed whereas leaves are the most general words that are related to the root. It must be noted that tree nodes have different branches and each branch represents a different word sense. Additionally, branches are ordered according to the frequency of that specific sense. We propose the following mechanism for Hyperonym rules: specific patterns are searched in the hyperonym tree. These patterns should be indicated by an expert in the area. If the rule matches, it produces a score SC_H that is equal to:

$$SC_H(w_i) = \frac{|S(w_i)| - (Order(s_i) - 1)}{\sum_{k=1}^{|S(w_i)|} k} \quad (2)$$

where $S(w_i)$ is the ordered set of senses of w_i, s_i is the sense where the pattern was found, and $Order(s_i)$ calculates the position of s_i in the ordered set $S(w_i)$.

Each IC agent tries to apply each of its own rules to the filtered service description. Then, each word/lemma has an associated score that is equal to the rule that matched with that specific word/lemma, and produced the highest score. The score of the event description with respect to a specific category is the sum of the associated score of every word/lemma that is part of the event description. These statements can be formalized as follows:

$$SC(W) = \sum_{i=1}^{|W|} \max_{r_j \in R} SC(w_i, r_j) \tag{3}$$

$$SC(w_i, r_j) = \begin{cases} SC_{TS}(w_i) & \text{if } r_j \in \text{TS rule} \\ SC_H(w_i) & \text{if } r_j \in \text{Hyperonym rule} \end{cases} \tag{4}$$

where W is the set of filtered words/lemmas, R is the set of rules of the IC agent, r_j is a rule, and $SC(w_i, r_j)$ is the score produced by rule r_j when applied to w_i. All of the

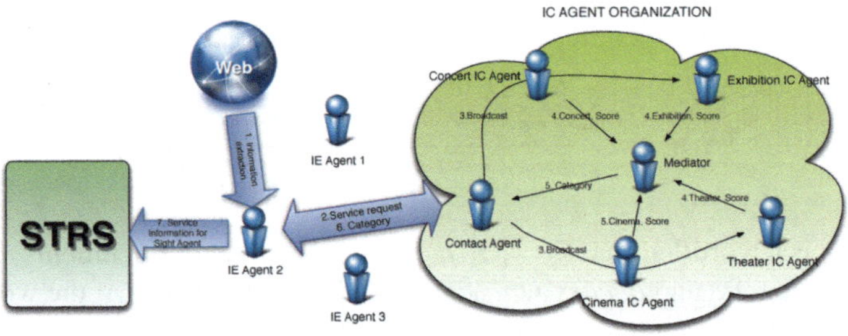

Fig. 1 1.An IE agent extracts a service description; 2.This agent requests a classification service; 3.The contact agent of the organization broadcasts the service call; 4. Each IC agent emits a vote/score and its specializing category to the trusted mediator; 5.The mediator agent decides which category to assign to the service description; 6.The classification result is sent back to the invoking IE agent; 7.This IE agent passes the service information to the STRS

IC agents form a team-based organization. The organization offers a classification service that takes an event description as an argument and returns the appropriate event category. A trusted mediator classifies the event description based on the opinion of the members of the team and their respective voting powers (adjusted by the mediator). The trusted mediator can adjust the voting power (vp) of each IC agent as follows:

$$Category(W) = \operatorname*{argmax}_{a_i \in IC} \; vp_{a_i} * SC_{a_i}(W) \tag{5}$$

where IC is the set of IC agents, and vp_{a_i} is the voting power that the mediator grants to the agent a_i. The complete architecture of the designed add-on can be found in Figure 1. It shows the whole process of information extraction, information classification and its integration in the STRS system.

3 Experiments

Two experiments were carried out in the implemented version of the extended STRS under the agent platform Magentix[1]. The goal of the first experiment was to test the classification accuracy of IC agents. Three service categories were selected as a testbed for this experiment: Concerts, exhibitions, and theater plays. A corpus of 600 service descriptions was built (200 event descriptions per category). The 70% of the corpus was used as training, whereas the other 30% was used for testing purposes. The voting power of each agent was fixed to w_{a_i} =1, and it remained static during the whole process. The test classification error using this approach was 11.11%, whereas a classic classification method such as term strength scored 17%. The second experiment was performed in order to observe how the IC Agent Organization deals with agents that have a bad behaviour, i.e. malicious agents or agents that have been badly designed. The three agents that were used in the first experiment were also used in this second experiment (music agent, theater agent, exhibition agent). Additionally, three malicious(or badly designed) agents that represent *music, theater, exhibition* categories were also introduced in the system. These malicious agents generate high scores with a high probability. The mediator checks agents' behaviours at intervals of 10 service calls. Then, it applies a decay on the voting power vp_{a_i} based on the behaviour of the agent in the past 10 service calls. The decay formula can be formalized as follows:

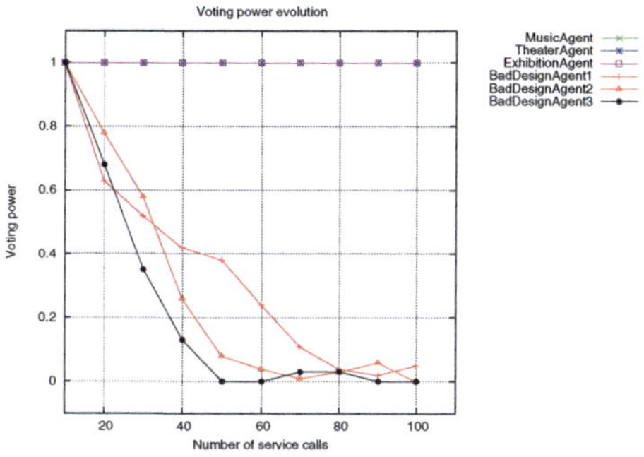

Fig. 2 Evolution of agents' voting power

$$vp_{a_i}^{t+1} = vp_{a_i}^{t} - \frac{FP_{a_i}}{|N_{other}|} + \frac{TP_{a_i}}{|N|} \quad (6)$$

where $vp_{a_i}^{t+1}$ is the new voting power, $vp_{a_i}^{t}$ is the voting power of agent a_i in the last check, FP_{a_i} is the number of times where the system decision was given by agent a_i and the correct service category was not the one a_i represents, TP_{a_i} is the number of times where the system decision was given by agent a_i and the correct service category was the one a_i represents, $|N|$ is the total number of service calls (10 in this case), and $|N_{other}|$ is the total number of service calls whose associated service category is not the one agent a_i represents. The experiment was run for 100 random service calls. The results of this experiment can be observed in Figure 2. This Figure shows the evolution of agents' voting power as the number of service calls increases. It can be observed how agents that were badly designed (malicious agents) had their voting power reduced to values close to zero, whereas the other agents's voting power remained at the maximum. Therefore, the agent organization was capable of reducing the voting power of those agents that introduced error in the system.

4 Conclusions

In this work, an add-on for the Social-Net Tourism Recommender (STRS) was presented. The designed add-on keeps information up-to-date and retrieves information about types of services (not found in the system) that are offered by third parties. The Web is used as a source of information for third party and system user services. In order to extract the desired information, two types of agents were created: information extraction agents (IE agents) and information classification agents (IC agents). IE agents are based on wrapper technology, and their goal is to extract the information that is required in order to keep the system up-to-date or extract information from third parties. IC agents are capable of scoring/voting the extracted information with respect to a leisure service category they are specialized in. They use natural language processing techniques to accomplish such task. All IC agents form a team-based organization where a trusted mediator governs a voting process where the category of the service is decided. The trusted mediator is capable of adjusting the voting power of each agent. Some experiments were carried out in order to test the performance of the add-on. First, the classification accuracy of IC agents was tested . Results show that the system classified 11.11% of the cases in test phase. Additionally, an experiment was carried out where a simple update rule for the voting power of each agent was used. Results show that badly designed agents have their voting power almost nullified by the mediator.

Acknowledgements. This work is supported by TIN2008-04446, TIN2009-13839-C03-01 and PROMETEO/2008/051 projects of the Spanish government, CONSOLIDER-INGENIO 2010 under grant CSD2007-00022, and FPU grant AP2008-00600 awarded to V.Sánchez-Anguix.

References

1. Alberola, J.M., Such, J.M., Espinosa, A., Botti, V., García-Fornes, A.: Scalable and efficient multiagent platform closer to the operating system. In: Artificial Intelligence Research and Development, vol. 184, pp. 7–15. IOS Press, Amsterdam (2008)
2. Fesenmaier, D., Ricci, F., Schaumlechner, E., Wober, K., Zanella, C.: Dietorecs: Travel advisory for multiple decision styles. In: Proc. of the ENTER 2003, pp. 232–242 (2003)
3. Kushmerick, N., Thomas, B.: Adaptive information extraction: Core technologies for information agents. In: Klusch, M., Bergamaschi, S., Edwards, P., Petta, P. (eds.) Intelligent Information Agents. LNCS, vol. 2586, pp. 79–103. Springer, Heidelberg (2003)
4. Loh, S., Lorenzi, F., Saldana, R., Litchnow, D.: A tourism recommender system based on collaboration and text analysis. Information Technology and Tourism 6, 157–165
5. Lopez, J.S., Bustos, F.A., Julian, V., Rebollo, M.: Developing a multiagent recommender system: A case study in tourism industry. International Transactions on Systems Science and Applications 4, 206–212 (2008)
6. Ricci, F., Werthner, H.: Case-based querying for travel planning recommendation. Information Technology and Tourism 4(3-4), 215–226 (2002)
7. Yang, Y.: Noise reduction in a statistical approach to text categorization. In: Proc. of the SIGIR 1995, pp. 256–263 (1995)

An Architecture for the Design of Context-Aware Conversational Agents*

David Griol, Nayat Sánchez-Pi, Javier Carbó, and José M. Molina

Abstract. In this paper, we present a architecture for the development of conversational agents that provide a personalized service to the user. The different agents included in our architecture facilitate an adapted service by taking into account context information and users specific requirements and preferences. This functionality is achieved by means of the introduction of a context manager and the definition of user profiles. We describe the main characteristics of our architecture and its application to develop and evaluate an information system for an academic domain.

1 Introduction

Ambient Intelligence systems usually consist of a set of interconnected computing and sensing devices which surround the user pervasively in his environment and are invisible to him, providing a service that is dynamically adapted to the interaction context, so that users can naturally interact with the system and thus perceive it as intelligent. To ensure such a natural and intelligent interaction, it is necessary to provide an effective, easy, save and transparent interaction between the user and the system. With this objective, as an attempt to enhance and ease human-to-computer interaction, in the last years there has been an increasing interest in simulating human-to-human communication, employing the so-called conversational agents [7]. Conversational agents have became a strong alternative to enhance computers with intelligent communicative capabilities with regard to the use of traditional interfaces, as speech is the most natural and flexible mean of communication among humans.

David Griol · Nayat Sánchez-Pi · Javier Carbó · José M. Molina
Group of Applied Artificial Intelligence (GIAA), Computer Science Department, Carlos III University of Madrid
e-mail: {david.griol,nayat.sanchez,
javier.carbo,josemanuel.molina}@uc3m.es

* Funded by projects CICYT TIN2008-06742-C02-02/TSI, CICYT TEC2008-06732-C02-02/TEC, SINPROB, CAM MADRINET S-0505/TIC/0255 and DPS2008-07029-C02-02.

Y. Demazeau et al. (Eds.): Advances in PAAMS, AISC 70, pp. 41–46.
springerlink.com

Adaptivity refers to several aspects in speech applications. In speech-based human-computer interaction users have diverse ways of communication. Novice users and experienced users may want the interface to behave completely differently, for example to have system-initiative instead of mixed-initiative. An example of the benefits of adaptivity in the interaction level can be found in [6]. The processing of context is essential in conversational agents to achieve this adapted behaviour and also cope with the ambiguities derived from the use of natural language [8].

In this paper we present an architecture for the design of conversational agents that provide personalized services by taking into account context information. In our architecture, agents handling different information sources and models interact to respond to the user's requests. Context information is used to provide a service that is then adapted to his location, geographical context, communication context, preferences, and needs. We provide a preliminary evaluation for the provisioning of personalized services to users in an academic information domain.

2 Our Architecture to Design Context-Awareness Conversational Agents

To successfully manage the interaction with the users, conversational agents usually carry out five main tasks: automatic speech recognition (ASR), natural language understanding (NLU), dialog management (DM), natural language generation (NLG) and text-to-speech synthesis (TTS). These tasks are usually implemented in different agents. Figure 1 shows a typical modular architecture of a conversational agent. Speech recognition is the process of obtaining the text string corresponding to an acoustic input. It is a very complex task as there is much variability in the input characteristics. Once the conversational agent has recognized what the user uttered, it is necessary to understand what he said. Natural language processing is the process of obtaining the semantic of a text string. It generally involves morphological, lexical, syntactical, semantic, discourse and pragmatical knowledge. The dialog management module updates the dialog context, provides a context for interpretations, coordinates the other modules and decides the information to convey and when to do it. Natural language generation is the process of obtaining texts in natural language from a non-linguistic representation. Finally, text-to-speech synthesizers transform the text into an acoustic signal.

As described in the introduction, context information is very valuable in order to enhance oral communication. For this reason, we have incorporated a context manager in the architecture of the designed conversational agents, as shown in Figure 1. This module deals with context information provided by the user and external positioning agents, and communicates this information to the different modules at the beginning of the interaction.

Kang et al [5] differentiate two types of context: internal and external. The former describes the user state (e.g. communication context and emotional state), whereas the latter refers to the environment state (e.g. location and temporal context). Most studies in the literature focus on the external context. In our case, context

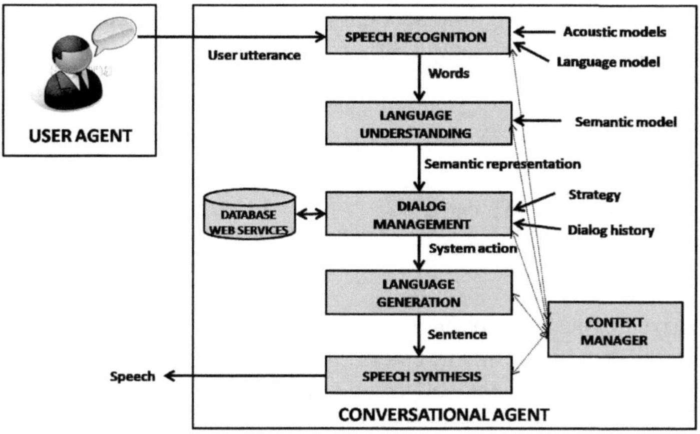

Fig. 1 Schema of the architecture of a conversational agent

information that is managed by the context manager includes both internal (user's information and preferences) and external information (user's identification and location into the environment provided by a positioning agent).

Once the context information has been obtained, it must be internally represented within the conversational agent so that it can be handled in combination with the information about the user interaction and the environment. Different models have been proposed in the literature to define this representation [4, 1]. In our proposal, context information is represented by means of user profiles that are managed by the context manager. The information included in the user profiles can be classified into three different categories: general user information (user's name, gender, age, current language, skill level when interacting with dialog systems, possible pathologies or speech disorders, user's preferences detected during previous dialogs, etc.), general statistics (number of previous dialogs and dialog turns, their durations, the date of the last interaction with the system, etc.), and usage statistics and user privileges (counts of each action over the system that a user performs, and a mark of user clearance for each possible action).

3 Domain Application

We have applied our context aware architecture to design and evaluate an adaptive system that provides information in an academic domain [2]. This information can be classified in four main groups: subjects, professors, doctoral studies and registration. The system must have gathered some data by asking the user about the name of the subjects, degrees, groups names, professors, groups, semesters, name of a doctoral program, name of a course, or name of a registration deadline. The way in which the user is queried for this information follows in most cases a system-directed initiative. A set of four different scenarios has been used to evaluate our proposal for this task, taking into account the four different queries that a user can perform to the system. An example of a scenario to obtain the office location of a professor is as follows:

```
User name: Patricia López
Location: Campus University, Main Building,
Side A, First Floor
Date and Time: 2009-11-03, 9:00am
Device: PDAQ 00-1C-41-32-0A-59
Objective: To know the location of the office of the professor
David Smith.
```

As summarized in the description of our architecture, a positioning agent is used to determine user's position while he enters the Wifi network in the University campus. Then, the conversational agent that provides the academic information (and has been previously detected) asks the user agent about information for his identification to provide the personalized service. Once this identification is received by the context manager included in the conversational agent, it loads the specific context profile characteristics. This information is then consulted by the rest of the modules in the conversational agent to personalize the provided service. For this example of scenario, this profile contains the following information:

```
Name: Patricia López
-------------------
Gender: Female | Age: 21 | Language: Spanish | Skill level:
High | Pathologies: None...
-------------------
Preferences: Computer Science, Tutoring Information...
-------------------
Current_Location: Main Building, Side A, First Floor
```

To evaluate our architecture, we have acquired a set of dialogs for each one of the four scenarios defined for the task, whether considering the introduction of the context manager in our architecture or not. A dialog simulation technique has been used to acquire a total of 300 successful dialogs. In this technique we automatically acquire a dialog corpus by means of the introduction in our architecture of a dialog manager simulator and a user simulator [3]. Both modules use a random selection of one of the possible answers defined for the semantic of the task. A user request for closing the dialog is selected once the system has provided the information defined in the objective(s) of the dialog. The dialogs that fulfill this condition before a maximum number of turns (defined after a study of real human-human dialogs for each specific task) are considered successful. Instead of building the user profiles automatically during several dialog sessions, we build them manually in order to demonstrate the goodness of our context management approach.

We defined seven measures for the comparison of the dialogs acquired using or not context information: the percentage of successful dialogs, the average number of turns per dialog, the percentage of different dialogs, the number of repetitions of the most seen dialog, the number of turns of the most seen dialog, the number of turns of the shortest dialog, and the number of turns of the longest dialog. Using these measures, we tried to evaluate the success of our approach with regard to the incorporation of context information in the dialog manager as well as its efficiency with regard to the different objectives specified in the scenarios.

Figure 2 shows a dialog for this task acquired without incorporating the context manager, and a dialog for the same scenario acquired using our approach. It can be observed how the different system answers can be personalized by taking into account the information provided by the context manager. The results showed that the system shows a tendency of answering with higher agility and more natural than without including the context manager.

S1: Welcome to the university information system. How can I help you?	*No prompts personalization.*
U1: I want to know information about the tutoring sessions of Professor David Smith.	
S2: Can you tell me the name of the subject?	*There is not information about the specific subject.*
U2: Language Processors.	
S3: The tutoring sessions of Professor Smith for Language Processors are Monday from 14:00 to 16:00 and Wednesday from 11:00 to 14:00. Anything else?	
U3: Can you tell me the office number?	
S4: The office number is 2.1.B.06 in the Main Building. Anything else?	*No information about the current position of the user.*
U4: No, thank you.	
S5: Thank you for using the university information system. Have a nice day!	

S1: Good evening Patricia, do you want to know tutoring information?	*Prompt personalization including the user's name. The system knows the preference of the user about consulting tutoring sessions.*
U1: Yes, I want to know the tutoring sessions of Professor Smith.	
S2: The tutoring sessions of Professor Smith for Language Processors are Monday from 14:00 to 16:00 and Wednesday from 11:00 to 14:00. Anything else?	*There is information about the specific subject.*
U2: Can you tell me the office number?	
S3: The office number is 2.1.B.06. You just have to walk along the Side A corridor and turn left. The office is the second door on the right. Anything else?	*The system uses current location of the user to personalize the answer.*
U3: No, thank you.	
S4: Thank you Patricia. Have a nice day!	*Prompt personalization including the user's name.*

Fig. 2 An example of a dialog for the academic domain without adding the context manager (above) and including this module in our architecture (below)

Table 1 shows the results of the comparison of the different measures for the academic information system. The first advantage of our approach is the number of dialogs that was necessary to simulate in order to obtain the total of 300 dialogs for the task. While, only a 3.7% of successful dialogs is obtained without using context information, this percentage increases to 17.5% when the context manager is introduced. The second improvement is the reduction in the number of turns. This reduction can also be observed in the number of turns of the longest, shortest and most seen dialogs. For this reason, the number of different dialogs is also lower using the context information.

Table 1 Results of the high-level dialog features defined for the comparison of the two kinds of dialogs for the academic information system

	Without Context Infor.	**Using Context Infor.**
Percentage of successful dialogs	3.7%	17.5%
Average number of user turns per dialog	4.99	3.75
Percentage of different dialogs	85.71%	67.52%
Number of repetitions of the most seen dialog	5	16
Number of turns of the most seen dialog	2	2
Number of turns of the shortest dialog	2	2
Number of turns of the longest dialog	14	12

4 Conclusions

In this paper, we have presented a multiagent architecture to develop context aware conversational agents. This allow us to deal with the increasing complexity that the design of this kind of systems requires, adapting the services that are provided by taking into account context information and user requirements and preferences by means of the introduction of a context manager. The results of the application of our architecture to evaluate an academic information system show how the main characteristics of the dialogs can be improved by taking into account context information. As a future work, we want to evaluate our system with real users and also carry out a study of the user rejections of system-hypothesized actions. Finally, we also want to apply our technique to carry out more complex tasks.

References

1. Doulkeridis, C., Vazirgiannis, M.: CASD: Management of a context-aware service directory. Pervasive and mobile computing 4(5), 737–754 (2008)
2. Griol, D., Callejas, Z., López-Cózar, R.: Acquiring and Evaluating a Dialog Corpus through a Dialog Simulation Technique. In: Proc. of the 9th SIGdial Workshop on Discourse and Dialogue, London, UK, pp. 326–332 (2009)
3. Griol, D., Hurtado, L., Sanchis, E., Segarra, E.: Acquiring and Evaluating a Dialog Corpus through a Dialog Simulation Technique. In: Proc. of the 8th SIGdial Workshop on Discourse and Dialogue, Antwerp, Belgium, pp. 39–42 (2007)
4. Henricksen, K., Indulska, J.: Developing context-aware pervasive computing applications: models and approach. Pervasive and mobile computing 2, 37–64 (2006)
5. Kang, H., Suh, E., Yoo, K.: Packet-based context aware system to determine information system user's context. Expert systems with applications 35, 286–300 (2008)
6. Litman, D.J., Pan, S.: Designing and evaluating an adaptive spoken dialogue system. User Modeling and User-Adapted Interaction 12, 111–137 (2002)
7. López-Cózar, R., Araki, M.: Spoken, Multilingual and Multimodal Dialogue Systems: Development and Assessment. John Wiley Sons, Chichester (2005)
8. Truong, H.L., Dustdar, S.: A survey on context-aware web service systems. International Journal of Web Information Systems 5(1), 5–31 (2009)

A Computational Model on Surprise and Its Effects on Agent Behaviour in Simulated Environments

Robbert-Jan Merk

Abstract. Humans and animals react in recognizable ways to surprising events. However, there is a lack of models that generate surprise intensity and its effects on behaviour in a realistic way, leading to impoverished and non-humanlike behaviour of agents in situations where humans would react surprised. To fill in this gap in agent-based modelling, a computational model is developed based on psychological empirical findings and theories from literature with which agents can display surprised behaviour. We tested this model in a simulated historical case from the domain of air combat and evaluated three behavioural properties against these simulated runs. The conclusion is that the model captures aspects of surprised behaviour and thus can help make agents behave more realistic in surprising situations.

1 Introduction

Surprise is considered a adaptive, evolutionary-based reaction to unexpected events with emotional and cognitive aspects (see for example [2], [14], [12]). Experiencing surprise has some effects on human behaviour, for example, expression through facial expressions [2] and the interruption of ongoing action [5]. However, there is little attention to the phenomenon of surprise in agent research and few agents have human-like mechanisms for generating surprise intensity and surprised behaviour (one exception for example is [7]) . This leads to impoverished and unrealistic behaviour of agents in situations where humans would react surprised.

In training simulations, there is an increasing demand for realistic computer-controlled actors for a number of reasons, such as cost effectiveness and the ability of having larger amounts of actors. One way to make these agents more realistic is to augment then with realistic models of surprise and the behaviour it influences.

The phenomenon of surprise has a more specific relevance in the military domain, besides its general importance as a basic human emotion. The element of surprise is considered an important factor in military operations by many military experts. Strategists such as Sun Tzu, F.C. Fuller and John Boyd have stressed the advantages of surprising the enemy [11].

Robbert-Jan Merk
National Aerospace Laboratory (NLR), Training, Simulation & Operator Performance. Anthony Fokkerweg 2, 1059 CM Amsterdam, The Netherlands
and
Vrije Universiteit Amsterdam, Department of Artificial Intelligence. De Boelelaan 1081, 1081 HV Amsterdam, The Netherlands

Y. Demazeau et al. (Eds.): Advances in PAAMS, AISC 70, pp. 47–57.
springerlink.com

From the previous paragraphs we can conclude that having realistic surprise models that are useable in simulation agents is useful. Because of this we propose in this paper a computational model that can be used in agents operating in training simulation that makes their behaviour more humanlike in surprising situations. The model is based on psychological empirical studies and is verified in a simulated scenario from the domain of military aviation against a number of properties.

2 Theory

One of the more influential models that explain the mechanisms behind how surprise intensity is generated in human is the *expectancy-disconfirmation* model [15]. According to the expectancy-disconfirmation theory, the main contributing factor to surprise is expectancy disconfirmation. In this view, people create expectations on how events in the world unfold. If they subsequently encounter an event that does not fall within their expectations, they will be surprised. This leads to an attribution process, a form of causal reasoning which leads to an attribution of the situation to certain causes in order to make sense of the situation. The duration of this causal attribution process depends on not only the surprise intensity but also other factors such as importance and valence of the surprising event. A number of experiments show that expectancy disconfirmation is indeed an important factor for surprise [15].

Several criticisms have been raised on the expectancy-disconfirmation model. They mainly contest the claim that expectancy-disconfirmation is the only factor that determines surprise intensity.

The experimental results shown in [4] show in a number of experiments that unexpected events that are seen as more important by a subject are experienced as more surprising. Also, failures are seen as more surprising than successes, establishing a correlation between the valence of an event and the intensity of surprise the event evokes. Further research confirms these findings.

Other research [8] shows that an unexpected events is seen as less surprising if the surprised person is offered a reasonable explanation that more or less justifies the occurence of the surprising event. This is explained by the authors as several experiments [13] show that amongst other factors, events that are familiar are less surprising. In the experinments, participants are more surprised if result contrast with earlier experiences. In other words, the event was more novel to the surprised person.

In conclusion, we have the following factors that influence the intensity of surprise:

1) expectation disconfirmation, 2) importance of observed event, 3) whether the observed event is seen as positive or negative (valence), 4) difficulty of explaining / fitting it in schema and 5) novelty (contrast with earlier experiences).

Besides the intensity of surprise, The effects of surprise on behaviour are explored in psychological research. Resulting from this research is that one of the main effects of surprise that is interesting for agents in training simulations is that it interrupt current activity and slows down the response to the surprising events because of the causal attribution process [5][15]. A consequent of this is that less time remains for actual decision making. As some studies have shown, decision making quality suffers under time pressure [9]. Especially in military tactical situations, decisions are made under considerale time pressure. It is reasonable to assume that a surprise indirectly leads to less quality in responding behaviour in time-critical situations such as military tactical situations. So we have two possible effects of surprise on behaviour that could be incorporated in agents: 1) slower response to surprising event compared to unsurprising event and 2) quality of response to surprising event is lower compared to unsurprising event.

3 Model

The model has been defined as a set of temporal relations between properties of states. A state property is a conjunction of atoms or negations of atoms that hold or do not hold at a certain time. The exact choice for what atoms to use depends on the actual model and domain and is defined by an ontology for that model. To model dynamics, transitions between states are defined.

In order to obtain an executable formal model, the states and temporal relations between them have been specified in LEADSTO [1], a temporal language in which the dynamic relations can be defined in the form of temporal rules that can be executed. Let α and β be state properties. In LEADSTO specifications the notation $\alpha \twoheadrightarrow_{e, f, g, h} \beta$, means:

if state property α *holds for a certain time interval with duration* g, *then after some delay (between* e *and* f*) state property* β *will hold for a certain time interval* h.

As all of the temporal relations used in the model are of the form $\alpha \twoheadrightarrow_{0,0,1,1} \beta$, the notation $\alpha \twoheadrightarrow \beta$ will be used instead. Intuitively, the symbol $\twoheadrightarrow$ can be read as an if-then rule, where the consequent holds at the next moment in time.

3.1 Model Overview

The surprise model can be divided into four parts: event evaluation, surprise generation, the sensemaking process and the effects of sensemaking on behaviour. In figure 1 an overview of the causal relations between the various states of the model is shown.

In the model, events in the environment are continually monitored and evaluated. This evaluation consists of determining the degree of expectation disconfirmation, how important the event is to the subject and how novel the event is.

This evaluation is then used to generate the surprise intensity. As the evaluation happens continually, this means that there is a surprise intensity value at any moment.

Based on the surprise intensity, the sensemaking process can be initiated, continued or halted. The sensemaking process is roughly analogous to the causal attribution process in the expectation-disconfirmation theory [15]. Its purpose is to revise the agent's beliefs on the current situation that have been invalidated by the surprising event. The sensemaking process has a feedback influence on surprise intensity, lowering it over time. This feedback represents the idea that the functional role of surprise is to regulate the sensemaking process. As sensemaking proceeds, the need for sensemaking decreases and likewise surprise intensity.

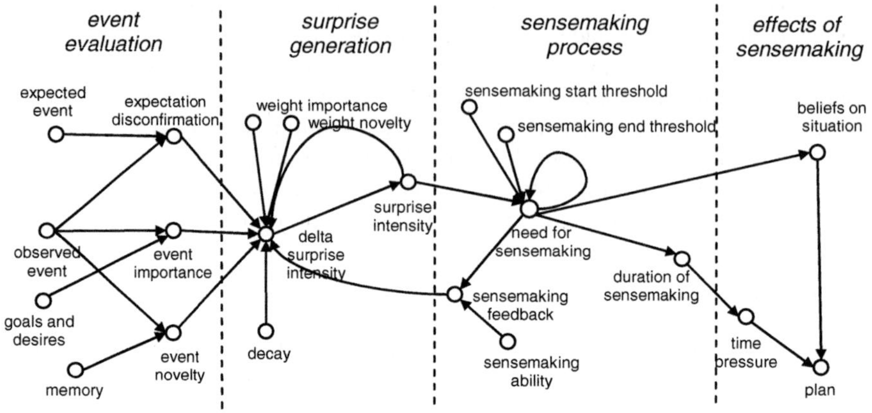

Fig. 1 Overview of the surprise model

The last part of the model deals with the effects of sensemaking on behaviour, represented by plans. The type and quality of the behaviour is determined by the beliefs the agent has and the time pressure, which rises with a longer sensemaking duration.

3.2 Event Evaluation

The three outcomes of the evaluation in our model are expectation disconfirmation, event importance and event novelty. These outcomes are represented by a real value between 0 and 1. We have not formalised the process that generate the evaluation outcomes as the focus of this paper is on generating surprise intensity and resulting behaviour. In this section we give some guidelines and ideas on how to interpret and generate these values.

The function of expectation disconfirmation in the model is to measure the degree of discrepancy between the expectations of the agent and the actual observed events. The higher this value, the more unexpected the event is to the agent.

Event importance measures the perceived impact the event has on the agent. A higher importance indicates that the event has relative farreaching consequences for the agent. Calculating the event importance can be done on basis of the goals, plans and desires the agent has, as well as other subjective aspects.

Event novelty gives an indication of how familiar an event is, how often the agent has experienced this situation before. A mechanism that links the agent's episodic memory on similar previous experiences with the observed event could be used for generating the value for event novelty.

3.3 Surprise Generation

Surprise intensity is represented as a real value between 0 and 1. In the model, the surprise intensity is not direcly calculated. Instead, the rate of change or derivative is calculated and this rate of change is then added to the current surprise intensity value. This rate of change is called the delta surprise intensity in the model.

The calculation of surprise intensity can then be informally described as follows:

> if currently the surprise intensity has value *si* and the delta surprise intensity has value *dsi*, in the next moment the surprise intensity will have the value *si* + *dsi*

More formally,

> surprise_intensity(si) & delta_si(dsi) →→ surprise_intensity(si + dsi)

The influences that determine surprise intensity that we identified in the previous section are used in the calculation of delta surprise intensity. The expectation disconfirmation, event importance and event novelty are the factors that increase surprise intensity. Two factors decrease surprise intensity, sensemaking feedback and decay. The idea behind the sensemaking process reducing surprise intensity is that the process in the model represents a cognitive effort to reduce surprise by explaining the event and fitting it in an revised view of the situation. The sensemaking feedback value represents the degree of succes of explaining the surprising event. In contrast to this, the decay factor represents the non-cognitive factors that reduce the intensity of emotions like surprise over time. Informally, the calculation of the delta surprise intensity occurs as follows:

> If currently the surprise intensity has value *si*, there is an expectation disconfirmation with value *ed*, the importance and novelty of the currently observed events have respectively the values *i* and *n*, the weights for importance and novelty have values *wi* and *wn* and th e decay parameter has value *d* and the sensemaking feedback has value *sf*, the delta surprise intensity for the next time step is determined by the formula

$$dsi = (1 - si) \cdot ed \cdot (wi \cdot i + wn \cdot n) - (si \cdot (d + sf)) \quad (1)$$

More formally, in LEADSTO format:

> surprise_intensity(si) & expectation_disconfirmation(ed) & weight_importance(wi) & importance(i) & weight_novelty(wn) & novelty(n) & sensemaking_feedback(sf) & decay(e) →→ delta_si((1 - si) · ed · (wi·i + wn·n) - (si · (d + sf)))

As formula (1) is an important part of the model, we will examine it in more detail. Formula (1) consists of the addition of two formulas, (2) and (3).

$$(1 - si) \cdot ed \cdot (wi \cdot i + wn \cdot n) \quad (2)$$

$$- (si \cdot (de + sf)) \quad (3)$$

Formula (2) is about the factors that increase surprise intensity while formula (3) represents the decreasing factors. Formula (3) is negated so that sensemaking and decay can be represented by positive values.

In formula (2) the expectation disconfirmation is multiplied with the sum of the importance and novelty factors that are themselves multiplied with their weight values. The reason for this construction consists of two assumptions: first the assumption that without expectation disconformation, there is no surprise. Second, the assumption that importance and novelty have a different effect in that they alone do not lead to surprise. For example, observing an important event that has been expected should not lead to surprise. These two assumptions are captured with formula (2). The weights *wi* and *wn* add up to 1, so that the outcome of *ed* ·

(wi · i + wn · n) alway lies between 0 and 1. With these weights, the relative influence between the two factors can be tuned.

We multiply *ed · (wi · i + wn · n)* with *(1-si)* in order to keep the value of surprise intensity below 1. As this value increases, the value of *(1-si)* decreases, reducing formula (2) and thus reducing the increase of formula (1). Including *(1-si)* ensures that the surprise intensity value changes smoothly over time.

In formula (3), the value obtained from the sensemaking process feedback and the decrease parameter are simply added. We multiply this addition with *si* for a similar reason as with the inclusion of the term *(1-si)* in formula (2): to keep the value of surprise intensity above zero.

3.4 The Sensemaking Process

The process of sensemaking is abstracted in our model. It is represented by two dynamic properties, the need for sensemaking and the sensemaking feedback.

The first property, the need for sensemaking, is represented by a boolean variable that is used to control the sensemaking process. If its value is true, the process is active and if false the process is inactive. The sensemaking process has two direct effects: it lowers surprise intensity by means of the sensemaking feedback and it causes beliefs the agent has on the situation to be revised.

Two parameters, the sensemaking start threshold and end threshold, deterimene wheen the need for sensemaking becomes true. The sensemaking process start as the surprise intensity rises above the start threshold, which causes the surprise intensity to drop. The sensemaking process continues until the surprise intensity falls below the end threshold. This mechanism represents the idea that the computationally costly process of sensemaking only takes place if a considerable surprise takes place and that this process endures until the feedback from sensemaking has reduced the surprise sufficiently.

Formally, the rules for determining the need for sensemaking are as follows:

```
surprise_intensity(si) & sensemaking_start_threshold(start_thr) & start_thr <= si
→ need_for_sensemaking(true)

surprise_intensity(si) & sensemaking_end_threshold(end_thr) & si <= end_thr
→ need_for_sensemaking(false)

need_for_sensemaking(currentValue) & surprise_intensity(si) & sensemak-
ing_start_threshold(start_thr) &
sensemaking_end_threshold(end_thr) & end_thr < si &si < start_thr
→ need_for_sensemaking(currentValue)
```

The second property, the sensemaking feedback, can only have two values: zero if the need for sensemaking is false and a value equal to the sensemaking ability parameter if the need for sensemaking is true. As we have no empirical support on the precise dynamics of surprise intensity, we have kept the mechanism for sensemaking feedback as simple as possible. In this mechanism, the sensemaking ability is a parameter than indicates how well sensemaking progresses. With this parameter it is possible to differentiate between skilled, experienced pilots and less experienced pilots.

3.5 The Effects of the Sensemaking Process

As explained in section 2, the occurrence of sensemaking has two effects on behaviour: delay in response and decline in response quality. The delay in reponse is implictly modelled in the model because the sensemaking process takes a number of time steps to complete. Only after the sensemaking process finishes can a response to the event be made. Time pressure is calculated by dividing the duration of sensemaking by the maximal possible duration, resulting in a value between 0 and 1. A higher time pressure results in a lower quality plan.

4 Case Study

In order to test the model, we constructed a case study loosely based on a historical event in air combat, Operation Bolo [6][10] Operation Bolo was a US Air Force (USAF) offensive operation during the Vietnam War against the North Vietnamese Air Force (NVAF). It is considered to be one of the most succesful surprise attacks in air combat history. The NVAF continually attacked the USAF bomber[1] missions in hit-and-run strikes, disengaging before the Americans could mount a counterattack. In response to this, colonel Robert Olds, and experienced fighter pilot in the USAAF, planned a deception. A number of fighters would fly in a bomber-like formation, have detectable bomber equipment with them, flew the standard route the bombers took and would carry anti-air missiles instead of bombs with them. Using these ruses, the Americans hoped that they could lure the NVAF fighters into open combat. The plan did indeed work. At January 2, 1967, 28 American fighters engaged 16 NVAF fighters, destroying 7 Vietnamese airplanes with no losses on the American side [6].

Although detailed information on the mission is hard to find, there is evidence that confirms that the Vietnamese pilots behaved surprised and that this had effects on the situation. One of the American pilots has stated that the NVAF pilots appeared to be confused[2]. Also, a military report on the operation states that "...however, the NVN air Force apparantly did not expect a strike and their reaction to Operation Bolo was much slower than anticipated" [3]. This is in line with the psychological research indicating that surprise leads to a slower response.

Based on this event, we constructed a case against which we can test our model. In this case, the agent plays the role of a Vietnamese fighter pilot. It expects to intercept an enemy bomber formation with no air-to-air missiles. Instead, it will encounter fighters well equiped for air combat. What our model should show is that an agent using the model will react in a worse way to this unexpected situation compared to a expected situation.

---

[1] Technically, most of the bombing runs were run with the F-105 Thunderchief, a fighter-bomber which was also capable of air-to-air combat. However, its air-to-air combat performance was inferior to the Mig-21, against which Bolo was aimed. The F-4 Phantom, the fighter that was used for Bolo, was more or less comparable in air-to-air combat to the Mig-21.

[2] In the History Channel documentary "Dogfights", season 1, episode 2 ("Air Ambush", 11/10/06), Robin Olds says that "They [the NVAF pilots] realized that we were not Thuds [nickname for F-105]...Mass confusion".

5 Simulation

The model described in the previous sections has been used to run a number of simulations, using the LEADSTO software environment as described in [1]. Within this software environment simulation traces (i.e., sequences of states) can be visualised. An example of such a simulation trace can be seen in Figure 2 and 3. Here, time is on the horizontal axis, the state properties are on the vertical axis. A dark box on top of the line indicates that the property is true during that time period, and a lighter box below the line indicates that the property is false.

An environment and scenario for the agent has been implemented based on the case described earlier. We programmed a simple mechanism for expectation generation in the agent. It is given a list of events that should occur after each other, representing a script or prototypical chain of events. As we based the case on Operation Bolo, this script mimics the historical events. The script is take_off, reach_interception_point, detect_aircraft and aircraft_recognition_bomber. The agent generate an expectation based on the first element in this list and generates an expectation based on the next element every time it observes an event.

The behaviour of the agent is represented by a plan. There are are three possible plans in this scenario: offensive_tactics_high_quality, offensive_tactics_medium_quality and offensive_tactics_low_quality. These represents the same offensive tactics that the Vietnamese displayed in the historical case, with different levels of quality[3] so that

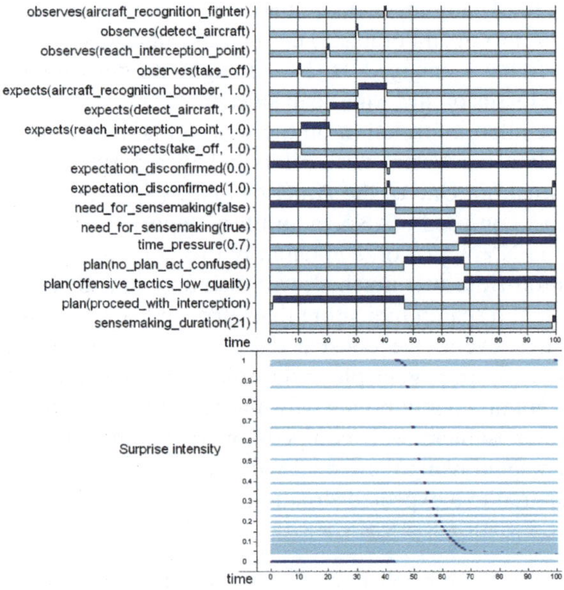

Fig. 2 Partial trace of the model reacting to a surprising event.(from LEADSTO software)

[3] Quality of response in this context is a measure of the lethality, survivability and resource controll of the behaviour. These three measurements are a standard way of evaluation military effectiveness of tactics and such.

the effect of surprise can be shown. In this simulation, the importance and novelty of events are parameters, as is sensemaking ability. An agent representing an experienced pilot has a low value for novelty (he has seen it all before) and a high value for sensemaking ability (experience improves situational evaluation).

To give an impression of how the model behaves, figure 2 shows the trace of a simulation as run in LEADSTO in which an inexperienced agent (high novelty, low sensemaking ability) encounters a surprising event: fighters instead of bombers. At the moment the agent sees that the enemy aircraft is a fighter, the expectation disconfirmation becomes 1 and coupled with a high importance and novelty values of 1.0, the surprise intensity rises. Coupled with a low sensemaking ability, the duration of the sensemaking is quite high (21 time steps out of a maximum of 30), so time pressure is quite high, lowering the quality of the response plan.

6 Evaluation

Three behavioural properties have been identified to evaluate the behavior of the proposed model. In order to test whether the model satisfies these properties, 8 differently configured simulations of the model have been run, the results of which can be seen in table 1. The first property is that if an agent observes an event which it did not expect, it will react slower and with a response of lower or equal quality than if the event was expected. As table 1 shows, trace 1 and 3 have identically configured agents. With surprise, there is a considerable delay and lower quality in response. Likewise with traces 2 and 5, a delay occurs with surprise.

The second property is that an agent representing a more experienced pilot will react faster and with a higher quality response to the same unexpected event than an agent representing a less experienced pilot. Traces 3, 4 and 5 illustrate this. Duration and quality level decrease with higher sensemaking ability and lower novelty.

Table 1 Simulation results. Expectation disconfirmation, event importance and event novelty refer to the aircraft_recognition events

trace	configuration description	sense-making ability	expectation disconfirmation	event importance	event novelty	duration of sense-making	quality of response[4]
1	no surprise, inexperienced agent	0.1	0.0	1.0	1.0	0	high
2	no surprise, very experienced agent	0.3	0.0	1.0	0.1	0	high
3	important surprise, inexperienced agent	0.1	1.0	1.0	1.0	21	low
4	important surprise, medium exp. agent	0.2	1.0	1.0	0.5	10	medium
5	important surprise, very experienced agent	0.3	1.0	1.0	0.1	6	high
6	unimportant surprise, inexperienced agent	0.1	1.0	0.1	1.0	16	medium
7	unimportant surprise, medium exp. agent	0.2	1.0	0.1	0.5	0	high
8	unimportant surprise, very experienced agent	0.3	1.0	0.1	0.1	0	high

[4] The quality of response is the quality of the agent's plan at the end of the trace.

The third property is that unimportant unexpected events do not result in a sensemaking process (and thus are effectively ignored by the agent). This holds for the medium experienced (trace 7) and very experienced configurations (trace 8), but not for the inexperienced configuration (trace 6). Further testing showed that no sensemaking takes place if the event novelty in trace 6 was lower than 0.9, which is still a reasonable parameter choice for representing an inexperienced pilot.

7 Discussion

This paper introduces a computational model for surprise generation and its effect on behaviour. A number of psychological theories and empirical studies found in the literature have been integrated into a single model. Verification shows that the model does indeed generate different behaviour in surprising events and with different representations of experience and importance evaluations.

There have been some research on computational models of surprise for use in agents. A notable example is S-EUNE [7], an agent architecture which uses surprise intensity to enable agents to explore unknown environments. S-EUNE differs from the model in this paper in many ways, most notably in the mechanism of surprise generation (in S-EUNE only expectation disconfirmation is used for calculating surprise intensity) and the lack of a sensemaking process.

The model presented in this paper can be used as part of an agent in simulated environments so that its behaviour is enriched with differentiated behaviour in case of surprising events. Additionally, the model incorporates the effects of experience and differences in personal capabilities in sensemaking, so that the generated surprised behaviour is further differentiated. While the model is relatively simple and there is room for improvement in for example the representation of the sensemaking process, it adresses the current lack of realistic models of surprise and its effect on behaviour in agent research.

Acknowledgements

This research has been conducted as part of the "SmartBandits" PhD project which is funded by the National Aerospace Laboratory (NLR) in the Netherlands. Furthermore, the author would like to thank his supervisors Mark Hoogendoorn, Jan Joris Roessingh and Jan Treur for their support and advice, and the anonymous reviewers for their comments that helped to improve the paper.

References

[1] Bosse, T., Jonker, C.M., Van der Meij, L., Sharpanskykh, A., Treur, J.: Specification and Verification of Dynamics in Agent Models. International Journal of Cooperative Information Systems 18, 167–193 (2009)

[2] Ekman, P., Friesen, W.V.: Unmasking the face. Englewood Cliffs, Prentice-Hall (1975)

[3] Futrell, R.F., et al.: Aces & Aerial Victories The United States Air force in Southeast Asia 1965-1973. The Albert F. Simpson Historical research center Air University (1976)
[4] Gendolla, G.H.E., Koller, M.: Surprise and Motivation of Causal Search: How Are They Affected by Outcome Valence and Importance? Motivation and Emotion 25(4), 327–349 (2001)
[5] Horstmann, G.: Latency and duration of the action interruption in surprise. Cognition and Emotion 20(2), 242–273 (2006)
[6] Isby, D.C.: Fighter Combat in the Jet Age. HarperCollins Publishers, London (1997)
[7] Macedo, L., Cardoso, A., Reisenzein, R.: A surprise-based agent architecture. In: Trappl, R. (ed.) Cybernetics and Systems, vol. 2. Austrian Society for Cybernetics Studies (2006)
[8] Maguire, R., Keane, M.T.: Surprise: Disconfirmed Expectations or Representation-Fit? In: Proceedings of The 28th Annual Conference of the Cognitive Science Society, pp. 1765–1770 (2007)
[9] Mann, L., Tan, C.: The Hassled Decision Maker: The Effects of Perceived Time Pressure on Information Processing in Decision Making. Australian Journal of Management 18, 2 (1993)
[10] Michel, M.L.: Clashes: Air Combat over North Vietnam 1965-1972. US Naval Institute Press (1997)
[11] Niederhauser, G.A.: Defeating surprise? Naval War College Newport, R. I. Report number AD-A279 463 (1994)
[12] Plutchik, R.: Emotions: A general psychoevolutionary theory. In: Scherer, K.R., Ekman, P. (eds.) Approaches to emotion, pp. 197–219. Erlbaum, Hillsdale (1984)
[13] Teigen, K.H., Keren, G.: Surprises: low probabilities or high contrasts? Cognition 87(2), 55–71 (2002) (March 2003)
[14] Scherer, K.R.: Emotion as a multicomponent process: A model and some cross-cultural data. In: Shaver, P. (ed.) Review of personality and social psychology, pp. 37–63. Sage, Beverly Hills (1984)
[15] Stiensmeier-Pelster, J., Martini, A., Reisenzein, R.: The role of surprise in the attribution process. Cognition and Emotion 9, 5–31 (1995)

[illegible]

Enhanced Deliberation in BDI-Modelled Agents

Fernando Koch and Frank Dignum

Abstract. Applications that operate in highly dynamic environments must deal with real-time changes of circumstances in order to be consistent and coherent. In this work, we propose an extended deliberation cycle for BDI-modelled agents that takes advantage of environmental events to coordinate the deliberation process. Our argument is that is it possible to optimise the deliberation performance by exploiting features of windows of opportunity. The proposed model let us define precisely the expected application behaviour that promotes the balance between reactive and proactive applications, leading to solutions with improved computational performance.

1 Introduction

Computing systems that operate in highly dynamic environments are growingly ubiquitous. The requirements to support mobility, operate in open environments, and rich end-user interface impose new levels of complexity in software design. This is the case for solutions for mobile services, robotics, and ambient intelligence for instance. They share, in different degrees, the issues of: highly dynamic environments, end-user interfacing, and constraint resources.

Our goal is to outline a model for intelligent services that operate in these environments. This model will help to classify, understand, and anticipate services' behaviours. We apply the *Belief-Desire-Intention model* [Rao and Georgeff(1995)] as the base technology suggesting that this model provides desirable features, such as: language for knowledge representation; outlines an enhanced inference system that inherently provides responsiveness and adaptivity, and; it is designed to develop systems equipped with sociability and interaction [Koch and Rahwan(2005)]. The BDI technology has been applied for solutions that require responsiveness balanced to some degree of proactive behaviour, such as robot control and factory equipment coordination. We propose to improve the behaviour of applications developed upon this technology even further, by equipping the deliberation process with the

Fernando Koch · Frank Dignum
Department of Information and Computing Sciences, Universiteit Utrecht, The Netherlands
e-mail: fkoch@acm.org, dignum@cs.uu.nl

Y. Demazeau et al. (Eds.): Advances in PAAMS, AISC 70, pp. 59–68.
springerlink.com

structures to observe and make sense of contextual information. We aim to optimise the deliberation performance by exploiting features of the "windows of opportunity" [Koch(2009)]. The solution considers environment information, provides a flexible deliberation cycle, and exploits the trade-off between quality of information and resource efficiency.

The paper is organised as follows. Section 2 details our motivation. Next, section 3 introduces our proposed model and section 4 presents a proof-of-concept implementation and discusses the results from simulations. Section 5 summarises our conclusions and future work.

2 Motivation

Our research investigates architecture and methods for building intelligent mobile services using agent-based technology. These are end-user support applications that collect, represent, and process the required information autonomously and implement user-friendly interfaces that hide the complexity of the underlying computational process [Maes(1994)]. Let us consider the problem scenario depicted in Figure 1 as an illustrative example.

> A user has a personal mobile assistant that helps him with task and time management activities. The application must operate in a fairly autonomous manner with regards to the user behaviour, but interacts to solicit necessary information, and to inform and confirm decisions.
>
> Assume a meeting about Project A is booked for 10am at office A110. At 9:53am, when the user is within two minutes walking distance from office A110, the application starts to retrieve information about the project, intending to notify the user of any updates before they step into the meeting (position (i)). However, enroute to the meeting room, the user detours to a sideline meeting with a colleague (position (ii)). The colleague's application interacts with the user's application to exchange relevant information about the project. Depending on how this conversation evolves, the user

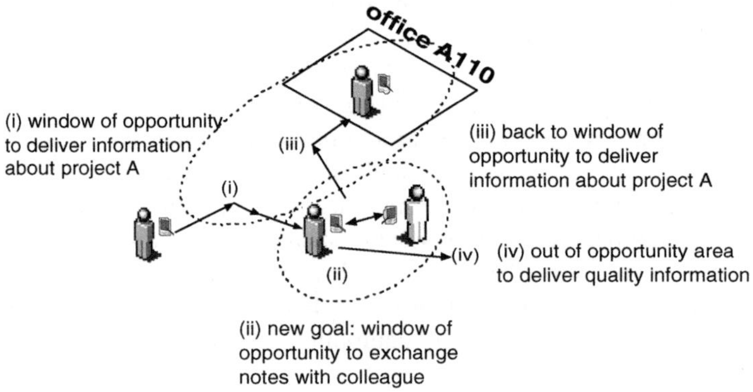

Fig. 1 Conceptual Problem Scenario

might decide to cancel his participation in the original 10 am meeting, in which case the application no longer needs to deliver information for Project A.

To fulfil this scenario, the application must be able to sense and represent the environment, keep a representation of the user's preferences, and adapt its behaviour to the evolving scenario. If the application is not able to adapt its processing line, then it will behave incoherently, such as delivering information when the user is out of the window of opportunity. For instance, what should the application do when the user moves into position (iv)? We consider two cases:

- over-committed applications that do not reconsider their processing – such as *bold agents* [Pollack and Ringuette(1990)] – would disregard the change of circumstance and deliver the information once retrieved; at position (iv), that information would be perceived as incoherent and useless.
- over-reactive applications analyse the environment and reconsider their goals continuously – such as *cautions agents* – will act correctly at the expense of computing processing.

In this example, the window of opportunity is represented in terms of time and space. However, the same concept extends to other dimensions. [Graham and Kjeldskov(2003)] proposed to segment context information into eight dimensions, which we translate in "questions" that can be used to assert the current situation: (i) *time*:. i.e. is the task still feasible in time?; (ii) *absolute location*, i.e. is the position still valid?; (iii) *relative location*, i.e. is the position still valid?; (iv) *objects present*, i.e. are the objects $o_1,\ldots,o_n$ present?; (v) *activity*, i.e. is the user engaged in the activities $a_1,\ldots,a_n$?; (vi) *social setting*, i.e. is the user part of the social setting segments $ss_1,\ldots,ss_n$?; (vii) *environment*, i.e. are the conditions $c_1,\ldots,c_n$ valid?, and; (viii) *culture*, i.e. is the user immersed in the cultural setting $cs_1,\ldots,cs_n$?

That is, *proximity* can be derived from the conditions inferred based on the questions above. For example, the application can infer that the user is *in* th window of opportunity if the user is physically near some position (absolute location) and the application can access some files (objects present) for a meeting book in a few minutes (activity). We suggest that the combination between observation and action at a lower-level improves reactiveness, saves resources, and preserves its proactive features. We present and discuss this approach next.

3 Proposed Solution

Figure 2 depicts the proposed solution. We highlight the presence of two main components, which sets the agenda for this presentation:

- (i) *Context Observer* (left box), which is the module specialised in observing environmental events, filtering relevant changes, and coordinating the deliberation process.
- (ii) *Planning Module* (right box), which implements the deliberation process *per se*; in our case, we proposed a BDI-modelled deliberation engine enhanced with:

parallel intention deliberation, checkpointing (i.e.. pause/resume deliberation), prioritisation, conflict resolution, and other features.

First, let us assume that the operational logic of BDI-modelled agents is based on the semantics provided by the 2APL Agent Programming Language, described in [Dastani et al(2007)Dastani, Hobo, and Meyer]. An *agent* is the tuple $\langle i, \mathscr{B}, \mathscr{D}, \mathscr{I}, \Pi \rangle$, where: i is the agent's identifier; $\mathscr{B}$ is the *belief base*; $\mathscr{D}$ is the *desire base* composed of desire tuples $\langle \varphi, C, redo \rangle$, where $\varphi \in \mathsf{Goal}$, C is the *commitment condition*; the parameter *redo* marks if the goal should be reprocessed after the goal is achieved; the set of goals $\mathsf{Goal} \subset \mathscr{L}$, with $\mathscr{L}$ the sentences in a standard propositional logic; $\mathscr{I}$ is the *intention base*, where intention is the tuple $\langle \varphi, C, p \rangle$, $\varphi \in \mathsf{Goal}$, C is the supporting condition, and p is the planning thread, detailed below, and; Π is the set of practical reasoning rules, composed of plan selection rules *PS* and the set of planning rules in the format $\varphi \leftarrow C \mid \pi$, $\varphi \in \mathsf{Goal}$, C is the *triggering condition*, $\pi \in \mathsf{Plan}$. and; *PR* is a set of planning rules in the format $\pi \leftarrow C \mid \pi'$, $\pi, \pi' \in \mathsf{Plan}$, C is the *reviewing condition*.

We represent intention execution in *planning threads* aiming to support parallel intention deliberation. A thread is a tuple $\langle st, \pi, \Theta \rangle$ where *st* is the operation status as a value in the set $\{run, pause\}$, $\pi \in \mathsf{Plan}$ is the current plan, and Θ is the local unification stack. The unification stack Θ contains the set of substitutions used to resolve the terms. The issues on parallel intention execution – i.e. conflict resolution, prioritisation, etc. – are discussed along with intention deliberation.

Context Observer. As mentioned, the *context observer* implements the functionality to observe the environment and signalise the plan deliberation process to reconsider the execution of current plans. It works as follows. Let us consider that β is an update of the belief base configuration due to the execution of a basic action; that is, if the execution of the action a upon the belief base $\mathscr{B}$ results in the new configuration $\mathscr{B}'$, then $\beta = \mathscr{B}' \backslash \mathscr{B}$. The relevance filters ρ_i are stored in $\mathscr{R}$. The implementation provides the *function to extract relevance filters*, which

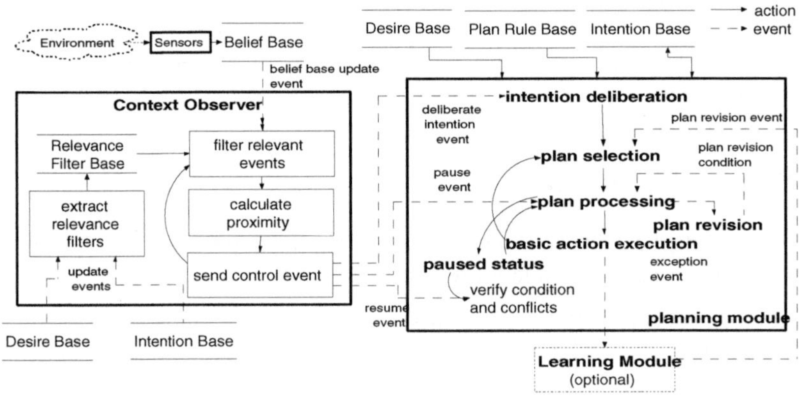

Fig. 2 Design Model

populates the relevance filter base by processing events related to assertion of desires and intentions, that is:

- when a desire $d = \langle \varphi, C, Redo \rangle$ is asserted, it processes: $\forall c_i \in C : \langle c_i, d \rangle$ to $\mathscr{R}$
- when an intention $i = \langle \varphi, C \rangle$ is asserted, it processes: $\forall c_i \in C : \langle c_i, i \rangle$ to $\mathscr{R}$.

The function to *filter relevant events* processes operation events. That is, for every update of a belief base entry β, if $\langle \beta, E \rangle \in \mathscr{R}$, then calculate the impact of β over the deliberation of the element E. Let us consider that the implementation provides a *function to calculate proximity* as the meta-operators: $in(d)$ and $in(i)$ are true if the condition of the desire d or the intention i evaluates as *in* the window of opportunity, and; $out(d)$ and $out(i)$ is true if the condition of the desire d or the intention i evaluates as *out* of the window of opportunity. The meta-operator $\mathsf{impact}(\beta, \varphi, V_r)$ that assesses the impact of the belief β on the processing of goal φ, as:

- $\mathsf{impact}(\beta, \varphi, false)$ if the belief β has no impact on the processing of φ;
- $\mathsf{impact}(\beta, \varphi, start) \iff \langle \beta, \langle \varphi, C, R \rangle \rangle \in \mathscr{R} \wedge in(\langle \varphi, C, R \rangle)$; that is, the configuration β leads to the execution of the desire $\langle \varphi, C, R \rangle$ iff the condition of the desire evaluates as *in* the window of opportunity.
- $\mathsf{impact}(\beta, \varphi, pause) \iff \langle \beta, \langle \varphi, C \rangle \rangle \in \mathscr{R} \wedge \langle \varphi, C, \langle run, \pi, \Theta \rangle \rangle \in \mathscr{I} \wedge out(\langle \varphi, C \rangle)$; that is, the configuration β leads to pausing the execution of the running planning thread iff the condition of the intention evaluates as *out* of the window of opportunity.
- $\mathsf{impact}(\beta, \varphi, resume) \iff \langle \beta, \langle \varphi, C \rangle \rangle \in \mathscr{R} \wedge \langle \varphi, C, \langle pause, \pi, \Theta \rangle \rangle \in \mathscr{I} \wedge in(\langle \varphi, C \rangle)$; that is, the configuration β leads to resuming the execution of the paused planning thread iff the condition of the intention evaluates as *in* of the window of opportunity.

Intention Deliberation. The agent can commit to multiple goals at the same time, as long as they are not conflicting. Let us assume that there is a meta-operator $A \nleftrightarrow B$ that reads "*element A conflicts with element B*". Let us also assume that there is a meta-operator $a >_p b$ that reads "*a has a higher priority than b*". Then, there are three possible operations for intention deliberation:

1. if there are no conflicts between the desired goal φ and any other goals in the intention base, then the intention is asserted and a new planning thread is created:

$$\frac{\langle \varphi, C, redo \rangle \in \mathscr{D}, \mathsf{impact}(\beta, \varphi, start),\ \mathscr{B} \models C\theta,\ \mathscr{I} \not\models \varphi\theta,\ \nexists \varphi_i : \mathscr{I} \models \varphi_i \wedge (\varphi \nleftrightarrow \varphi_i)}{\langle \mathscr{B}, \mathscr{D}, \mathscr{I} \rangle \rightarrow_\theta \langle \mathscr{B}, \mathscr{D}, \mathscr{I} \cup \{\langle \varphi\theta, C, \langle run, \varepsilon, \Theta \rangle \rangle\} \rangle}$$

2. if there is a conflict between φ and any φ' in the intention base, and φ has a lower priority than φ', then the intention is asserted and a new planning thread is created with paused status:

$$\frac{\langle \varphi, C, redo \rangle \in \mathscr{D}, \mathsf{impact}(\beta, \varphi, start),\ \mathscr{B} \models C\theta,\ \mathscr{I} \not\models \varphi\theta,\ \exists \varphi_i : \mathscr{I} \models \varphi_i \wedge (\varphi_i \nleftrightarrow \varphi) \wedge (\varphi_i >_p \varphi)}{\langle \mathscr{B}, \mathscr{D}, \mathscr{I} \rangle \rightarrow_\theta \langle \mathscr{B}, \mathscr{D}, \mathscr{I} \cup \{\langle \varphi\theta, C, \langle pause, \varepsilon, \Theta \rangle \rangle\} \rangle}$$

3. if there is a conflict between φ and any φ' in the intention base, and φ has a higher priority than φ', then the process switches the running planning threads:

$$\frac{\langle \varphi, C, redo\rangle \in \mathscr{D},\ \mathsf{impact}(\beta,\varphi,start),\ \mathscr{B} \models C\theta,\ \mathscr{I} \not\models \varphi\theta,\quad \exists \varphi_i : \mathscr{I} \models \varphi_i \wedge (\varphi_i \not\leftrightarrow \varphi) \wedge (\varphi >_p \varphi_i)}{\begin{array}{c}\langle \mathscr{B},\mathscr{D},\mathscr{I} \cup \{\langle \varphi_i, C_i, \langle run, \pi_i, \Theta_i\rangle\rangle\}\rangle \rightarrow_\theta \\ \langle \mathscr{B},\mathscr{D},\mathscr{I} \backslash \{\langle \varphi_i, C_i, \langle run, \pi_i, \Theta_i\rangle\rangle\} \cup \{\langle \varphi_i, C_i, \langle pause, \pi_i, \Theta_i\rangle\rangle \langle \varphi\theta, C, \langle run, \varepsilon, \Theta\rangle\rangle\}\rangle\end{array}}$$

If $\mathsf{impact}(\beta, \varphi, pause)$, then the deliberation cycle must pause the processing of φ:

$$\frac{\mathsf{impact}(\beta,\varphi,pause), \langle \varphi, C, \langle run, \pi, \Theta\rangle\rangle \in \mathscr{I}}{\langle \mathscr{B},\mathscr{D},\mathscr{I}\rangle \rightarrow \langle \mathscr{B},\mathscr{D},\mathscr{I} \backslash \{\langle \varphi, C, \langle run, \pi, \Theta\rangle\rangle\} \cup \{\langle \varphi, C, \langle pause, \pi, \Theta\rangle\rangle\}\rangle}$$

If $\mathsf{impact}(\beta, \varphi, resume)$, then the deliberation cycle must resume the processing of φ:

$$\frac{\begin{array}{c}\mathsf{impact}(\beta,\varphi,resume), \mathscr{B} \models C\theta,\ \forall \varphi_i : \mathscr{I} \models \varphi_i \wedge \neg(\varphi_i \not\leftrightarrow \gamma), \\ \forall \langle \varphi_j, C_j, \langle pause, \pi_j, \Theta_j\rangle\rangle \in \mathscr{I}, \varphi \neq \varphi_j : (\varphi_j <_p \varphi),\ \mathsf{revise}(\Theta, \mathscr{B}, \Theta')\end{array}}{\langle \mathscr{B},\mathscr{D},\mathscr{I}\rangle \rightarrow \langle \mathscr{B},\mathscr{D},\mathscr{I} \backslash \{\langle \varphi, C, \langle pause, \pi, \Theta\rangle\rangle\} \cup \{\langle \varphi, C, \langle run, \varepsilon, \Theta'\rangle\rangle\}}$$

This operation considers that the implementation provides the meta-operator $\mathsf{revise}(\Theta, \mathscr{B}, \Theta')$ to revise the contents of the unification stack Θ with regards to the current situation $\mathscr{B}$ resulting Θ' with revised values. This operation can be implemented via programming constructs in the supporting logic.

Plan Processing. Next step is to process the plans. The function for *plan selection* selects the best plan rule from the set or practical reasoning rules Π whose planning thread's goal implies the plan rule's header and the belief base implies the plan rule's condition. Let us assume that there is a meta-operator $\mathsf{findprs}(t, P)$ that selects the planning rules in Π whose head match the element t and a meta-operator $\mathsf{mincost}(P, \pi*)$ that selects the plan rule $\pi* \in P$ with minimum cost, as:

$$\frac{\mathsf{findprs}(\varphi,P),\ \mathsf{mincost}(P, \langle \varphi_h \leftarrow C \mid \pi_b, \eta, \theta\rangle),\ \mathsf{unify}(\varphi,\varphi_h,\eta),\ \mathscr{B} \models C\theta}{\langle \mathscr{B},\mathscr{D},\mathscr{I}\rangle \rightarrow \langle \mathscr{B},\mathscr{D},\mathscr{I} \backslash \{\langle \varphi, C, \langle run, \varepsilon, \Theta\rangle\rangle\} \cup \{\langle \varphi, C, \langle run, \pi_b\eta\theta, \Theta'\rangle\rangle\}\rangle}$$

The unifier θ contains the bindings produced by the evaluation of the plan rule's guard C' using the substitutions in η from the unification of the head. The evaluation of the guard may compute new bindings θ for free variables in $\varphi_h\eta$. The operation adjusts the unification stack to Θ' by adding the set of substitutions.

Next, the operation for *plan processing* continues to execute the plan by resolving the "next step" of current plans. The term t_h in the plan's head can be either $t_h \in \mathsf{AbstractPlan}$ or $t_h \in \mathsf{BAction}$. For the first situation, if $\langle \varphi, C, \langle run, t_h; \pi, \Theta\rangle\rangle \in \mathscr{I}$ is a planning thread:

$$\frac{\begin{array}{c}t_h \in \mathsf{AbstractPlan},\ \mathsf{findprs}(t_h,P),\ \mathsf{mincost}(P, \langle \pi_h \leftarrow C \mid \pi_b, \eta, \theta\rangle), \\ \mathsf{unify}(t_h,\pi_h,\eta),\ \mathscr{B} \models C\eta\theta\end{array}}{\langle \mathscr{B},\mathscr{D},\mathscr{I}\rangle \rightarrow \langle \mathscr{B},\mathscr{D},\mathscr{I} \backslash \{\langle \varphi, C, \langle run, t_h; \pi, \Theta\rangle\rangle\} \cup \{\langle \varphi, C, \langle run, (\pi_b; \pi)\eta\theta, \Theta'\rangle\rangle\}\rangle}$$

The unifier θ contains the bindings produced by the evaluation of the plan rule's guard C using the substitutions in η from the unification of the head. This operator may compute new bindings θ for free variables in $\pi_h \eta$. The substitutions apply to both parts of the plan, thus $(\pi_b;\pi)\eta\theta$. The operation adjusts the unification stack to Θ' by adding the set of substitutions.

For the second situation, if $t_h \in \mathsf{BAction}$, the execution of basic actions update the belief base through the partial function $\mathscr{T}$, where $\mathscr{T}(a,\mathscr{B}) = \mathscr{B}'$ returns the result of updating the belief base by performing the action a. In this case, the problem of unexpected events in dynamic environments cannot be overlooked. It is possible that basic actions fail to execute due to instabilities or unexpected situations. The application can detect the error condition either via an external logic or programming constructs.

Plan Revision. Finally, *plan revision* is a required component to support the deliberative behaviour of BDI-model agents. The rationale is as follows: agents are situated in some environment which can change during the execution of the agent, requiring flexibility of the problem solving behaviour, i.e. the agent should be able to respond adequately to changes in its environment. The operation is defined below.

$$\frac{\pi_h \leftarrow C \mid \pi_b \in PR,\ \pi = \pi_h\theta,\ \mathscr{B} \models C}{\langle \mathscr{B},\mathscr{D},\mathscr{I}\rangle \rightarrow \langle \mathscr{B},\mathscr{D},\mathscr{I}\backslash\{\langle \varphi,C,\langle run,\pi;\pi',\Theta\rangle\rangle\}\cup\{\langle \varphi,C,\langle run,(\pi_b;\pi')\theta,\Theta'\rangle\rangle\}\rangle}$$

It works as follows: if there is a plan revision rule whose head π_h is implied by the head of the current plan $\pi;\pi'$, resulting in the set of substitutions in θ. Then, the body of the plan revision rule replaces the head of the current plan, and; the substitutions in θ apply to the new plan. The unification stack is adjusted to Θ' by adding the substitutions in θ.

We argue that these structures equip the application to better operate in multi-purpose, open environments. Next, we describe the result of simulation case studies to demonstrate this allegation.

4 Case Study

In this section, we introduce a simulation environment and run a case study to demonstrate the performance enhancement obtained with the proposed model. The experimental system is based on a proof-of-concept implementation of the proposed operation logic (discussed in the previous sections), called *2APL-M*. We built a simulation environment that works as an extension of the well-known *TileWorld applications*, called “the MobileWorld”.

We use the following agent configurations: *bold agents* never reconsiders its options before the current plan is executed; *cautious agents* considers the impact of environmental changes each timer cycle, and; *model agents* implements the deliberation model introduced in this work.

Figure 3 presents the behaviour of the different reconsideration strategies. The (A) bold agent fails to provide convenient information, delivering the notification for goal $M1$ before the notification for $M2$ and out of the window of opportunity. Both the *cautious* and *model agents* respect the window of opportunity by switching from $M1$ to $M2$ respecting the window of opportunity. Hence, these agents deliver similar qualitative results.

The *computational performance* can be inferred by the number of operations required to resolve the game. We opted for a solution based on counting the number of unifications executed by the agent's internal logic. The number of unifications can be directly mapped to the number of CPU instructions executed providing a realistic measure of the application's performance.

Figure 4(A) provides the average number of unifications required to resolve the game for variations of degree of dynamism. We concluded that *cautious agent* present the worst computational performance, with resource utilisation growing proportionally to the degree of dynamism. This can be explained as cautious agents are calculating the conditions of the existing elements for every environmental update. On the other hand, *model agent* presents a performance in between the other two, as it is able to filter the relevant updates before processing the elements' conditions. Figure 4(B) depicts the relative performance gain showing that the advantage grows proportionally to the degree of dynamism.

Figure 5(A) presents the average number of unifications required to resolve the games for variations of degree of relevance, which is the number of environmental changes that actually impact the agent's elements. We concluded that *cautious agents* provide steadily measures. This happens because these configurations are not taking relevance in consideration, processing every update even if not relevant. However, the performance of *model agents* reduces proportionally to the degree of

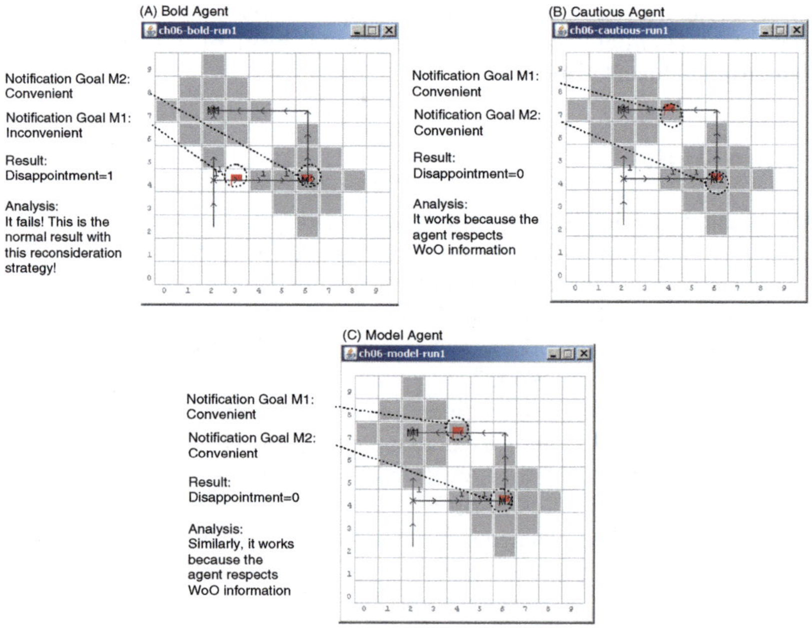

Fig. 3 Effect of Reconsideration Strategy

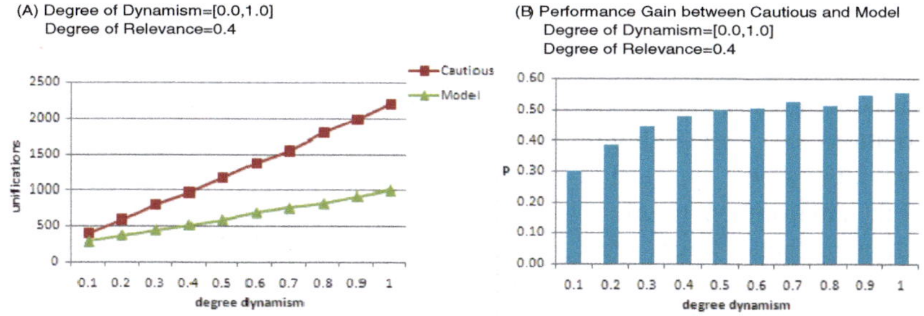

Fig. 4 Performance for Different Degrees of Dynamism

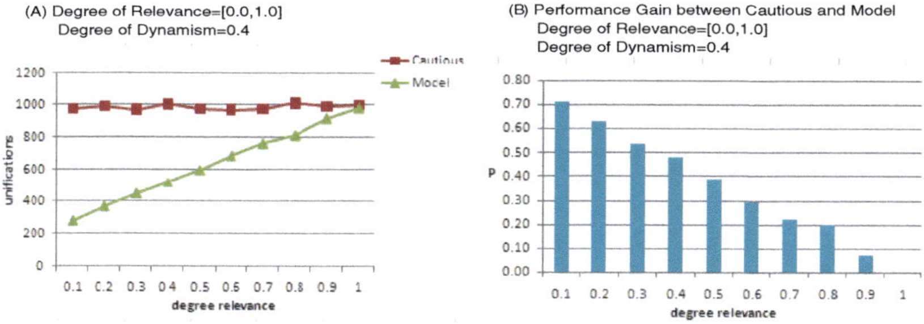

Fig. 5 Performance for Different Degrees of Relevance

relevance. This is because the more relevant the environmental updates, the more likely they will affect the elements' conditions. Figure 5(B) presents the relative performance gain between the *cautious* and *model agents* for variations in degree of relevance.

The results underline the gain in computation performance provided by the proposed model. We conclude that *model agents*' performance gain is more prominent in highly dynamic environment with more relevant events. Moreover, applications operating using this model (1) save computing power and (2) acts coherently by avoiding executing action at inopportune moments, as the execution is paused when the agent moves out of the window of opportunity.

5 Conclusion

Our research sought a model for intelligent services that operate in highly dynamic environments aiming to classify, understand, and anticipate services' behaviours. We claim that applications built upon the proposed model will behave better when implementing solutions for these environments, such as mobile services. The optimisation in deliberation performance means that less CPU-cycles are required to run the application, which reflects in the application's overall performance. In the case of mobile services, it means for instance reduction in battery utilisation. The solution contributes to the theory by:

- Introducing an extended BDI-model of agent computing that provide inherent solutions to operate in highly dynamic environments.
- Providing a new approach to allow agent-based applications to exploit features of the environment to infer when to revise current processing lines.
- Allowing that different agent platforms could be adapted to support the proposed design to reduce CPU cycles and increase overall application performance.

Moreover, the technology can be applied to implement applications in different knowledge domains that involve dynamic environment, use context information for decision making, and mix between reactive and proactive behaviour. That is the case of robotics, ambient intelligence, and others.

In future work we intend to exploit this proposal in different domain: mobile social networks. We will be looking into: how to extend mobile services' functionality by using social-interation information, and; how the proposed model contributes to support this development.

References

Dastani, M., Hobo, D., Meyer, J.J.: Practical Extensions in Agent Programming Languages. In: Press, A. (ed.) Proceedings of the Sixth International Joint Conference on Autonomous Agents and Multiagent Systems, AAMAS 2007 (2007)

Graham, C., Kjeldskov, J.: Indexical representations for context-aware mobile devices. In: Proceedings of The IADIS e-Society Conference, pp. 373–380. IADIS Press, Lisbon (2003)

Koch, F.: An agent-based model for the development of intelligent mobile services. PhD thesis, Utrecht University (2009)

Koch, F., Rahwan, I.: The role of agents in intelligent mobile services. In: Barley, M.W., Kasabov, N. (eds.) PRIMA 2004. LNCS, vol. 3371, pp. 115–127. Springer, Heidelberg (2005)

Maes, P.: Agents that reduce work and information overload. Communications of the ACM 37(7), 31–40 (1994)

Pollack, M., Ringuette, M.: Introducing the Tileworld: Experimentally Evaluating Agent Architectures. In: Dietterich, T., Swartout, W. (eds.) Proceedings of the Eighth National Conference on Artificial Intelligence, pp. 183–189. AAAI Press, Menlo Park (1990)

Rao, A.S., Georgeff, M.P.: BDI-agents: from theory to practice. In: Proceedings of the First International Conference on Multiagent Systems, San Francisco, USA (1995)

Cooperative Behaviors Description for Self-* Systems Implementation

Hiroyuki Nakagawa, Akihiko Ohsuga, and Shinichi Honiden

Abstract. Agent platforms have recently attracted attention for use as a basis for self-* systems development because it provides a mechanism for autonomous functionalities. From among these platforms, JADE allows developers to describe concurrent behaviors on it, and this can be a foundation of constructing the multi-processes on self-* systems. This paper shows an overview of our approach for implementing highly collaborative behaviors by introducing a component style behavior model and its life cycle on the basis of the agent platform.

1 Introduction

Agent technologies are one of the more effective approaches for dealing with the recent increase in software complexity because of its autonomy. In particular, self-* systems such as self-managed [4], self-adaptive [1], and self-healing [2] systems, require autonomous mechanisms and some researches make use of agent platforms [7, 5, 8]. Among the several existing agent platforms, JADE provides behavioral descriptions that enable the running of concurrent processes on individual agents, and this could be a fundamental approach to the construction of self-* systems that usually contain multi-processes, such as for monitoring their environments and internal states, for managing their goal achievement states, and for executing their functionalities. However, the current JADE does not provide enough methods for controlling and combining behaviors, and therefore it is still difficult to construct self-* systems.

Hiroyuki Nakagawa · Akihiko Ohsuga
The University of Electro-Communications, Tokyo, Japan
e-mail: nakagawa@is.uec.ac.jp, ohsuga@is.uec.ac.jp

Shinichi Honiden
National Institute of Informatics, Tokyo, Japan
e-mail: honiden@nii.ac.jp

Y. Demazeau et al. (Eds.): Advances in PAAMS, AISC 70, pp. 69–74.
springerlink.com

Our approach builds upon the idea of mapping behaviors into components. We introduce a component style behavior class and it helps to enforce a uniform interface for the behavior cooperation and autonomous activations. These features allow developers to effectively construct self-* systems.

2 Background

Self-* systems consist of processes with various functions, such as monitoring their environments and internal states, managing their goal achievement states, and executing their functionalities, and these processes interact together inside the systems. To realize these processes, we define the following requirements for implementing self-* systems.

Self-* systems process implementation requirements:

- **Concurrent execution:** The execution environment should execute multiple processes concurrently.
- **Process cooperation:** The programming framework should provide rich APIs for process interaction and control.
- **Autonomous activation:** The programming framework should provide an API that processes can automatically activate by checking the designated conditions.

We try to realize self-* systems implementation framework based on JADE [13]. JADE is an agent platform that helps developers to construct agents described in Java language. One of the main features of JADE-based development is that JADE enables developers not only to describe multiple agents but also to describe multiple concurrent tasks for individual agents. JADE provides some extended behavior classes inheriting the `Behaviour` class, and developers can implement their own behaviors by inheriting one of these classes. Fig. 1 shows an example of the JADE behavior description. In many cases, developers using JADE override the *action* methods, which repeatedly execute, and the *done* methods, which express at-end conditions. In this example, the implemented class inherits SimpleBehaviour (*line 8*), which provides the standard API set, increments a variable as an iterative task (*lines 16–19*), and checks the variable for the at-end condition (*line 20*). An agent is assigned two instances of the behavior (*lines 3–4*), in this example, and these two behavior instances are executed concurrently until each at-end condition is satisfied.

Even though JADE provides behavior descriptions that enable the running of concurrent processes on individual agents and satisfies the *concurrent execution* requirement, it is insufficient for description of the behaviors of self-* systems that have to interact with each other and react to their environment. In particular, we can identify the following two difficulties with the description of behaviors.

Behavior cooperation: While the JADE framework provides the mechanism for describing multiple behaviors and assigning them to an agent, it does not provide a way to reference the other behaviors' states. Moreover, it does not allow for safe state transitions. While JADE provides the *block* method that mandatorily suspends the behavior, the currently running behavior cannot execute suitable evacuation

```
public class SampleAg extends Agent{
  protected void setup(){
    addBehaviour(new Prt(this,"p1"));
    addBehaviour(new Prt(this,"p2"));
  }
}

class Prt extends SimpleBehaviour{
  int n=0;
  String name;

  public Prt(Agent a, String st){
    super(a);
    name = st;
  }
  public void action(){
    System.out.println("name: "+name);
    n++;
  }
  public boolean done(){return n>=5;}
}
```

Fig. 1 Example of behavior description in JADE

processes before being suspended. As a result, certain problems may occur when the behaviors have to interact together. For instance, when several behaviors use a service provided by the same behavior, conflicts or inconsistencies may occur in the service providing behavior.

Autonomous activation: While JADE provides the *done* method to specify an at-end condition, for the activation, it just provides an activation mechanism by specifying the delay time or by explicitly invoking in an agent setup method or in other behaviors' action methods. Developers cannot explicitly specify the activation conditions.

3 Constructing Behaviors Using Component Model

We aim to establish a development framework for building self-* systems satisfying the requirements described in Section 2. In our framework, developers build an architecture model representing components and the connections between them first, and then implement these components by inheriting our extended behavior class in order to construct an agent corresponding to the self-* system. This section explains our framework by using the development of a cleaning robot as an example, the descriptions of which can be found in previous studies such as [7].

In order to describe the relationship between behaviors, our development framework uses an architecture model that consists of several components and their connections. In particular, we use the updated Darwin model [3] as the component model. The Darwin Model is also used in the lowest layers of the three-layer architecture [4] for self-* systems. The main characteristic of this Darwin model is that components have two port types, one for providing services and the another for using other component services, and the system architecture is expressed by connecting these ports. In addition, the components have an externally visible state, called a *mode*. We utilize this mode for visualizing the behavior states and component connections for grasping the dependencies between behaviors.

Fig. 2 is an architecture model for a cleaning robot. For example, this model shows that the "DealWithDust" component uses two services provided by the "ApproachDust" and "CleanField" components in order to provide its own service. This model is used in several studies for developing self-* systems, and we have

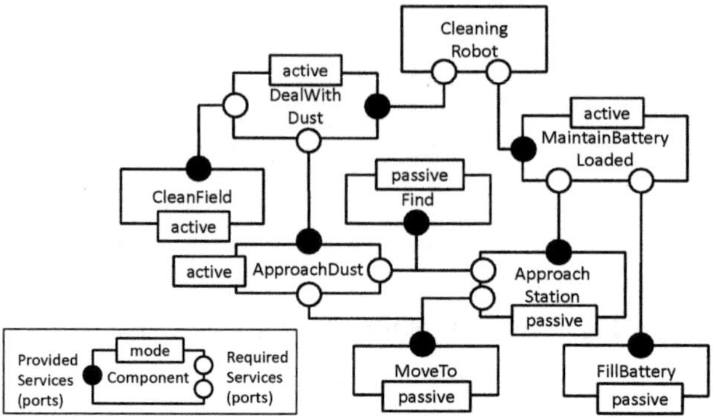

Fig. 2 Example architecture model: architecture for cleaning robot

proposed a method for determining such architecture models from the requirements descriptions in [9]. In our framework, we map each component in the architecture model into each behavior to be implemented one-to-one.

Developers implement behaviors on the basis of the architecture model. We introduce *ComponentBehaviour* class, which is inherited SimpleBehaviour class, in order to support behavior implementation. Fig. 3 shows the class diagram of this behavior class. Developers are able to implement behaviors that satisfy the requirements described in Section 2 by just inheriting from the ComponentBehaviour class.

The ComponentBehaviour class has the mode variable and reveals the behavior state through this variable. This can provide a mechanism to observe the behavior states; however, we are confident that the introduction of a mode variable by itself is insufficient because various state sets of individual behaviors do not guarantee the grasp of other behaviors' states. Moreover, the implementation of state changes by substitution into the mode variable may cause some discrepancies between the expected state and the mode variable because JADE concurrently executes multiple behaviors.

Therefore, first, we limit the available mode states to *waiting*, *active*, *achieved*, and *not achieved*, in order to provide a uniform state reference. Fig. 4 shows the state transitions in the ComponentBehavior class. *Waiting* represents the service in standby, and *active* represents the service is being provided. Achieve-type behaviors can transition to the *achieved* state when the goal is achieved or to the *not achieved* state when it can no longer be achieved. Next, in order to appropriately control behaviors, we combine mode states with control methods, i.e. *activate()*, *passivate()*, and *finalize()*, as described in Fig. 4. These extensions allow for a sufficient degree of processes cooperation.

We also add an autonomic activation mechanism into this behavior class. In generic JADE development, developers describe the execution task in the *action()* method. We define a new method for the description of the execution task, called the *perform()* method, and embed the condition branch following the mode state and the invocation of the perform method in the ComponentBehaviour class. We

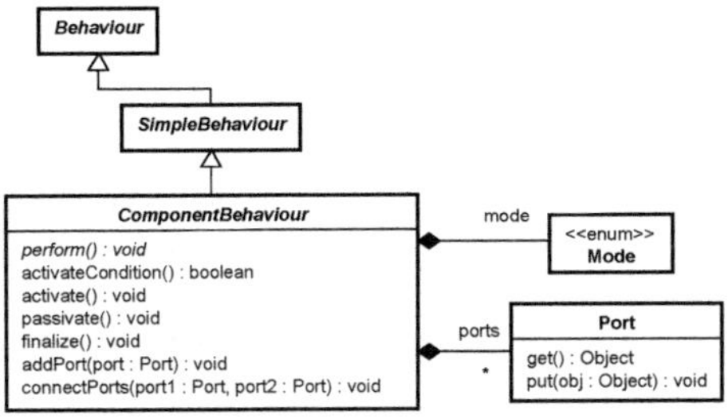

Fig. 3 Structure of ComponentBehaviour class

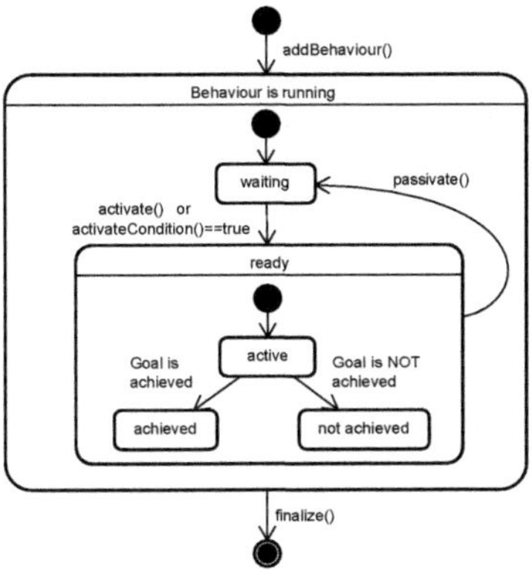

Fig. 4 Life cycle of ComponentBehaviour

also add the *activateCondition()* method for describing the activation conditions, and the class invokes the perform method when the method returns *true*. These additions enable autonomous activation by just describing activation conditions in the activateCondition method.

4 Related Work

While JADE provides the multi-threaded programming style for constructing agents, the implementation on the basis of the belief-desire-intention (BDI) [12] model is another agent building style. This model provides an agent architecture that provides a reasoning mechanism with the concepts of belief, desire (or goal) and plan. Jadex [11] is one of these frameworks for providing the BDI programming style. This framework is also used as a platform for self-* systems [7]. Since developers can describe plans and these activating conditions in Jadex, it satisfies the *autonomous activation* requirement; however, it requires developers to describe the relationship between plans and goals, and this makes it difficult to understand the concurrent processes. Therefore, the *process cooperation* requirement is difficult to achieve.

Several development processes have been proposed for constructing agents on JADE. Moraïtis [6] proposed a way of implementing JADE agents by using the design model of the Gaia methodology [14], and Nikraz [10] proposed a way of supporting whole agent development processes. However, these processes only use the existing behaviors, e.g. FSMBehaviour, SimpleBehaviour, and SequentialBehaviour, and they do not satisfy the requirements identified in this paper.

5 Conclusion

We focused on the implementation of self-* systems in this paper, and presented our approach to realizing a development process that makes use of an agent programming framework, JADE. We also added an extended behavior class into JADE to allow developers to describe multiple behaviors for implementing self-* systems. This behavior class allows for efficient implementation of self-* systems by supporting behavior cooperation and autonomous activation. In future work, we will evaluate our approach through self-* systems implementation experiments.

References

1. Cheng, B.H., de Lemos, R., Giese, H., Inverardi, P., Magee, J., et al.: Software engineering for self-adaptive systems: A research road map. In: Dagstuhl Seminar Proceedings (2008)
2. Garlan, D., Kramer, J., Wolf, A. (eds.): Proceedings of the first workshop on Self-healing systems (WOSS 2002). ACM, New York (2002)
3. Hirsch, D., Kramer, J., Magee, J., Uchitel, S.: Modes for software architectures. In: Gruhn, V., Oquendo, F. (eds.) EWSA 2006. LNCS, vol. 4344, pp. 113–126. Springer, Heidelberg (2006)
4. Kramer, J., Magee, J.: Self-managed systems: an architectural challenge. In: Future of Software Engineering (FOSE 2007), pp. 259–268 (2007)
5. Lorenzoli, D., Tosi, D., Venticinque, S., Micillo, R.A.: Designing multi-layers self-adaptive complex applications. In: Fourth international workshop on Software quality assurance (SOQUA 2007), pp. 70–77. ACM, New York (2007)
6. Moraïtis, P., Petraki, E., Spanoudakis, N.I.: Engineering JADE agents with the Gaia methodology. In: Kowalczyk, R., Müller, J.P., Tianfield, H., Unland, R. (eds.) NODe-WS 2002. LNCS (LNAI), vol. 2592, pp. 77–91. Springer, Heidelberg (2003)
7. Morandini, M., Penserini, L., Perini, A.: Towards goal-oriented development of self-adaptive systems. In: Proc. of the International Workshop on Software Engineering for Adaptive and Self-managing Systems (SEAMS 2008), Leipzig, Germany, pp. 9–16 (2008)
8. Nafz, F., Ortmeier, F., Seebach, H., Steghofer, J.-P., Reif, W.: A generic software framework for role-based organic computing systems. In: International Workshop on Software Engineering for Adaptive and Self-Managing Systems, pp. 96–105 (2009)
9. Nakagawa, H., Ohsuga, A., Honiden, S.: Constructing self-adaptive systems using a KAOS model. In: Proc. of Second IEEE International Conference on Self-Adaptive and Self-Organizing Systems Workshops (SASOW 2008), Venice, Italia, pp. 132–137. IEEE, Los Alamitos (2008)
10. Nikraz, M., Caire, G., Bahri, P.A.: A methodology for the analysis and design of multi agent systems using JADE (2006), `http://jade.tilab.com/doc/tutorials/JADE_methodology_website_version.pdf`
11. Pokahr, A., Braubach, L., Lamersdorf, W.: Jadex: Implementing a BDI-infrastructure for JADE agents. EXP - in search of innovation (Special Issue on JADE) 3(3), 76–85 (2003)
12. Rao, A.S., Georgeff, M.P.: BDI agents: From theory to practice. In: ICMAS, pp. 312–319. The MIT Press, Cambridge (1995)
13. Telecom Italia. JADE: Java agent development framework, `http://jade.tilab.com/`
14. Zambonelli, F., Jennings, N.R., Wooldridge, M.: Developing multiagent systems: The Gaia methodology. ACM Trans. on Software Engineering and Methodology 12(3), 317–370 (2003)

Collaborative Dialogue Agent for COPD Self-management in AMICA: A First Insight

Mario Crespo, Daniel Sánchez, Felipe Crespo, Sonia Astorga, and Antonio León

Abstract. This work presents the dialogue interface that is being developed for the AMICA project[1], aimed at self-management and follow-up of patients with Chronic Obstructive Pulmonary Disease (COPD). This platform tries to overcome possible problems of interaction of patients by extending the traditional mouse and keyboard with new information-communication technologies such as natural dialogue. Since most of these patients are elderly, these technologies can help suit their users' needs and overcome their impairment by providing more natural ways of interacting with technology. The first analyses suggest that dialogue models must be flexible enough to be adapted to requirements of natural language.

Keywords: Agent cooperation, natural dialogue processing, medical platforms.

1 Introduction

Home telemonitoring presents an alternative for the close follow-up of patients by ensuring timely transmission of clinical and physiological data and by supporting prompt medical intervention before deteriorations occur in patients' conditions (Jaana et al. 2009). Several studies have emphasized that home telemonitoring has the potential to improve health results and reduce cost of health systems (Kun 2001) (Giorgino et al. 2004). This is the reason why new methods for health care are currently arousing a great interest.

Telemonitoring solutions deployment have been on the rise since the early nineties (Jaana et al. 2009) varying in the use of medical instruments. However, only few cases have focused on an actual interaction system that allows for consultation at home (Guy et al. 2007) (Young et al. 2001). Most COPD patients are older than 60

Mario Crespo · Daniel Sánchez · Felipe Crespo
Biomedical Engineering and Telemedicine Lab. University of Cádiz, Spain
e-mail: mario.crespo@uca.es

Sonia Astorga · Antonio León
Puerta del Mar University Hospital, Cádiz, Spain

[1] This work was supported in part by the Ambient Assisted Living (AAL) E.U. Joint Programme, by grants from Instituto de Salud Carlos III and the European Union under Project AAL-2008-1-176 http://www.amica-aal.com/

Y. Demazeau et al. (Eds.): Advances in PAAMS, AISC 70, pp. 75–80.
springerlink.com

(Devereux 2006), thus, platforms used to monitor them must be easy to use. This work presents some challenges and draws some conclusions about the development of a dialogue model system within the Autonomy Motivation & Individual Self-Management for COPD Patients (AMICA) project, aimed at self-management and follow-up of patients with Chronic Obstructive Pulmonary Disease (COPD).

2 The Challenge of Home Telemonitoring in AMICA

The major goal of COPD treatment is to ensure that the patient's overall health is improved. A critical step in the disease management is to obtain from the patient reliable and valid information on the impact of COPD on their health status (Jones et al. 2009). To achieve this, a series of physiological signals are obtained daily by means of an ad-hoc sensor (Morillo et. al 2009). This information is then extended by that provided by the patient interacting with the interface. By combining both pieces of information the system is expected to be able to set off medical alarms, modify small aspects of the patients' treatment program or lifestyle, or even suggest hospitalization.

Each user will have a personal profile in which all relevant information about their individual health conditions will be stored in an electronic patient record. When patients connect the platform, it loads their profile and gets ready for dialogue. This will allow the dialogue between the patient and the interface to change according to the physician's requirements and the clinical conditions of the user.

2.1 Medical Interview and COPD

Generally speaking, the AMICA dialogue platform tries to emulate a medical consultation by inquiring patients about their health conditions. The goal of a medical interview is to assess a list of symptoms to diagnose patients. During consultation, physicians usually ask patients about symptoms such as 'pain', 'dizziness', 'shortness of breath', etc., which vary depending on the disease. The course of COPD involves a decline in general health conditions and its most common symptoms are breathlessness, excessive sputum production or chronic cough (WHO 2008). The need for a validated and simple instrument to quantify COPD has motivated the development of questionnaires or health status measures such as the St. George's Respiratory Questionnaire (SGRQ) (Jones et al. 1991), the COPD Clinical Questionnaire (CCQ) (Molen et al. 2003) or the COPD Assessment Test (CAT) (Jones et al. 2009) aiming at:

- **Approaching the detection of COPD exacerbations** or deteriorations of patient's respiratory symptoms.
- **Assessing the severity of COPD exacerbations.**
- **Establishing a possible bacterial etiology**. Infection is the most common cause for acute deteriorations.

2.2 Dialogue Modeling

Dialogue systems require the implementation of dialogue models to analyze and structure conversation. Conversation is usually interpreted in terms of dialogue games or tasks that people must accomplish together (Clark 1996). Participants work together to complete levels of execution and attention, presentation and identification, signaling and construing, and more extended dialogue problems. Possible tasks vary from one platform to another. Generally the physician talks with their patients about a list of medical items to determine a patient's health status. Medical agents that interact with patients must be able to infer such items from a conversation as physicians would do.

In the creation of dialogue agents, a corpus in which dialogue is thoroughly analyzed is required so that linguistic and probabilistic conversational models are defined (Alcacer et al. 2005). On this basis, a corpus of medical dialogues between physicians and COPD patients from University Hospital of Puerta del Mar of Cádiz is being collected, as for similar projects (Cattoni R. et al., 2002) (Leech & Weisser 2003) (Alcácer N. et al 2005) (Allen et al. 2001) (Allen et al. 2007). The corpus is based on medical interviews conducted by physicians who ask COPD hospitalized patients about their symptoms. This corpus shall consist of a collection of 50 medical dialogues of approximately 300 minute-physician-patient conversation, which will be a valuable resource for the implementation of dialogue models. This corpus is being codified in XML according to current trends in dialogue processing. On the basis of the DAMSL annotation (Allen & Core, 1997), a new layer of information is provided with those tasks /goals achieved in conversation. Annotation is organized in five graded levels of generalization:

1. Level of intelligibility: uninterpretable, abandoned or self-talk.
2. Task or communication management: presentation, meaning, cancellation, etc.
3. Type of Discourse Act: assert, command, invite, offer, commit, answer, etc.
4. *Conversational games* or set of speech acts with a similar purpose (e.g. to know quantity of sputum, regularity of cough, color of sputum, etc.)
5. *Transactions or* set of conversational games that accomplish one major step in the participants' plan for achieving the task (e.g. degree of phlegm or cough).

3 COPD Medical Dialogue: First Analysis

10% of this corpus was extracted and analyzed by a linguist with the aid of a physician of Puerta del Mar University Hospital in Cádiz. The goal of this analysis was to observe how medical items were presented naturally in dialogue. This excerpt was composed of 6 conversations, 28 minutes and 1218 participant interventions and 1236 speech acts. From this set of interventions 6 general items were identified, each of them including different subtopics:

SEMANTIC DOMAIN	SEMANTIC DOMAIN	CONVERSATION FIGURES
Personal information	Age	*Dialogues*: 6
	Profession	*Interventions*: 1218
	Smoking	*Dialogue acts*:1236
Medical history	Hospitalization date	*Communication management dialogue acts*: 306
	Operations / surgery	
	Medical treatment	*Dialogue acts aimed at task-management*: 105
Symptoms	Cough	
	Phlegm	*General medical items*: 6
	Fever	*Subtopics*: 11
Functionality	Level of breathlessness on performing different actions	*Mean of Interventions for subtopic*: 12

Fig. 1 Medical items analyzed in 10% of AMICA corpus

The mean of interventions for each of these 11 subtopics is 12. From the 1236 dialogue acts, 306 (25%) were aimed at communication management (e.g., 'Uh huh', 'well, or 'ok') and 105 (9%) aimed at task-management. Figure 7 shows an excerpt of this small corpus in which the physician inquires a patient about the degree of coughing. As it can be seen, speakers cooperate to complete levels of information and ensure that his interlocutor understands what it was meant. The number of interventions in this fragment is 7 including 8 different dialogue acts. In this sense, the third intervention contains 2 different speech acts: the first is aimed at communication-management and the second one presents a new question to the patient. It is worth pointing out the three last interventions aimed at making clear a piece of information that is required by the physician. The last intervention is used to point out the patient that the physician has understood properly.

ACQUITING INFORMATION ABOUT COUGHING	
Physician: <<Normalmente, ¿tos le da usted?>>	*[Physician: <<Do you usually get coughs?>>]*
Patient: <<toso, pero muy poco>>	*[Patient: <<I do, but seldom>>]*
Physician: <<ajá…>> <<¿diariamente y por las noches le da la tos?>>	*[Physician: <<*Uh huh…*>> << do you get coughs daily or just at night?>>]*
Patient: <<por la noches, por las noches, me da... pero muy poco>>	*[Patient: <<I get coughs at night, but rarely>>]*
Physician: <<¿poquito?>>	*[Physician: <<¿rarely?>>]*
Patient: <<sí>>	*[Patient: yes]*
Physician: <<vale>>	*[Physician: <<ok>>]*

Fig. 2 Example of conversation in AMICA corpus

4 Discussion

Current trends in dialogue modelling define each possible conversation as Finite State Automata (Lewin 2000) as a 5-tuple (S,Σ,T,s_0,f) where S is a set of states, Σ is an alphabet, T is a transition function $T:Sx\Sigma \rightarrow S$, s_0 is the initial state and f the set of final states. A possible conversation γ of consists of an automata$_\gamma$ whose alphabet (usually depicted as arcs) stands for possible dialogue acts, and the states of such an automata as conversation stages associated to actions within the system. When a final state of a certain automata is achieved, a certain conversation has concluded. Figure 3 depicts a simple automaton:

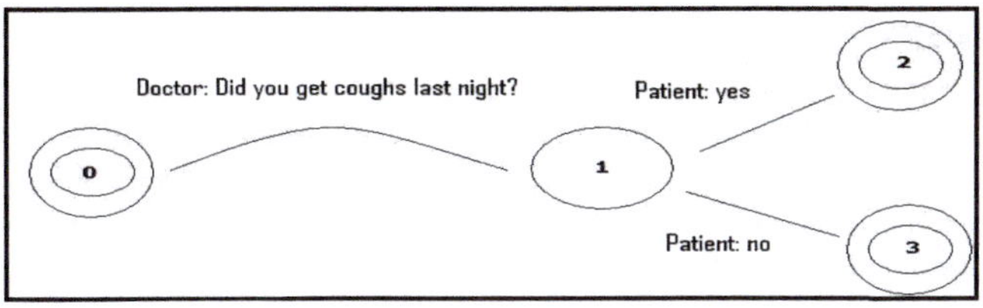

Fig. 3 Dialogue-based Finite State Automata

The system must contain a big amount of these automata representing possible dialogues or speech variations arising in new conversations. The corpus above referred will provide the dialogue model with an important source to create such automata whose arcs will be filled in with the interventions registered. However, speech *acts* cannot account for the frequent acknowledgements and occasional repairs that occur in a dialogue. Conversation is more than sequences of utterances produced in turns. In most cases, speech acts require several utterances in order to be realized. (Clark and Schaefer 1989) states that speech acts are usually the result of contributions in two senses: 1) *presentation phase* and 2) *acceptance phase*. Conversation can unfold on the belief that partners have understood what the contributor meant. (Traum & Allen 1992) (Traum 1999) call this new level as *grounding acts* that allows for the creation of more flexible linguistic models.

5 Conclusion

This work presents the dialogue interface that is being developed for the AMICA project along with a first insight into the issues arising in the creation of dialogue models. These linguistic models should be flexible enough to be adapted to all requirements of natural dialogue. The development of this kind of platforms will hopefully try to improve patients' quality of life and help spread the use of technology for health care.

Acknowledgements. The authors would like to thank the European Commission, the Spanish Ministerio de Industria, Turismo y Comercio and the Instituto de Salud Carlos III for the financial support provided to the AMICA project. We are also grateful to the medical staff at the University Hospital Puerta del Mar, Cádiz, Spain.

References

Alcácer, N., et al.: Acquisition and Labelling of a Spontaneous Speech Dialogue Corpus. In: Procs. of 10th International Conference on Speech and Computer (SPECOM), Patras, Greece, pp. 583–586 (2005)

Allen, J., Core, M.: DAMSL: Dialog Act Markup in Several Layers. Draft (1997)

Allen, J., et al.: Towards Conversational Human-Computer Interaction. AI Magazine 22(4), 27–38 (2001)

Allen, J., et al.: PLOW: A Collaborative Task Learning Agent. In: Proceedings of the AAAI Conference on Artificial Intelligence: Special Track on Integrated Intelligence (2007)

Cattoni, R., et al.: ADAM: The SI-TAL Corpus of Annotated Dialogues. In: Procs. of LREC (2002)

Clark, H.H.: Using Language. Cambridge University Press, Cambridge (1996)

Clark, H.H., Schaefer, E.F.: Contributing to discourse. Cognitive Science, 13 (1989)

Devereux, G.: ABC of chronic obstructive pulmonary disease. Definition, epidemiology, and risk factors. BMJ 332, 1142–1144 (2006)

Giorgino, T., et al.: The HOMEY Project: a telemedicine service for hypertensive patients. In: Dialogue Systems for Health Communication, pp. 32–35. AAAI Press, Menlo Park (2004)

Jaana, M., et al.: Home Telemonitoring for Respiratory Conditions: A Systematic Review. The American Journal of Managed Care 15(5), 313–320 (2009)

Jones, P.W., et al.: The St. George's Respiratory Questionnaire. Respir. Med. 85 (1991)

Jones, P., et al.: Development and first validation of the COPD Assessment Test. Eur. Respir. J. 34, 649–654 (2009)

Kun., L.: Telehealth and the global health network in the 21st century. From homecare to public health informatics. Comput. Meth. Prog. Bio. 64(3), 155–167 (2001)

Leech, M., et al.: Generic speech act annotation for task-oriented dialogues. In: Procs. of the 2003 Corpus Linguistics Conference. Centre for Computer Corpus Research on Language Technical Papers, pp. 441–446. Lancaster University (2003)

Lewin, I.: A formal Model of Conversational Game Theory. In: Procs. of the 4th workshop on the semantics and pragmatics of dialogue (2000)

Young, M., et al.: A telephone-linked computer system for COPD care. Chest 119, 1565–1575 (2001)

van der Molen, T., et al.: Development, validity and responsiveness of the Clinical COPD Questionnaire. Health Qual Life Outcomes 1, 13 (2003)

Morillo, S., et al.: Accelerometer-Based Device for Sleep Apnea Screening. IEEE transactions on information technology in biomedicine (2009)

Traum, D.R., Allen, J.F.: A Speech Acts Approach to Grounding in Conversation. In: Proc. 2nd International Conference on Spoken Language Processing, pp. 137–140 (1992)

Traum, D.R.: Computational models of grounding in collaborative systems. In: Brennan, S.E., Giboin, A., Traum, D. (eds.) Working Papers of the AAAIFall Symposium on Psychological Models of Communication in Collaborative Systems, pp. 124–131. American Association for Artificial Intelligence, MenloPark, California (1999)

WHO World Health Organization, Chronic obstructive pulmonary disease, COPD (2008), `http://www.who.int/respiratory/COPD/en/`

Application of Model Driven Techniques for Agent-Based Simulation

Rubén Fuentes-Fernández, José M. Galán, Samer Hassan, Adolfo López-Paredes, and Juan Pavón

Abstract. Agent-based simulation is being recognized as a useful tool for the study of social systems. It is based on the idea that agents can be used as a good abstraction of members of a society, and by simulating their interactions, observe the emergent behavior. Usually, agent-based models for social simulation are rather simple, but there are more and more works that try to apply this technique for rather complex systems, both in the typology of the agents and their relationships, and in scalability. For this reason, software engineering techniques are required to facilitate the development of agent-based simulation systems. In this sense, it can be useful to consider agent-oriented methodologies, as they cope with these requirements. This work explores the application of a model-driven methodology for the development of MAS, INGENIAS. There are several advantages of this approach. One is the possibility to define specific modeling languages that are conceptually close to the domain-expert, in this case for simulation of social systems. This will facilitate the communication of multidisciplinary teams and, by generating code from models, the social scientist is alleviated from the concerns of programming for a simulation platform. This is illustrated with a case study of urban dynamics.

Keywords: Agent-based modeling, model driven techniques, social simulation.

1 Introduction

Agent-based simulation (ABS) is being recognized as a useful tool for the study of social systems, especially in Social Sciences [1]. One of the main advantages of

Rubén Fuentes-Fernández · Samer Hassan · Juan Pavón
GRASIA, Universidad Complutense de Madrid, Spain
e-mail: {ruben,samer,jpavon}@fdi.ucm.es

José M. Galán
INSISOC, Universidad de Burgos, Spain
e-mail: jmgalan@ubu.es

Adolfo López-Paredes
INSISOC, Universidad de Valladolid, Spain
e-mail: adolfo@insisoc.org

Y. Demazeau et al. (Eds.): Advances in PAAMS, AISC 70, pp. 81–90.
springerlink.com

modeling a social system with agents, and in our opinion the one that distinguishes it from other paradigms, is that it facilitates a direct correspondence between the entities in the target system, and the parts of the computational model that represent them (i.e. the agents) [2].

The process of abstraction to transform the real target system into a model for simulation is complex. This process involves different subtasks and roles, which need diverse backgrounds and competences in the design, implementation and use of an archetypical agent based simulation. Drogoul et al.[3] identify three different roles in the modeling process: the *thematician*, the *modeler*, and the *computer scientist*. This classification has been expanded by Galán et al. [4], including an additional role, the *programmer*.

The role of the thematician, which ideally would be performed by an expert in the domain, aims at producing the first conceptualization of the target system. This task entails: defining the objectives and the purpose of the modeling exercise, identifying the relevant components of the system and the relations between them, and describing the most important causal dependencies. The modeler's job is to produce formal requirements for the models starting from the thematician's ideas. These requirements allow the computer scientist to formulate a feasible model that can run in a computer. However, not all the formal specifications can be directly implemented in a computer. The computer scientist role finds a suitable approximation to the modeler's formal model that can be executed in a computational system with the available technology. Finally, the programmer's role is to implement the computer scientist's model to a target simulation platform.

In practical terms, modeling in Social Sciences faces two problems. The first problem appears when the same person plays all the roles of the process [4], as it is not common being an expert in software engineering and social sciences at the same. Besides, scientists may find difficult understanding the detailed behavior of the underlying software, since doing it would imply a full understanding of its implementation.

On the other hand, many problems require a multidisciplinary perspective, involving members with specialized roles. The second problem arises because effective communication between experts of fundamentally different domains (e.g. sociology and computer science) is not trivial. In most cases, it is difficult to grasp how the social features have been mapped to program constructions. Thus, there are difficulties to assure that the program really implements its conceptual model.

To address these problems, our research promotes the creation of a set of high-level tools, methods and languages, to assist the transfer of models between different roles in the modeling process. These tools should work with modeling languages that include, ideally, concepts close to the thematician's background, but at the same time representing ideas from a software engineering point of view. We must take into account that any mismatch between the specifications and the actual model passed to the next stage, will end up producing an error.

Moreover, a high-level communication tool may also help in the validation. Model validation is the process of determining that the model behavior represents the real system to satisfactory levels of confidence and accuracy, which are determined by the intended model application and its application domain. When dealing with complex systems, as it is frequent on ABS, the traditional methods used in model validation are not widely accepted [5]. In such cases, a good option for the validation of the conceptual model is to check whether the theoretical foundations and assumptions are reasonable within the context of the objectives of the simulation. This structural validation is sometimes performed on the basis of participatory methods with experts in the modeled domain and stakeholders [6]. Again, these expert panels do not usually have a software engineering background. Intermediate languages between the different roles endowed with a high-level descriptiveness facilitate the communication, modification and criticism of the models in the validation stages.

This paper shows how the application of metamodeling and transformation techniques can facilitate the definition of specific modeling languages for agent-based social simulation, taking as starting point the INGENIAS agent-oriented methodology, which provides tools for agent-based modeling, model transformations, and code generation. The role of metamodels and how to use them for agent-based social simulation modeling is introduced in the next section. Subsequently, the methodological process is illustrated step-by-step by means of a case study on urban dynamics. The paper ends with a discussion and concluding remarks concerning the approach.

2 Metamodels for Agent-Based Modeling and Simulation

An approach for addressing the mentioned problems is the use of domain-specific languages (DSL) [7] to produce intermediate models between the thematician's abstract non-formal models and the final program. A DSL explicitly defines its concepts, attributes, relationships, and potential constraints applicable to the models specified with it. These elements are "domain-specific" because they are extracted from the target domain. Thus, researchers working with a DSL specify their models using the language of their discipline. Moreover, since the language elements are clearly defined, mapping them to software constructions is significantly easier and more reproducible than in the case of an arbitrary set of elements, which is the situation with non-formal descriptions.

Metamodels [8] allow the specification of DSLs. Such specification corresponds to a graph with nodes (i.e., entities or concepts) linked by arcs (i.e., relationships or references), and both of them can have properties (i.e., features or attributes). A metamodel for a given DSL defines the types of nodes and arcs that correspond to the domain concepts. It indicates their names and attributes, and the rules and constraints that they satisfy, for instance the concepts a relationship can link or its cardinality.

Metamodels have two key advantages for defining DSLs. First, they can be extended to satisfy specific modeling needs: if the current form of the DSL is not enough to model a given problem, new elements can be introduced as extensions or specializations of the existing ones. These new elements provide an accurate representation of the domain notions according to the definition of the thematician. Second, there is a wide range of support tools to work with them. This facilitates the production of modeling tools for specific domains.

INGENIAS [9] has been chosen as it provides tools for model-driven development. Its modeling language supports the specification of agents' organizations, as well as agent intentional behavior, characteristics that are present in social systems. This is a feature that general purpose modeling languages lack. Modeling with this language is supported by the INGENIAS Development Kit (IDK) with a graphical editor, which can be extended to work with new modeling concepts. Second, the INGENIAS model-driven engineering approach facilitates the independence of the modeling language with respect to the implementation platform. This is especially important here in order to abstract away programming details and concentrate on modeling and analysis of social patterns. With this purpose the IDK supports the definition of transformations between models and code for a range of target platforms.

Table 1 summarizes the main concepts of INGENIAS used in the rest of the paper. Note that this table considers concepts but not their relationships. The relationships used in the case study have self-explanatory names. For instance, this is the case of the relationship "*WFProduces*" from a "*Task*" to a "*Frame Fact*". The first two letters indicate the main type of an INGENIAS diagram where the type of relationship appears. In this case, "*WF*" stands for "*WorkFlow*" diagram. The rest of the name provides the meaning of the relationship. "*Produces*" shows that the fact is the result of the execution of the task.

Table 1 Main concepts of the INGENIAS modeling language

Concept	Meaning	Icons
Agent	An active concept with explicit goals that is able to initiate some actions involving other elements of the simulation.	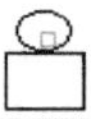
Role	A role groups related goals and tasks. An agent playing a role acquires the goals and tasks of such role.	
Environment application	An element of the environment. Agents act on the environment using its actions and receive information through its events.	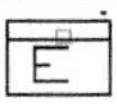
Goal	An objective of an agent. Agents try to satisfy their goals executing tasks. The satisfaction or failure of a goal depends on the presence or absence of some elements (i.e. frame facts and events) in the society or the environment.	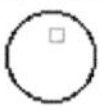
Task	A capability of an agent. In order to execute a task, certain elements (i.e. frame facts and events) must be available. The execution produces/consumes some elements as result.	
Frame Fact	An element produced by a task, and therefore by the agents.	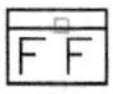
Event	An element produced by an environment application.	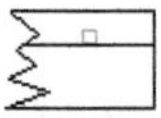
Interaction	Any kind of social activity involving several agents.	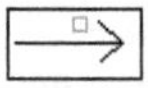
Group	A set of agents that share some common goals and the applications they have access to.	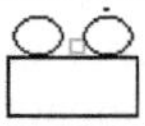
Society	A set of agents, applications and groups, along with some general rules that govern the agent and group behavior.	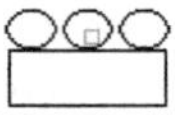

It is possible to define an agent-based model in a process with the following activities:

1. *Domain analysis*. Thematicians consider the concepts that are required to express their hypotheses and the related information in the group or society.
2. *Determine interactive concepts*. Among the domain concepts, some of them focus on the analysis, and thus they are considered as decision makers that follow certain rationality. Besides, some concepts represent elements that initiate interactions with others, for instance asking for some services. All these concepts are candidate agents in the agent based model and are represented as subclasses of Agent. Any element that engages in interactions with other elements of the system should be modeled also as an agent.
3. *Determine non-interactive concepts*. Passive elements that do not take decisions are regarded as part of the environment. In INGENIAS, these elements are modeled as subclasses of the environment application.
4. *Determine specialization hierarchies between concepts*. The elements introduced for a problem usually share some features. In order to highlight the common aspects of elements and encourage their reusability, these new elements are arranged in inheritance hierarchies. A super-concept contains all the elements/attributes and participates in all the relationships common to its sub-concepts. Sub-concepts only modify their own specific features, constraining or adding some features of the super-concept.
5. *Determine groups and societies*. In case that several agents share common global goals or environment applications, they can be gathered in groups. If they also share common rules of behavior, they constitute a society.
6. *Determine interactions*. Agents can use environment applications and communicate with other agents. A group of interconnected activities aimed at satisfying a global goal constitutes an interaction.
7. *Assign roles, objectives and capabilities to agents*. INGENIAS refines agent definitions with the tasks they are able to do, which correspond to their capabilities and goals. A goal is linked to the tasks which are able to satisfy it. These tasks produce some elements (frame facts). The presence or absence of some frame facts and events (produced by environment applications) make the evidence for the satisfaction or failure in the achievement of the goal.
8. *Refine interactions*. A refined interaction indicates the agents and environment applications that participate in it, the tasks that agents execute, the goals they pursue with those executions, and the elements produced and consumed in it.
9. *Validate ABS*. These agent-based models are a refinement of the thematician's non-formal models. They are expected to represent it accurately, while providing additional details that facilitate the transition to the running system.

These activities do not need to be performed sequentially. For instance, if there is a precise idea of the existing interactions, modelers can begin with activity 5 and then use this information to discover the agents and environment applications. Besides, this presentation of the process is necessarily simplified given the space limitations of the paper, but more detailed activities are required to provide a full modeling guideline. Also, note that this process can be used to describe models or metamodels. If the concepts have a wide application for a domain, they become part of a metamodel; if they are specific for a problem, they remain at the model level.

This process can be regarded as iterative. That is, a given model can be the abstract model for a new and more platform-oriented model. This allows a transition from the abstract non-formal description to the program in several steps, improving the traceability of the process. This approach is in the basis of model driven engineering, and INGENIAS fully implements it. The advantage of this decomposition is the possibility of using specific guidelines and support tools for each step, which crystallize and automate the expertise of thematicians and modelers and helps novel researchers.

3 Case Study: Urban Dynamics in Valladolid

In order to illustrate the usefulness of metamodeling with INGENIAS in social simulation contexts, an urban dynamic model has been selected. Several complementary theories from fields as Sociology, Geography, Political Science or Economics attempt to explain the complex problem of the dynamic spatial occupation [10]. One of the most descriptive models of urban dynamics applied to real systems is the Yaffo-Tel Aviv model developed by Benenson et al. [11]. This model has been adapted to the Valladolid metropolitan area (Spain) and, together with other socioeconomic models, is used for exploring the dynamics of urban phenomena [12]. In order to adapt it to this new context and develop the different layers of the model, extensive discussions with domain experts (thematicians) were needed. The description presented in this section is a result of the intensive communication between modelers and those thematicians.

The model comprises two layers. The first layer is retrieved from a vectorial GIS. This GIS explicitly represents every block with households in the studied geographical region and characterizes them by their spatial and socioeconomic characteristics. In the second layer, the computational agents representing the families that live in the area are spatially situated.

The main assumption of the model is that agent's selection of residence is influenced by intrinsic features of the candidate households and by the similarity of the agents' socioeconomic factors with those of their neighborhoods. Thus, some of the rules of behavior of the agents in this model are based on the concept of neighborhood, defined taking into account the centroids of their blocks and the Voronoi tessellation [13]. The variable "residential dissonance" quantifies the dissimilarity between an agent, its neighborhood and its household. The probability that an agent leaves a residence is considered proportional to such residential dissonance. This variable may be influenced by differences in terms of nationality or education level, or by imbalances between an agent's wealth and the value of the house where it lives.

Once each agent has calculated the dissonance, the opportunity to change its current residence is modeled through a stochastic process that transforms dissonance into probability of change. Those selected agents are included in a set M of potential internal migrants. If external immigration is enabled, immigrant agents are also included into the set M. Subsequently, each agent A in M estimates the attractiveness (one minus the dissonance) of a number of candidate empty households H_A.

Finally, agents in set M are assigned to empty households. Each potential migrant chooses the household found with highest attractiveness. If the dwelling is already empty, the agent occupies it with a probability depending on its attractiveness; if it is not, the agent removes the household from its H_A list. The process is iterative until H_A is empty for all the agents in M. In each iteration, the order of the agents is randomly selected to avoid bias in the agent selection. The agents that have not been able to find a suitable house leave the city with probability L_A and remain in its current house with probability $1-L_A$. The immigrant population without household leaves the city as well.

Another submodel dynamically updates the prices of the households [14]. The value of a household depends on the wealth of its family and neighbor families, and the value of the surrounding empty houses. The value of the empty households decreases at a constant rate considered as an exogenous parameter of the model.

4 Following the Process

The agent-based model for the case study is the result of the activities that were described in section 2. Activity 1 corresponds to the descriptive analysis realized by the thematician before the discussions which resulted on the previous section, already from the modeler's point of view.

Activities 2 and 3 are made in parallel to identify the entities in the system. In this case, there are at least two active elements, the families and the households. Families initiate interactions to keep low their dissonance level; households perform an interaction with other households to update their value. Thus, both elements are modeled as two different classes of agents.

According to the specification of the problem, when a family wants to migrate, it receives a list of potential candidate households for migration, chosen randomly all over the city. However, communication constraints only allow a family to communicate directly with their neighbors, not all over the city. In order to allow families to get this information about other parts of the city, we introduce a *map* environment application, which is a non-interactive concept as defined in Activity 3. The map is part of the environment and it provides methods to get the list of empty households and retrieve the house data (such as if it is empty).

Activities 4 and 5 are intended to create hierarchies and groups of elements that share features. At this step, there are only two types of agents and one environment application, so the hierarchies are flat. In order to represent the organization of the city, a society is introduced, together with three groups that can be seen in Fig. 1. The *city* society comprehends all the participants in the problem. They share rules of collaborative (e.g. all the agents are willing to provide the requested information) and non-violent behavior (e.g. no family can occupy the household of other family, and no household can eject its family). The three groups in this society are *families*, *households*, and *neighborhoods*. The neighborhoods include the households and the families that are able to communicate among them. This information is used to make explicit the influence of the neighborhood in the calculation of the dissonance level and the price of households. Note also that this information is useful in the implementation of the model. As Repast (http://repast.sourceforge.net/) was chosen as target platform, an agent can only communicate with those placed in adjacent cells. Thus, the neighborhoods tell us which agents must be in which cells. This is a typical example of information with social basis refined to guide the implementation.

Activity 6 identifies the relevant interactions for the problem: the calculation of the household prices and the family dissonance levels. Due to space restrictions, the following diagrams focus on the second interaction.

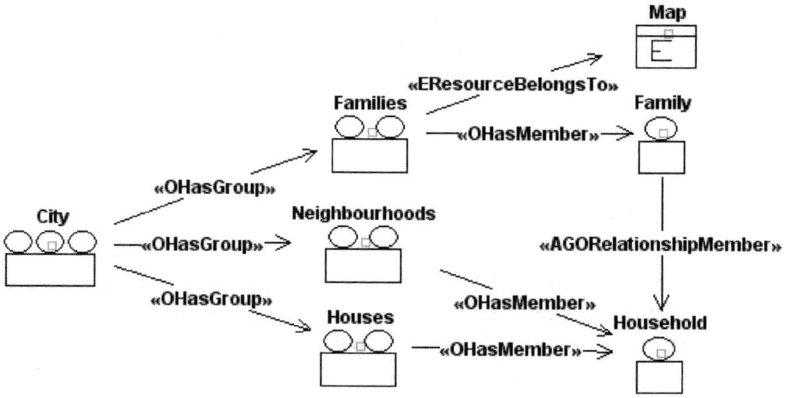

Fig. 1 The *city* society and its components. For the legend see Table 1

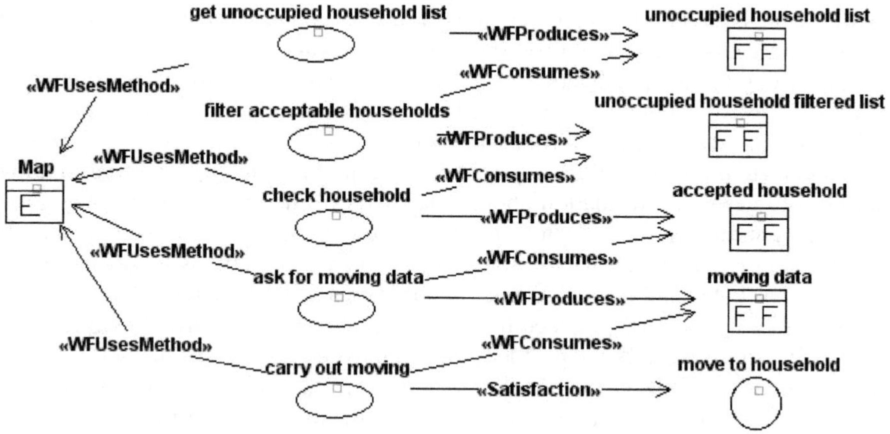

Fig. 2 *Migrate in the city* workflow for the *family* agent. For the legend see Table 1

Activities 7 and 8 identify the tasks and goals of agents, the elements exchanged and their relations to specify the interactions. Fig. 2 and Fig. 3 include part of the results of these activities. They focus on the workflow the family agent performs after it finds out that its dissonance level is too high with respect to its neighborhood.

Fig. 2 shows the part of the workflow where the family agent looks for a suitable and unoccupied household. It begins asking for H unoccupied households using the map. Afterwards, it filters the list to get only the households whose dissonance level is below its threshold. Then, the family uses the map to try to get one of the suitable households. Note that according to the specification of the non-formal description, all the families that are uncomfortable with their households try to make these tasks at the same time. First, all of them get the list, then they filter it, and afterwards each of them tries to get into a household. Given this order, it is possible that when a family tries to occupy an initially unoccupied household, another discontent family that chose first has already occupied it. For this reason, the task *check household* can fail.

5 Conclusions

The introduction of a specific modeling language, which is founded on well established agent concepts and close to the domain-expert in the form of diagrams, facilitates communication, specification, implementation and validation of agent-based models for the simulation of social systems. This has been validated by providing guidelines for agent-based modeling using the support of the default modeling tools of a specific agent-oriented modeling language, INGENIAS. Its use has been illustrated with a case study on urban dynamics.

This framework will allow the specification of social systems with a graphical modeling language, the simulation of the models of these systems by exploiting the capabilities of existing agent-based simulation tools, and the identification and analysis of social patterns (at a macroscopic or aggregate level) in terms of the atomic elements of the social system specification (at a microscopic or individual/interaction level). The advantages go further than usability. As it has been discussed in [15], this solution facilitates the replication of an experiment on different simulation engines, in order to contrast results. The availability of a graphical view of the system facilitates its understanding and improves the identification of patterns in the system too.

It has still to be evaluated the effort of learning a new language, but, in principle, a visual modeling language should be easier to use than a typical programming language. The main issue, however, is the effort that is required to adapt existing agent metamodels for creating domain specific languages. However, in the case of INGENIAS, this adaptation is feasible as both the language and the tool easily allow extensions to introduce new concepts and relations, together with graphical icons for them. A possible extension could be to differentiate the neighborhood from a standard agents group (such as "Families"), as it should be related with the space in some way.

Despite of these issues, this approach is considered a step forward in the search of more reliable and transparent agent-based models. The increase of formalization associated, together with the facilitation of replication, would restrain the typical criticism of complex models as obscure black-boxes.

Acknowledgements. We acknowledge support from the project "Agent-based Modelling and Simulation of Complex Social Systems (SiCoSSys)", supported by Spanish Council for Science and Innovation, with grants TIN2008-06464-C03-01 and TIN2008-06464-C03-02.

References

1. Gilbert, N., Troitzsch, K.G.: Simulation for the Social Scientist, 2nd edn. Open University Press, Stony Stratford (2005)
2. Gotts, N.M., Polhill, J.G., Law, A.N.R.: Agent-based simulation in the study of social dilemmas. Artificial Intelligence Review 19, 3–92 (2003)
3. Drogoul, A., Vanbergue, D., Meurisse, T.: Multi-Agent Based Simulation: Where are the Agents? In: Sichman, J.S., Bousquet, F., Davidsson, P. (eds.) MABS 2002. LNCS, vol. 2581, pp. 1–15. Springer, Heidelberg (2003)
4. Galán, J.M., Izquierdo, L.R., Izquierdo, S.S., et al.: Errors and artefacts in agent-based modelling. Journal of Artificial Societies and Social Simulation 12(1), 1 (2009)
5. Brown, T.N., Kulasiri, D.: Validating models of complex, stochastic, biological systems. Ecological Modelling 86, 129–134 (1996)
6. López-Paredes, A., Saurí, D., Galán, J.M.: Urban water management with artificial societies of agents: The FIRMABAR simulator. Simulation 81, 189–199 (2005)
7. Mernik, M., Heering, J., Sloane, A.M.: When and how to develop domain-specific languages. ACM Computing Surveys 37, 316–344 (2005)
8. OMG: Meta Object Facility (MOF) Core Specification, Version 2.0
9. Pavón, J., Gómez-Sanz, J., Fuentes, R.: Model driven development of multi-agent systems. In: Rensink, A., Warmer, J. (eds.) ECMDA-FA 2006. LNCS, vol. 4066, pp. 284–298. Springer, Heidelberg (2006)
10. Aguilera, A., Ugalde, E.: A Spatially Extended Model for Residential Segregation. Discrete Dynamics in Nature and Society 1, Article ID 48589 (2007)
11. Benenson, I., Torrens, P.M.: Geosimulation: automata-based modeling of urban phenomena. John Wiley and Sons, Chichester (2004)
12. Galán, J.M., del Olmo, R., López-Paredes, A.: Diffusion of Domestic Water Conservation Technologies in an ABS-GIS Integrated Model. In: Corchado, E., Abraham, A., Pedrycz, W. (eds.) HAIS 2008. LNCS (LNAI), vol. 5271, pp. 567–574. Springer, Heidelberg (2008)
13. Okabe, A., Boots, B.N., Sugihara, K.: Spatial tessellations: concepts and applications of Voronoi diagrams. John Wiley & Sons, New York (1992)
14. Benenson, I.: Modeling population dynamics in the city: from a regional to a multi-agent approach. Discrete Dynamics in Nature and Society 3, 149–170 (1999)
15. Sansores, C., Pavón, J.: Agent-based simulation replication: A model driven architecture approach. In: Gelbukh, A., de Albornoz, Á., Terashima-Marín, H. (eds.) MICAI 2005. LNCS, vol. 3789, pp. 244–253. Springer, Heidelberg (2005)

Using ICARO-T Framework for Reactive Agent-Based Mobile Robots

José Manuel Gascueña, Antonio Fernández-Caballero, and Francisco J. Garijo

Abstract. This paper describes the experience and the results of using agent-based component patterns for developing mobile robots. The work is based on the open source ICARO-T framework, which provides four categories of component patterns: agent organization pattern to describe the overall architecture of the system, cognitive and reactive agent patterns to model agent behavior, and resource patterns to encapsulate computing entities providing services to agents. The experimental setting is based on the development of a team of cooperating robots for achieving surveillance tasks. The approached development, illustrated with working examples of the utilization of the patterns, is detailed.

Keywords: Agent-oriented programming languages, Agent framework, Reactive mobile robots.

1 Introduction

There is a huge amount of work and valuable proposals about agent oriented programming frameworks and languages - see the works collected in [1] and the agent platforms overview that appears in web page [2]. JADE, JACK, Jadex, and Cougaar are just some examples. This paper analyzes the MAS development process using the ICARO-T framework [6]. The differentiating factor from ICARO-T is the use of component patterns for modeling MAS. These patterns are described in UML including static and dynamic aspects and Java code consistent with the design description. Development guidelines for creating application components using agent patterns are also provided. The main advantage of ICARO-T framework is that it provides to engineers not only concepts and models, but also customizable MAS

José Manuel Gascueña · Antonio Fernández-Caballero
Universidad de Castilla-La Mancha, Departamento de Sistemas Informáticos & Instituto de Investigación en Informática de Albacete, 02071-Albacete, Spain
e-mail: caballer@dsi.uclm.es

Francisco J. Garijo
Institut de Recherche en Informatique de Toulouse, Equipe SMAC,
Université Paul Sabatier, 118 route de Narbone, 31062 Toulouse Cedex 9, France
e-mail: fgarijo@gmail.com

Y. Demazeau et al. (Eds.): Advances in PAAMS, AISC 70, pp. 91–101.
springerlink.com

design and Java code fully compatible with software engineering standards, which can be integrated in the most popular IDEs. While other agent based platforms, such as FIPA, focus on communication standards, ICARO-T focuses on providing high level software components for easy development of complex agent behavior, agent coordination, and MAS organization.

An additional reason for choosing ICARO-T is cost-effectiveness of agent patterns in application development. Evaluation results in previous experiences [7] have reported significant reductions by an average of 65 percent of time and effort in the design and implementation phases. Cost reduction is achieved without minimizing or skipping activities like design, documentation and testing. In the phases of testing and correction the errors are also reduced. The article structure is as follows.

2 ICARO-T Fundamentals

The ICARO-T framework is the result from the cumulative experience in the development of agent-based applications in the last ten years. Therefore, the framework architecture and the underlying patterns have been elaborated, refined and validated through the realization of several agent-based applications. The first such system discovered patterns for building reactive and cognitive agents. It was a cooperative working system [8], which was refined with the development of a project management system for the creation of intelligent network services [11]. Scalability of the cognitive agent model was considered in a context with thousands of users, in a MAS that supported the personalization of web sites [10], and the reuse of this solution in an online discussion and decision making system [12], as well as a prototype to validate the MESSAGE methodology [3]. Refinements were applied in telecommunications company Telefonica for developing several voice recognition services [7]. In 2008 a new version of the framework was delivered as open source (`http://icaro.morfeo-project.org/`); then new teams started using ICARO as support for academic courses and for research projects (e.g. the e-learning project ENLACE [4]). Utilization as implementation support for a new integrative MAS methodology is also considered [5] in the domain of multisensory surveillance (e.g. [13], [9]).

ICARO-T offers three categories of reusable component models: agent organization models to describe the overall structure of the system, agent models, based on reactive and cognitive agent behavior, and resource models to encapsulate computing entities providing services to agents. The next three paragraphs are focused in describing these components.

The basic working cycle of a reactive agent is summarized in Fig. 1. The reactive agent is able to receive events from different components (agents and resources) through their use interface. The perception (1) provides interfaces to store events in a queue and to consume them, and (2) provides events to the control when requested via the perception consumption interface. The control performs the cycle that is described next.

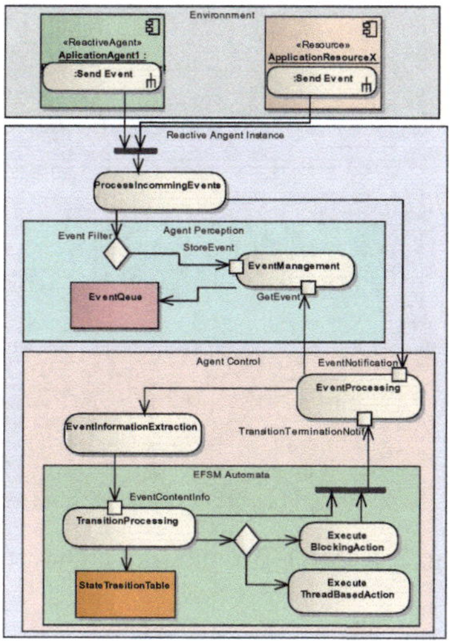

Fig. 1 Basic working cycle of a reactive agent

Firstly, when the agent is created, it starts a process to obtain and process perception events. It requests an event via the perception event consumption interface. If it does not get any event, it waits for an event notification to proceed for a new request. Otherwise, it starts an event processing cycle that consists of the following actions: (i) it extracts the event information - it obtains the action to execute and the object list that will be used as parameters for the action; (ii) it sends the information to the automaton in order to transit to another state; and (iii) it waits for transition termination to get and process a new event. When the control is in a wait state and receives a notification from the perception indicating that an event has occurred, it starts a process to obtain and process the event.

To achieve their goals, the agents need to interact with the computing entities in their environment. Agents view these entities as "resources". More formally, in agent-based applications developed using ICARO-T resources are those computing entities that are not agents, and are used by the agents to obtain information for achieving their objectives. As agents do, resources should offer standard interfaces in order to gain access easily to the provided functionalities. Examples of resources are persistent systems that provide object persistency through relational database management, visualization systems that provide user interface facilities such as presentation screens and user data acquisition for agents to interact with the users, text to speech translators, speech processors, syntactic analyzers, and so on.

An application in the ICARO-T framework is modeled as an organization made up of controller components, which are agents, and resources. Therefore, there are three layers in the organization (see Fig. 2): the control layer (CL), which is made up of controller components; the resource layer (RL), made up of the components that supply information or provide some support functionality to the agents to achieve

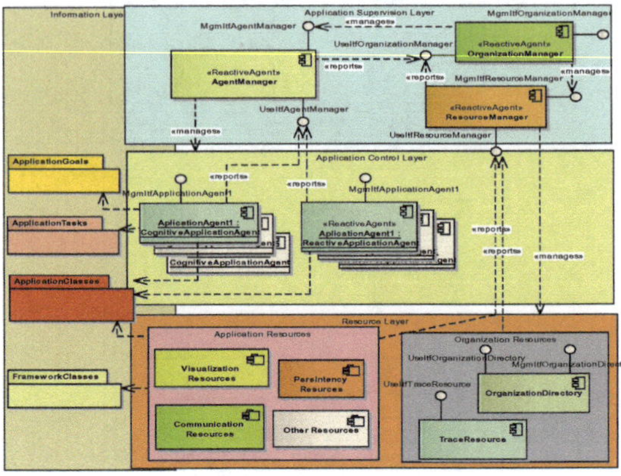

Fig. 2 ICARO-T framework architecture

their goals; finally the information layer contains ontology and/or information entities needed for modeling both the framework itself and applications.

3 Reactive Collaborative Mobile Robots

The selected case study, for illustrating the development process using ICARO-T, implements the collaboration among several mobile robots (a minimum of five) to carry out a common surveillance task in an industrial estate. The robots navigate randomly through pre-defined surveillance paths in a simulated environment. When there is an alarm in a building, a robot is assigned the role of the chief, three robots will be subordinated, and the other ones will be expecting in rearguard to receive orders from the chief (e.g. to replace a damaged robot). Failures are discovered by the robot itself when any of its mounted devices (e.g., sonar, laser, camera, etc) does not work in a right way. The robots perceive that an alarm has occurred through two mechanisms: (1) the security guard notifies robots that an alarm has occurred and where it has taken place, (2) the robot is equipped to perceive an alarm itself when it is close enough to the corner of a building; therefore it does not have to wait for the security guard announcement. The alarm is covered when a robot coalition (one chief and three subordinates) surrounds the building, that is, the robots that form a coalition are located at the four building corners where the alarm has occurred. In this case study three hypotheses are assumed: (1) several alarms cannot take place simultaneously, (2) the robots cannot collide as the streets are wide enough, and (3) the robots navigate from corner to corner.

3.1 Application Resource and Agent Identification

The first step undertaken by the developer, in order to implement a multi-agent system with ICARO-T, is to identify the application agents and resources from the established requirements. For the proposed application an agent (RobotApplication-Agent) and three resources (InterfaceRes, EnvironmentRes and RobotLocationRes) were identified. RobotApplicationAgent is a reactive agent that supports the robot functionalities. InterfaceRes is a visualization resource that allows the user to interact with the application (simulating an alarm rising in a given building, notifying that an alarm has occurred, simulating that a robot has detected a failure, and restarting the application), and to visualize what is happening in the simulated environment. EnvironmentRes is a resource that provides information about the simulated environment (industrial estate dimensions, building where the alarm has taken place, robots that do not work well, and robots initial locations). RobotLocationRes is a resource that stores the robots location and the moment in which they were updated - in this way any robot is able to know the location of the other ones.

In any ICARO-T application, an agent retrieves information contained in a resource and/or updates it via the resource use interface. Likewise, the resources send events to an agent via the agent use interface. Moreover in this application, InterfaceRes uses the EnvironmentRes use interface to update the information introduced by the user, being in this way accessible to RobotApplicationAgent agent instances.

3.2 The Application Agent Description

The reactive agent behavior is modeled with a finite state automaton, where the states represent concrete situations of the agent life cycle. The state diagram interpretation corresponding with the reactive agent that controls a robot is as commented next (see Fig. 3a). There are three kinds of states: initial (InitialState), final (FinalState) and intermediate ones (e.g., AlarmDetection, Rearguard, and so on). When an agent is in a given state and its event queue has an event, which belongs to the valid inputs (events) for transition, then the agent transits from the current state to the transition target state and executes the action associated with that transition. This mechanism is not repeated again in the new state until the execution of such an action has been completed. There is a kind of particular transition, the universal transition, which is valid for any automaton state. This transition takes place for a given input. Then, the action is executed and the automaton transits to the next state, regardless of the automaton's state. On the other hand, in Fig. 3a the following notation is adopted for the purpose of clarity. A transition that goes from a boundary to a state means that such transition is able to go from any state enclosed into the boundary to the specified target state (see for instance rolesReassignation / notifyPosition).

For the reactive agent to be capable of interpreting the graphically represented automaton it is necessary to express it in a textual way through an XML file. This file has to be in the same package as the class that implements the agent actions and named "automaton.xml". The file "StateTable.dtd" establishes restrictions about the structure and syntax necessary to define valid XML files to reactive agents. Let us highlight that in our application all actions are modeled as blocking - this feature is specified in the "automaton.xml" file. This means that if an action does not finish

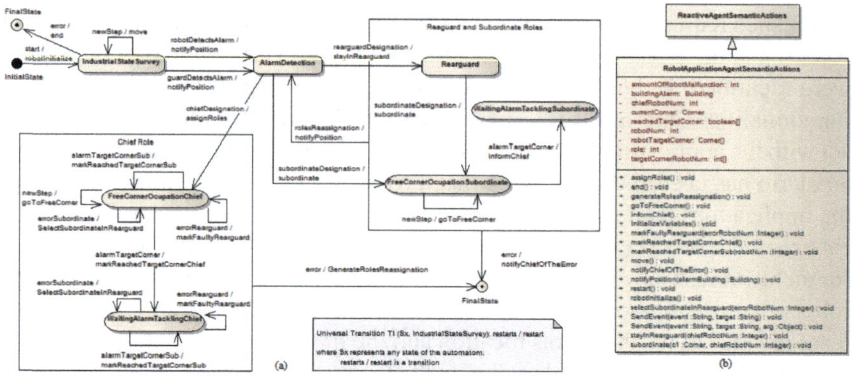

Fig. 3 (a) Graphical representation for robot automaton; (b) Semantic actions class

(it is blocked), then the agent will remain inactive as it is not able to process new events, and therefore the application will be out of service. ICARO-T also offers the possibility to define an action as non-blocking, which means that even if the action does not finish the agent will continue working.

Actions are part of the reactive agent automaton and are defined as methods of a semantic actions class. This class should be (1) named NAMEApplicationAgentSemanticActions, to be conforming to ICARO's naming rules, where NAME is the name chosen for the agent, and (2) created in a package named icaro.applications.agents.ReactiveNAMEApplicationAgent.behaviour. Moreover, that class should extend the ReactiveAgentSemanticActions class (see Fig. 3b). In robotInitialize action the agent sends itself the newStep event to start moving the robot. End action marks the agent as damaged. Restart action restarts the simulation process, and, afterwards the agent sends itself a newStep event. The rest of actions are described in Table 1, where the role played by the agent is mentioned when it is meaningful. The agent starts executing when the Agent Manager sends a start event, causing the agent to change to IndustrialStateSurvey state and to execute the robotInitialize action. The agent does not consult if there are events associated with transitions that go from IndustrialStateSurvey state until robotInitialize has not completed. Once the agent has been initialized, it will navigate randomly (cycling in IndustrialStateSurvey state) until an alarm appears (a robotDetectsAlarm or guardDetectsAlarm event arrives). In this case, notifyPosition action is executed to determine who the chief is. After that, assignRoles action produces the assignment of subordinate and rearguard roles to the other agents. The agent with chief role enters the states enclosed in the 'Chief Role' boundary, whereas the agents with other roles go to states enclosed into 'Rearguard and Subordinate Roles' boundary. The EnvironmentRes resource is responsible of sending error (the robot is damaged), restart, robotDetectsAlarm and guardDetectsAlarm events to the agents. The other events are sent by application agents. Table 1 together with Fig. 3a helps to a better understanding of how the modeled automaton works.

Table 1 Description of the automaton's actions

Action	Description
move	The agent sends itself a newStep event if there is no alarm.
notifyPosition	Determines if the agent becomes the chief. The chief is the agent closer to the alarm and the ties are solved in favor of the agent that has a lower index. The agent sends itself a ChiefDesignation event if it becomes the chief.
stayInRearguard	The agent updates its role as rearguard and learns who the chief is. The robot does not move while it has this role.
subordinate	The agent (1) updates its role as subordinate and learns who is the chief, and (2) sends itself a newStep event.
assignRoles	The agent (1) updates its role as chief, (2) assigns to what corner the chief should go, (3) assigns to what corners the three next closets agents to the alarm should go to, and sends them a subordinateDesignation event that contains the following information: the target corner that it should occupy and who is the chief robot, (4) send to the other agents a rearguardDesignation event, and finally, (5) sends itself a newStep event. The notifyPosition action may be consulted to know how the ties are solved.
goToFreeCorner	If the chief/subordinate agent is on the assigned target corner, then it sends itself an alarmTargetCorner event; otherwise it determines to what corner it will move next, and it sends itself a newStep event.
notifyChiefOfTheError	If the agent is a rearguard, then it sends an errorRearguard event to the chief; whereas if it is a subordinate agent, then it sends an errorSubordinate event, which contains its identification number to the chief. In both cases the agent marks the controlled robot as damaged.
informChief	The subordinate agent sends to the chief agent an alarmTargetCornerSub event, which contains the subordinate agent identification number.
markReachedTarget CornerSub	The chief agent (1) marks the target corner that the subordinate agent has occupied, (2) increases the number of occupied corners, and, (3) notifies the user when four target corners have been occupied.
markReachedTarget CornerChief	The chief agent (1) marks the target corner that it has occupied, (2) increases the number of occupied corners, and, (3) notifies the user when four target corners have been occupied.
selectSubordinate InRearguard	The chief agent (1) increases the number of damaged robots, (2) identifies the closest rearguard agent to the target corner to be occupied by the damaged subordinate agent, and, (3) sends a subordinateDesignation event that contains the target corner and the chief identification number.
markFailureyRearguard	The chief agent (1) marks the rearguard agent that sent an errorRearguard event as damaged, and, (2) increases the number of damaged robots.
generateRoles Reasignation	The chief agent (1) marks itself as damaged, (2) increases number of damaged robots, and, (3) sends to the rest of agents a rolesReassignation event that contains the location of the building where the alarm occurred.

3.3 *Application Resources Description*

Application resources inherit the management interface from the resource pattern. The developer should define the use interface and the class that implements the interface for each resource. In general, an interface NAMEUseItf is created in a package called icaro.applications.resources.NAME for a NAME resource. This interface defines methods callable from other components. On the other hand, a NAMEGeneratorClass class that extends the SimpleResourceImpl class and implements the interface in icaro.applications.resources.NAME.imp package is created. For example, Fig. 4 depicts the classes that take part to define the RobotLocationRes resource.

3.4 *How to Access a Resource and Send Events*

Fig. 5 shows how an agent accesses the RobotLocationRes resource. Firstly a variable, whose type is resource use interface, is defined (see line 1). After that, the resource use interface is retrieved from interface repository (notice in line 3 that

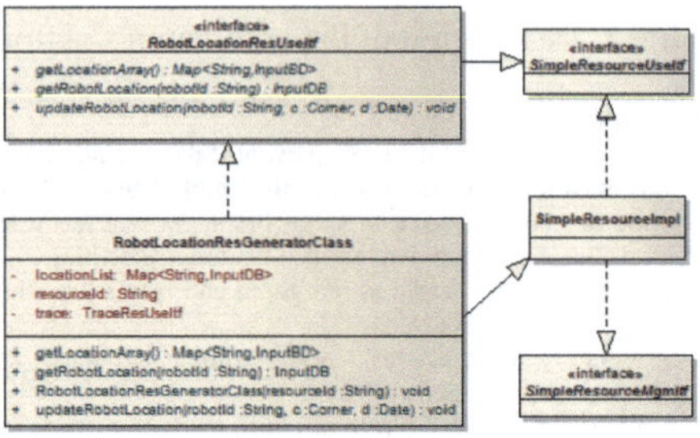

Fig. 4 RobotLocationRes resource description

the value passed into getInterface ends with the resource instance name given in the organization description XML file). Then, all is ready to access the methods offered by the interface (line 5). A resource accesses another resource by following a mechanism similar to the previously described one.

```
(1) RobotLocationResUseItf robotLocationDB;
(2) try {
(3)      robotLocationDB = (RobotLocationResUseItf) repositoryUseItf.getInterface(
                              PredefinedNames.USE_ITF + "RobotLocationRes1"); }
(4) catch(Exception e){ e.printStackTrace();}
(5) InputDB input = robotLocationDB.getRobotLocation("ReactiveRobotApplicationAgent" + j);
```

Fig. 5 Access to a resource from an agent

```
                    public void subordinate (Corner c, Integer chiefRobotNum)

(1) Object arg[] = new Object[2];    arg[0] = robotTargetCorner[i];    arg[1] = (Integer)robotNum;
(2) InterfaceRepositoryUseItf interfaceRepositoryUseItf = InterfaceRepositoryGeneratorClass.instance();
(3) try {
(4)      ReactiveAgentUseItf agentUseItf = (ReactiveAgentUseItf)
                        interfaceRepositoryUseItf.getInterface(PredefinedNames.USE_ITF +
                        "ReactiveRobotApplicationAgent" + i);
(5)      EventResAgent ev = new EventResAgent("subordinateDesignation", arguments,
                           this.getAgentName(), "ReactiveRobotApplicationAgent" + i);
(6)      agentUseItf.acceptEvent(ev);  }
(7) catch (Exception e1){ e1.printStackTrace(); }
```

Fig. 6 Sending an event from agent to agent

Next, an example is used to illustrate how an event containing information is sent by an agent (see Fig. 6). First, an array of objects is created with a size equal to the objects to be sent, and then it is initialized (a value of type Corner and another one of type Integer is assigned in line 1). Let us highlight that the order of assignment should be the same that appears in the parameters definition of the semantic action that will be executed upon receiving the event (in this example subordinate action). After that the interface repository use interface is retrieved (line 2) and it is used to get the use interface associated with the agent instance to which the event will be sent (line 4). Then, the event is created (line 5). The first value passed is the type of event to send, the second one is the information contained, and the third and fourth ones are the source and target, respectively. Finally, the event is sent using the use interface (line 6) retrieved in line 4. Notice that sending an event to an agent from a resource follows this same idea.

Fig. 7 (a) Organization description XML file; (b) A fragment of the execution interface

3.5 Organization File Description

The application organization is described through an XML file (in our case it is "RobotSimulation.xml") that conforms to "OrganizationDescription-Schema.xsd" file. Firstly, the XML file describes the organization components features: managers' behavior, application agents' behavior, and application resources description. Secondly, the application instances are defined (see Fig. 7a). The instance identifier is the value provided in id (observe the numbering used to distinguish among different instances), and the type is the value given into descriptionRef. Moreover, the instances managed for each manager are also specified (see values in idRef attribute). Finally, a script file in bin folder should be created to launch the developed application. This file will contain the following command: 'ant -buildfile=../build.xml run -DdescriptionPath= DESC' (in our case DESC is RobotSimulation). Fig. 7b depicts a fragment of the execution interface.

4 Conclusions

The study of ICARO-T framework, and the implementation of a case study using this framework, enables reaching the following conclusions. In order to cope with application development complexity, the availability of architectural framework ICARO-T facilitates the development of MAS in several ways. (1) The categorization of entities either as agents or resources implies a clear design choice for the developer. (2) Environment can be modeled as a set of resources, with clear usage and management interfaces. There are standard patterns and mechanisms in the framework to facilitate their access. (3) Management of agents and resources follows certain patterns and most management functionality is already implemented. This relieves the developer of a considerable amount of work, and guarantees that the component will be under control. (4) The framework enforces a pattern for system initialization, which is particularly important in MAS where multiple distributed entities have to be initialized consistently, and this turns out to be a complex issue in many systems. (5) Agents work as autonomous entities and encapsulate their behavior (reactive, cognitive) behind their interfaces.

On the other hand, two disadvantages have been found. (1) The developer needs to know and to manually create the structure of all files (xml and java). And, (2) the current version of ICARO-T can only be run on a computer. In the future an agent-oriented methodology will be chosen in order to model an ICARO-T application and generate code skeletons from the models. In this way, the developer just will need to learn how agents and resources send events and how agents access the methods provided by the resources. Moreover, we are thinking in extending this study case in order to propose more complex coordination protocols (e.g. what happens when there are simultaneous alarms) and checking the framework's suitability. After that, the next task will be to use ICARO-T in physical robots instead of simulated environments.

Acknowledgements. This work was partially supported by Spanish Ministerio de Ciencia e Innovación TIN2007-67586-C02-02, and Junta de Comunidades de Castilla-La Mancha PII2I09-0069-0994 and PEII09-0054-9581 grants.

References

1. Bordini, R.H., Dastani, M., Dix, J., El Fallah Seghrouchni, A.: Multi-Agent Programming: Languages, Platforms and Applications. Springer, Heidelberg (2005)
2. Braubach, L., Pokahr, A.: (2009), `http://jadex.informatik.uni-hamburg.de/bin/view/Links/Agent+Platforms`
3. Caire, G., Coulier, W., Garijo, F., et al.: Agent-oriented analysis using MESSAGE/UML. In: Wooldridge, M.J., Weiß, G., Ciancarini, P. (eds.) AOSE 2001. LNCS, vol. 2222, pp. 119–135. Springer, Heidelberg (2002)
4. Celorrio, C., Verdejo, M.F.: Adapted activity deployment and configuration in a pervasive learning framework. In: Pervasive Learning, pp. 51–58 (2007)
5. Fernández-Caballero, A., Gascueña, J.M.: Developing multi-agent systems through integrating Prometheus. In: INGENIAS and ICARO-T. CCIS, vol. 67, pp. 219–232 (2010)
6. Garijo, F., Polo, F., Spina, D., Rodríguez, C.: ICARO-T User Manual. Technical Report, Telefonica I+D (2008)
7. Garijo, F., Bravo, S., Gonzalez, J., Bobadilla, E.: BOGAR_LN: An agent based component framework for developing multi-modal services using natural language. In: Conejo, R., Urretavizcaya, M., Pérez-de-la-Cruz, J.-L. (eds.) CAEPIA/TTIA 2003. LNCS (LNAI), vol. 3040, pp. 207–220. Springer, Heidelberg (2004)
8. Garijo, F., Tous, J., Matias, J.M., Corley, S., Tesselaar, M.: Development of a multi-agent system for cooperative work with network negotiation Capabilities. In: Albayrak, Ş., Garijo, F.J. (eds.) IATA 1998. LNCS, vol. 1437, pp. 204–219. Springer, Heidelberg (1998)
9. Gascueña, J.M., Fernández-Caballero, A.: On the use of agent technology in intelligent, multi-sensory and distributed surveillance. Know. Eng. Rev. (2009) (to appear)
10. Gómez-Sanz, J., Pavón, J., Díaz-Carrasco, A.: The PSI3 agent recommender system. In: Cueva Lovelle, J.M., Rodríguez, B.M.G., Gayo, J.E.L., Ruiz, M.d.P.P., Aguilar, L.J. (eds.) ICWE 2003. LNCS, vol. 2722, pp. 30–39. Springer, Heidelberg (2003)

11. Gómez-Sanz, J., Pavón, J., Garijo, F.: Intelligent interface agents behavior modeling. In: Cairó, O., Cantú, F.J. (eds.) MICAI 2000. LNCS (LNAI), vol. 1793, pp. 598–609. Springer, Heidelberg (2000)
12. Luehrs, R., Pavón, J., Schneider, M.: DEMOS tools for online discussion and decision making. In: Cueva Lovelle, J.M., Rodríguez, B.M.G., Gayo, J.E.L., Ruiz, M.d.P.P., Aguilar, L.J. (eds.) ICWE 2003. LNCS, vol. 2722, pp. 525–528. Springer, Heidelberg (2003)
13. Pavón, J., Gómez-Sanz, J., Fernández-Caballero, A., Valencia-Jiménez, J.J.: Development of intelligent multisensor surveillance systems with agents. Rob. Aut. Syst. 55, 892–903 (2007)

REST-A: An Agent Virtual Machine Based on REST Framework

Abdelkader Gouaïch and Michael Bergeret

Abstract. This article presents a framework based on agent concepts and REST architectural style. Based on this framework, an agent virtual machine along with its operational semantics are introduced. The idea is to consider agents' actions as manipulation of resources within environments using only a limited set of primitives. This makes both the agent abstract machine and its operational semantics easy to comprehend and the implementation straightforward. Finally, performances of our implementation are evaluated using a simple benchmark application.

1 Introduction

Many efforts have been recently made within multi-agent systems (MAS) community to facilitate building large-scale distributed software systems using agents as building blocks. In fact, starting from fundamental works in the 80's MAS have been considered as an appropriate framework when both the control and data are inherently distributed. Despite its potential, MAS are not yet considered as a mainstream paradigm for large scale distributed software engineering. This can be partially explained by the lack of practical frameworks that tackle current distributed systems challenges such as reliability and distributed data management. We want to address this shortcoming by specifying a framework and an agent virtual machine with the following characteristics:

- Expressiveness and pragmatism: Most of works on agent oriented programming are mostly based on logic. These works offer solid theoretical basis to develop theories about agents and reason about their actions. However, the dominant culture of programming is imperative which makes adopting logic based agent-programming languages difficult. It would be interesting to have a hybrid

Abdelkader Gouaïch · Michael Bergeret
University of Montpellier, CNRS, LIRMM, Montpellier, France
e-mail: gouaich@lirmm.fr, bergeret@lirmm.fr

Y. Demazeau et al. (Eds.): Advances in PAAMS, AISC 70, pp. 103–112.
springerlink.com

approach using logic to express states of affairs and let the control expressed imperatively.
- Target resource-oriented systems: Within MAS many architectures have been proposed for building software systems [1, 11]. However, most of these frameworks are centered on the concept of agent or organization and have neglected persistence and data management. This contrasts with current large scale software systems, such as Grids [9], are resource centered. Our goal is to fill this gap by suggesting a framework that coherently encapsulates agents to represent activity aspects and resources that represent data/resource aspects.
- Efficiency: We want to propose a framework that is consistent with MAS and a virtual machine built with solid theoretical foundations. Meanwhile, the implementation must be efficient enough to ensure scalability.

The rest of the paper is organized as follows: section 2 presents the Agent-REST framework; section 3 presents the formal model the agent abstract machine; section 4 presents the implementation of the agent virtual machine (VM); section 5 presents a simple test case and a benchmark; section 6 presents related works; and finally section 7 concludes the paper.

2 REST-Agent Framework

REST (REpresentational State Transfer) is an architectural style introduced by T. Fielding [5] to facilitate building networked applications. The central concept of REST is *resource*: a resource is identified using a uniform resource identifiers (uri) and owns an internal state. Access to a resource is performed using HTTP operations. The semantics of these operations is expressed as follow:

- 'put' and 'delete' are used to respectively store and delete resources
- 'post' modifies the state of a resource
- 'get' is used to retrieve a representation of the resource's state

REST has gained popularity and is becoming a *de facto* architectural style of Interned based applications. This is explained by the flexibility, scalability and efficiency offered by REST architectures, which are consequences of encapsulation of states within resources rather than in processes. The other advantage of REST is its simplicity and expressiveness, which are very useful to diminish complexity and consequently ease maintenance of large scale systems.

We present REST-Agent framework (Figure 1) that combines both REST and MAS concepts as follow:

- Environment: an environment represents a container of resources and agents. The environment handles agents' commands (or influences) to either modify resources' state or read their representation using respectively 'post' and 'get' commands
- Resource: a resource encapsulates an internal state and is identified by a uri
- Agent: an agent is an autonomous process that reads resource representations as inputs and submit influences on the environment

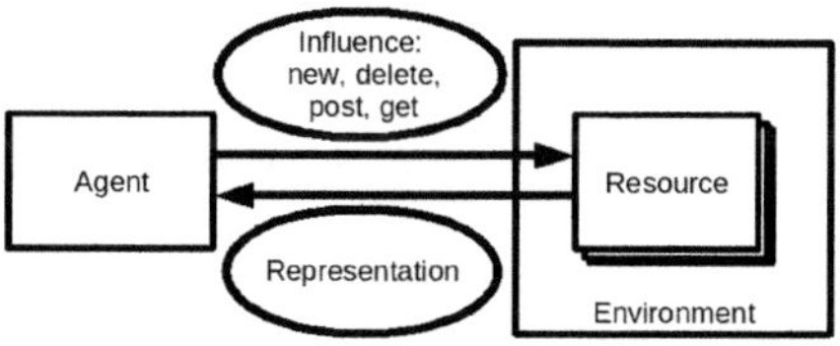

Fig. 1 The Agent-REST framework

- Influence: the term 'influence' is preferred to 'command' because an agent does not modify directly a resource's state. Thus, a resource state modification is always an attempt that has to be validated by the environment.
- Resource representation: as said previously, a resource is an abstraction and agents have access only to their representations. Resource representation are taken as perceptions within the agent's deliberation process.

3 Agent Abstract Machine

3.1 Modeling Influences and Representations

Influences and resource representations are expressed as syntaxic terms using the following definitions.

Definition 1. Let N be a countable set of names and V a countable set of values.

We build a set of influence terms $S_{\text{Influence}}$ as a syntaxic construction over N and V as follows (with $n \in N$, $v \in V$):

$$
\begin{aligned}
S_{\text{Influence}} ::= \ & \textbf{get}\ n \\
& \textbf{post}\ n\, v \\
& \textbf{new}\ n \\
& \textbf{delete}\ n
\end{aligned}
$$

Terms for resource representation is also built syntaxically over N and$V$ as follows (with $n \in N$, $v \in V$):

$$S_{\text{Representation}} ::= \textbf{repr}\ n\, v$$

3.2 Behavior Structure

Definition 2. Let C be a countable set of *capabilities*. The set of behaviors generated from C is defined by the power set $\mathscr{P}(C)$ equipped with set union as composition law. As a notation, we represent the structure $(\mathscr{P}(C), \cup)$ as $(C, +)$.

In the previous definition, the notation $\mathscr{P}(C)$ denotes the power set of C and $\cup$ denotes set union law.

Example 1. Let suppose a turtle agent [15] with the following capabilities:

$$C = \{\uparrow, \downarrow, \rightarrow, \leftarrow, \text{penup}, \text{pendown}, \text{canMove}\}$$

These capabilities are interpreted as follows: a turtle is able to move in four directions (up, down, right and left); put down a pen to start drawing, take off the pen to stop drawing, and finally check if the movement is allowed (for instance by checking if the turtle has enough energy).

From this set we are able to create behaviors using the composition law. For instance, $(\rightarrow + \text{pendown}) \in (C,+)$ expresses the willing to turn right and *simultaneously* putting down the pen.

Definition 3. We denote by $(C,+,.)$ the freely generated monoid having as generator the set $(T,+)$; . (dot) is the formal composition law.

Example 2. By introducing the . (dot) law, we are able to express a second level of composition that is not commutative. This could be useful for instance to express guards. For instance, one can construct formally the following element: $\text{canMove}.(\rightarrow + \text{pendown}) \in (T,+,.)$ to express the fact that 'canMove' must succeed before turning right *and* starting the draw mode.

To be evaluated, a behavior needs input representing agent's perceptions. To represent this unit of evaluation comprising a behavior and its associated set of perceptions we introduce the concept of *taskgram.*

Definition 4. A *taskgram* is defined as a couple (t,i) element of $(B,+,.) \times \mathscr{P}(S_{\text{Perception}})$ where t represents a behavior and i is its associated input data.

As an execution unit a taskgram is evaluated *as a whole* using an evaluation device to produce a set of influences:

Definition 5. An evaluation device M is considered as a black-box capable of producing a mapping between a taskgram and a set of influences. Formally, this device is modeled as a partial function:

$$\begin{aligned} M : (B,+,.) \times \mathscr{P}(S_{\text{Perception}}) &\rightarrow \mathscr{P}(S_{\text{Influence}}) \\ (t,i) &\mapsto M(t,i) \end{aligned}$$

with the following properties:

$$M(\{l\},i) = eval_M(l,i) \tag{1}$$

$$M(\emptyset,i) = \emptyset \tag{2}$$

$$M(u+v,i) = M(u,i) \cup M(v,i) \tag{3}$$

$$M(u.v,i) = \begin{cases} M(u,i) \cup M(v,i) & \text{iff}\, M(u,i) \neq \emptyset \\ \emptyset & \text{otherwise.} \end{cases} \tag{4}$$

In the previous definition, $\emptyset$ denotes the empty set and $eval_M$ denotes an evaluation function producing an influence given a capability label l and a perception i.

Equation (1) indicates that any singleton containing a capability label is directly evaluated by the device; (3) indicates that the evaluation process is linear on + law. (2) and (3) says simply that evaluation device acts as monoid homomorphism. (4) defines an evaluation of behaviors of the form $u.v$: whenever the evaluation of the first operand is an empty-set then the overall expression is evaluated as an empty-set. This models a conditional evaluation of task v according to u.

3.3 Fluents

Fluents have been introduced by situation and fluent calculus to represent facts about situations [6]. We use a simpler notion of fluents to express conditions that trigger behaviors. Conditions are expressed on resource representations using selectors. A selector is a pattern used to match some uri and select the value of an attribute. For conciseness, the full syntax of fluents is not presented; only notations illustrate how they are express and evaluated.

Definition 6. *Fluents* are expressed through syntactic constructs as follows:

$$P ::= exp \mid \rho_i \underbrace{(P,...)}_{i\,\text{elements}} \qquad exp ::= uri\_selector \mid \Omega(exp, exp)$$

Where ρ_i denotes usual logic operators of arity i such as : $\{\top, \bot, \neg, \wedge, \vee\}$ and Ω denotes usual comparison operators of arity 2 such as: $\{=, \neq, <, >\}$.

3.4 Environment

The presented framework follows ideas developed in [7, 10] to expose environments explicitly as a first-class entity within a multi-agent system. From an agent perspective, an environment is considered as a container of resources and interaction among agents holds through common resources. In fact, to exchange data agents have to perform simply post and get operations on common resources shared through uri. In the theoretical model, an environment is considered as a black-box with two primitives to: (i) send an influence to a resource, (ii) and get a resource representation. These operations are noted respectively as follow: $E(r)$ and $E \leftarrow p$ where r represents an influence of type 'get' and p a request of type 'post'.

3.5 Agent Abstract Machine

An agent program is given by describing a set of behaviors; a set of fluents and a set of triggers that relate fluents to behaviours.

Definition 7. Let C be a set of capabilities; an *C-agent* program is given by a 3-uplets (B, F, G) where:

- $B : \{b_i \in (C, +, .)\}$ is a set containing behaviors
- $F : \{f_i\}$ is a set containing fluents
- $G \subseteq F \times B$ is a trigger relation between fluents and behaviors

The operational semantics methodology suggests to specify formally all states or configurations of the abstract machine. The dynamics of this machines is then given as evolution rules linking configurations.

Definition 8. A configuration of the agent abstract machine is given by the 5-uplet (s, b, i, o, n) where:

- $s \in \mathbb{Z}/5\mathbb{Z}$ represents the state of the machine; 5 stages are needed for the complete evaluation cycle.
- $b \subseteq B$ represents the set of active behaviors
- $i \in \mathscr{P}(S_{\text{Perception}})$ represents the set of current perceptions
- $o \in \mathscr{P}(S_{\text{Influence}})$ represents the set of current influences
- $n \in \mathbb{N}$ represents the cycle number.

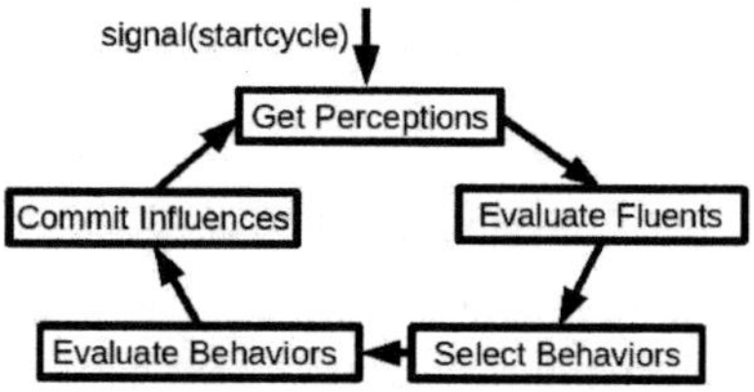

Fig. 2 Agent abstract machine phases

The agent abstract machine dynamics is characterized by five stages:

1. Get perceptions phase: this is the starting point of each cycle where agent's perceptions are gathered by retrieving requested resource representations
2. Evaluate fluents phase: during this phase all fluents are evaluated to determine their truth values
3. Select behaviors phase: the last phase has calculated the truth values of fluents which permit to select triggered behaviors
4. Evaluate behaviors phase: all triggered behaviors along with perceptions are used to construct taskgram that are submitted to the evaluation device
5. Commit influences phase: the evaluation of taskgrams produces a set of influences that are submitted to the environment. This ends the current cycle and a new cycle is started when a heartbeat signal is received.

For conciseness, formal rules are not presented in this article and could be found in [8].

4 Implementation of the Abstract Machine

The architecture of the MAS interpreter is described in Figure 3. Environments are responsible for: (i) updating resource states when getting new, delete and post influences and (ii) retrieving resource representations in case of a get influence. The evaluation device is responsible for computing taskgrams to produce influences. The current implementation takes advantage from taskgram properties: In fact, as demonstrated by functional programming [14], computations that are bounded to a static data set and that avoid side effects are efficiently parallelized. Taskgrams exhibit such property so a multi-threaded evaluation device has been used to enhance performances.

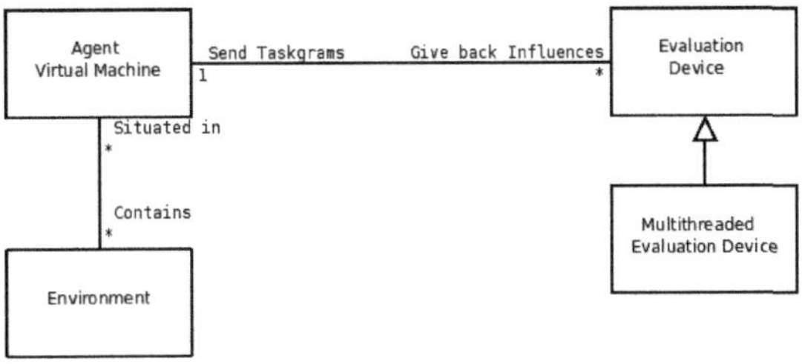

Fig. 3 Architecture of the MAS interpreter

5 Benchmark

The scenario of the benchmark application is inspired from predator-prey scenario. A predator agent that moves within a 2D grid tries to catch preys that flee. First, we present how this game has been modeled and implemented and then present some performance results. Three types of resources are defined for the game:

Resource	Description	URI
Arena	This is a 2D grid composed of cells. Each cell has a position (x,y) and can hold an agent	/arena/cell
Prey	Represents a prey agent. Attributes are: location, velocity, and angle.	/prey
Predator	Represents a predator. Attributes are: location.	/predator

The set of capabilities is as follow: (i) *flee* : according to the near position of the predator, this capability calculates a direction to flee; (ii) *restore* : no predator in sight, don't move and restore energy; and (iii) do nothing.

Two fluents are used: (i) *predatornear* : that holds when the predator is in vinicity; and (ii) *tired* : the prey agent is tired, and his velocity isn't in his maximal power.

Triggers are as follow:

- $predatornear \Longrightarrow flee$
- $\neg predatornear \wedge tired \Longrightarrow restore$
- $\neg predatornear \wedge \neg tired \Longrightarrow donothing$

5.1 Results

The game described in the last section was considered as a benchmark to test performances of the agent VM. In this case, the predator moves randomly. Results[1] of benchmark are presented in Figure 4. They show that despite the large number of agents that have been used, the implementation gives interesting performances: the duration time progresses linearly with the increase of agents. This is a interesting behavior that demonstrates that the implementation scales with the number of agents. This is due to the use of parallel evaluations and Stackless Python library to evaluate taskgrams.

#agents\#iteration	100	500	1000
100	0,097 s	0,311 s	0,549 s
500	0,272 s	1,491 s	2,817 s
1000	0,565 s	2,407 s	5,179 s
5000	2,565 s	12,828 s	34,022 s

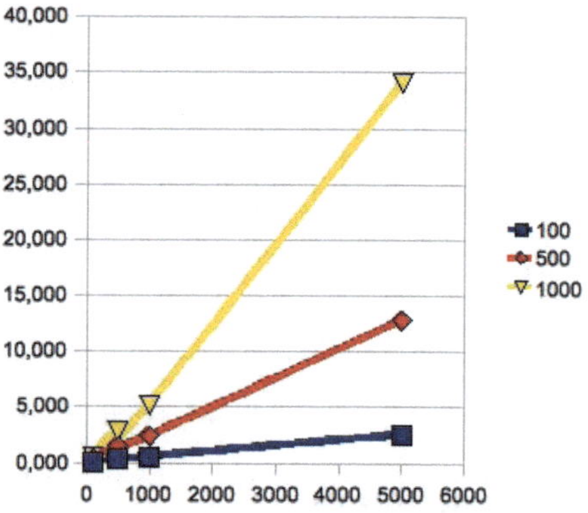

Fig. 4 Prey-Predator benchmark results (Y duration, X number of agents)

However, it is important to note that the presented scenario does not explore explicit collaboration between agents. Future works will explore collaborative scenarios that require agents to share and synchronize some resources.

6 Related Works

As mentioned in [3] works on agent programming fits in the classical decomposition of programming languages being either declarative or imperative. Declarative approaches are represented by languages based upon Prolog and BDI, like AgentSpeak [13] or 3APL [12]. These approaches often suffer from practices of software engineering and complexity of real world applications that cannot be captured at once as a set of first-order logical expressions. For imperative style one can mention

[1] Hardware and software configuration: Intel(R) Core(TM)2 Duo CPU E7300 @ 2.66GHz running under Linux Ubuntu 9.04.

works like Jack Intelligent System [17] or CLAIMS [4] to model mobile agents. This approach is syntactic and properties are expressed as rewriting rules. As consequence, the operational semantics is a big collection of rules that are not easy to comprehend. Our approach uses algebraic semantics to embed properties within algebraic structures. REST-A framework can also be related to MAS works on using environments as first class entities [16] These works have emphasized the central role played by the environment as an *artifact* of communication and coordination. However, artifacts are quite heterogeneous ranging from simple blackboards to complex tuple spaces. When building a MAS that involves different types of artifacts, agents have to use multiple adapters, which makes agent's implementation more difficult to maintain. To deal with this problem, REST-A can be regarded as a generic wrapper of artifacts: agents uniformly access resources located on heterogeneous artifacts using only a limited set of standardized primitives.

7 Conclusion

The framework presented in this paper defines a simple but yet powerful way to program agent oriented application. The dynamics of the abstract machine expresses a well specifies agent's cycle, with clear transition from a step to another. The evaluation device ensures computation of influences within a timely-bounded period. The framework has been implemented and the benchmarking shows some interesting and promising results. Future works will consider testing this framework with communicating agents to evaluate the performances when multiple agent exchange messages. Since exchanging messages is implemented as shared resource modification, the behavior of the implementation as described in this paper is not expected to change radically.

References

1. Bellifemine, F.L., Caire, G., Greenwood, D.: Developing Multi-Agent Systems with JADE. Wiley, Chichester (2007)
2. Bordini, R.H., Dastani, M., Dix, J., Fallah-Seghrouchni, A.E. (eds.): Multi-Agent Programming: Languages, Platforms and Applications. Multiagent Systems, Artificial Societies, and Simulated Organizations, vol. 15. Springer, Heidelberg (2005)
3. Bordini, R.H., Dastani, M., Dix, J., Seghrouchni, A.E.F.: Multi-Agent Programming: Languages, Tools and Applications. Springer Publishing Company, Heidelberg (Incorporated) (2009)
4. Fallah-Seghrouchni, A.E., Suna, A.: Claim and sympa: A programming environment for intelligent and mobile agents. In: Bordini et al (eds.) [2], pp. 95–122
5. Fielding, R.T.: Architectural Styles and the Design of Network-based Software Architectures. PhD thesis, University of California (2000)
6. Giacomo, G.D., Levesque, H.J.: An incremental interpreter for high-level programs with sensing. In: Logical Foundation for cognitive agents: contributions in honor of Ray Reiter, pp. 86–102 (1999)

7. Gouaich, A.: Requirements for achieving software agents autonomy and defining their responsibility. In: Nickles, M., Rovatsos, M., Weiss, G. (eds.) AUTONOMY 2003. LNCS (LNAI), vol. 2969, pp. 128–139. Springer, Heidelberg (2004)
8. Gouaich, A., Bergeret, M.: An operational semantics of a timely bounded agent abstract machine. Technical Report RR-09028, University of Montpellier, CNRS, LIRMM (2009)
9. Gouaïch, A., Cerri, S.A.: Movement and interaction in semantic grids: dynamic service generation for agents in the mic* deployment environment. In: Dimitrakos, T., Ritrovato, P. (eds.) LeGE-WG. Workshops in Computing. BCS, vol. 4 (2004)
10. Gouaïch, A., Michel, F.: Towards a unified view of the environment(s) within multi-agent systems. Informatica International Journal 29(4), 423–432 (2005)
11. Gutknecht, O., Ferber, J.: The madkit agent platform architecture. In: Infrastructure for Multi-Agent Systems, London, UK, pp. 48–55 (2001)
12. Hindriks, K.V., Boer, F.S.D.: Agent programming in 3apl. AAMAS Journal 2, 357–401 (1999)
13. Rao, A.S.: Agentspeak(l): Bdi agents speak out in a logical computable language, pp. 42–55. Springer, Heidelberg (1996)
14. Roe, P.: Parallel Programming using Functional Languages. PhD thesis, Glasgow University (1991)
15. Tisue, S., Wilensky, U.: Netlogo: A simple environment for modeling complexity (2004)
16. Weyns, D., Schumacher, M., Ricci, A., Viroli, M., Holvoet: Environments in multiagent systems. The Knowledge Engineering Review 20(2), 127–141 (2006)
17. Winikoff, M.: Jack intelligent agents: An industrial strength platform. In: Bordini, et al. (eds.) [2], pp. 175–193

Detection of Overworked Agents in INGENIAS

Celia Gutierrez and Ivan García-Magariño

Abstract. Overworking behaviors appear in multi-agent systems communication quite often. This occurs when an agent receives many messages in a short period of time. As the agent pays attention to the large amount of messages it worsens its performance, causing the system's performance to worsen as well. The reasons for this behavior are varied and depend on the nature of the messages: an agent that sends messages to the same agent when it should have sent it to a group of agents in a balanced way; an agent that does not receive a quick response becomes impatient and keep on sending the same message to the same agent until it gets a response, and so on. This article presents a technique that detects the overworked agents in INGENIAS with a new metric and efficiently measures it with a new version of the INGENIAS Development Kit. The experimentation of this technique advocates that the existence of overworked agents is strongly related with the quality of service of multi-agent systems.

Keywords: Metric, Multi-agent Systems, Overworking, Quality of Service, Response Times.

1 Introduction

Metrics are useful for designing Multi-agent System (MAS) communications [3] when systems are complex for varied reasons: the number of agents, the protocol communication, or the negotiator agent selection policy. There are some undesirable behaviors in MAS that can cause delays when executing a MAS, such as the *overworking situations*. In these situations, an agent is overworked because of receiving an excessive number of messages from other agents in a short period of time. The consequence of the misuse in this communication is usually that an agent overworks. This agent provides a worse response time because it is busy

Celia Gutierrez · Ivan García-Magariño
Departamento Ingeniería de Software e Inteligencia Artificial
Facultad de Informática
Universidad Complutense de Madrid
Prof. Jose Garcia Santesmases, s/n, 28040, Madrid, Spain
email: {cegutier,ivan_gmg}@usal.es

Y. Demazeau et al. (Eds.): Advances in PAAMS, AISC 70, pp. 113–118.
springerlink.com

paying attention to the continuous requests, causing the overall response time to be higher. Response time is a parameter to measure the Quality of Service (QoS). Therefore, it is necessary to detect overworking situations. This work presents a metric for detecting overworking situations.

The INGENIAS [5] Agent Oriented Software Engineering (AOSE) methodology can incorporate an stage in which the running system is tested with the presented metric. This methodology is selected because its model-driven approach allows designers to develop and modify MAS with a low effort. Thus, a system can be changed with a low cost if the metric reveals that a design is overworking a particular agent. The INGENIAS Development Kit (IDK) [2] is modified in order to write logs with the timestamps of messages. These logs facilitate the measurement of the presented metric by means of an external tool.

This article is structured as follows: section 2 describes the overworked situations in MAS in a formal way and the background; section 3 contains the experimental stage of this approach and the conclusions.

2 Metrics for Detecting Overworked Agents and Background

An agent suffers from an overworking situation if it receives an excessive number of messages in a short period of time computed during a MAS execution. For this reason, two thresholds are defined for respectively determining the amount of messages and the duration of the period:

- *OT* is the overload threshold.
- *DMW* is the duration of measurement window.

Current research uses the Foundation of Intelligent Physical Agents (FIPA) standard; hence the messages follow its notation. Taking this into account, it is possible to accurately detect the overworked situations suffered by an agent, by considering the nature of the messages that are received by this agent. For instance, the *Request* messages usually require more work than the *Inform* messages. Thus, the received messages are weighted regarding the types of the messages. An overworking situation is defined as follows:

Def: An agent A_i suffers an *overworking situation* between two time instants t_i and t_f if both of the following conditions are satisfied, as in Equations (1) and (2):

- $$|t_i - t_f| \leq DMW \quad (1)$$
- $$\sum_{x \in R(A_i, t_i, t_f)} w(x) \geq OT \quad (2)$$

where:

- $R(A_i, t_i, t_f)$ is the set of the messages that are received by the agent A_i between the t_i and t_f instants.
- $w(x)$ is the weight assigned to a message *x* according to its type.

From the experience, this work recommends some values for the constants DMW and OT and some weights for some message types, all of which are presented in Table 1.

Table 1 Recommended Values for detecting Overworking Situations

Constants		Weights of Messages	
Constant	Value	Type of Message	Weight
OT	25	Inform	0.5
DMW	2 seconds	Request	1.0

In order to determine the grade in which an agent overworks, this work defines the *Overworked (O)* metric that counts the number of overworking situations for an agent per unit of time, as in Equation (3):

$$O(A_i)= S(A_i) / T_E \quad (3)$$

where:

- $S(A_i)$ is the number of overworking situations of agent A_i that are not overlapped in time.
- T_E is the time of the MAS execution.

From the experience, this work recommends MAS not to have any agent whose Overworked measurement exceeds *10 times per minute* (*t.p.m.*). In other words, agents are not recommended to approximately receive more than 25 messages in 2 seconds, and if this situation occurs, at least it should not occur more than 10 t.p.m.

It is possible to detect the origin of the some overworking situations by keeping track of the content of the field messages. In this case, one can notice if a group of messages belonging to a certain agent and type is causing damage by overworking another one.

In the literature, several works take overworking into account when designing MAS architectures. [7] presents a distributed MAS architecture to assist users to arrange meetings, so that the solution must satisfy the users' meeting requirements and preferences. This works presents advantages over the current meeting scheduler systems. The use of TCP/IP sockets speeds up the scheduling process for inter-agent messages and communication with the server, because it avoids the situation of overworked email servers, which often take a lot of time to deliver email messages. The mentioned work does not measure the overworked agents; however, it just present architectures or techniques that try to avoid having them.

Furthermore, [4] is closer to our work. That work describes a performance measurement system built into a MAS architecture. Message transport metrics capture statistics on the messages for run-time adaptation and for post-mortem analysis. They show no evidence of an identification of the overworked agents, although these metrics can be internally used for load-balancing agents.

Works related with QoS in MAS appear several times in the literature. In many applications there are designs of MAS that reach a high QoS level but they differ on the types of solutions. In [6], they have used machine-learning techniques to build adaptive multimedia applications. A highly satisfactory QoS is provided as the system adapts the settings to the network capacity. In another work [1], a system was designed to allocate resources for cellular data services in such a way that it meets both customer satisfaction and cost effectiveness. In this case, a high level of QoS is achieved by building the agents within three modules. In contrast to those works, our approach detects the origin of a low QoS by considering the misuse of communications in terms of overworked agents.

3 Results and Conclusions

A cinema case study is selected because little differences in its design can imply noticeble differences in the number of overworking situations. As a brief introduction to the cinema case study, it is designed for the acquisition of cinema tickets of the user's preferences, i.e. the film, the starting time, the row and column in the cinema, and the price limit. This case study has been implemented with IDK, so that interactions between agents are defined. The measurement of the metric has been made with a greedy algorithm built into the running system. This algorithm checks the logs of the messages generated in the interactions between agents.

In the experiment there are three kinds of agents who have been depicted in this way:

- *BuyerAgent*, who represents the expert in purchasing the cinema tickets, with 100 instances.
- *InterfaceAgent*, who represents the user, with 100 instances.
- *SellerAgent*, who represents the cinema, with 7 instances.

The execution begins when a user requests a movie. It finishes when it is obtained: the ticket is delivered, through the different representatives, to the user.

Two configurations of this case study are measured with the presented metrics and the recommended constant values.

3.1 Cinema with an Overworked Agent

This version is characterized by the default selection of the BuyerAgent by the InterfaceAgent. This mainly causes two problems:

- In the interaction InterfaceAgent-BuyerAgent, of Request type and depicted with the IDK, the same BuyerAgent is selected: this implies that an agent overworks in several periods of times.
- In the interaction SellerAgent-BuyerAgent, of Inform type also depicted with the IDK, the same BuyerAgent receives all the messages: this causes that an agent suffers several overworking situations.

Tables 2 and 3 collect the results for this case:

Table 2 Overworked metric for the first configuration of cinema case study

Agent name	Overworked Metric (t.p.m)
BuyerAgent_0	0
BuyerAgent_1	48
...	(the same results as in uyerAgent_0)

Table 3 Time intervals and number of each type of requests for the overworking situations

Time Intervals of Overworking situations	REQUEST	INFORM
(0 – 1.23)	10	30
(1,24 – 2.47)	10	30
(2,48 –3.72)	10	30
(intervals of similar length)	...	...

These results reflect that there are overworking situations because the selection process has not been equitable. The messages that create the overworking situations revealed that the senders were all the Interface agents while the receiver was always the same Buyer agent, although there were others Buyer agents. These facts reveal that the cause of these overworking situations was a non-balanced selection mechanism in the Interface-Buyer interactions.

3.2 Cinema without Overworked Agents

This version is characterized by its random selection process of the BuyerAgent, whose results are collected in Table 4:

Table 4 Overworking metric for the second configuration of the cinema case study

Agent name	Overworking metric
BuyerAgent_0	0
...	(the same results as in BuyerAgent_0)

These results reflect that there are neither overworked agents nor overworking situations because the selection process has been equitable.

3.3 Review of Response Times

The response times of both configurations of the cinema case study are shown in Table 5. As one can observe, the configuration without overworked agents has shorter time responses and, consequently, provides a higher quality of service. These experiments advocate that there is a relationship between the existence of overworked agents (the Overworked metric) and a low QoS (in terms of high response times).

In conclusion, this work describes a metric, called Overworked, to detect overworking situations in MAS. This metric takes types of messages into consideration for being more accurate in detecting overworked agents. The presented metric considers short periods of times instead of the whole execution times. This metric is used to detect overworked agents in the INGENIAS methodology. The experiments of this work shows the relationship between the Overworking metric and the QoS in terms of response times.

Table 5 Response times in milliseconds for both cases

Number of iterations	Response time for the cinema case study with an overworked agent	Response time for the cinema case study without overworked agents.
10	6755	3572
20	8106	3992
30	8994	4435
40	9511	5412
50	10166	6419
60	11099	7540
70	11746	9306
80	12035	10203
90	12463	11488

Acknowledgements. This work has been done in the context of the project Agent-based Modelling and Simulation of Complex Social Systems (SiCoSSys), supported by Spanish Council for Science and Innovation, TIN2008-06464-C03-01.We also acknowledge support from Programa de Creación y Consolidación de Grupos de Investigación UCM-BSCH GR58/08.

References

1. Chen, J.L.: Resource allocation for cellular data service using multiagent schemes. IEEE Transactions Syst. Man Cybern, Part B 31(6), 864–869 (2001)
2. Gomez-Sanz, J., Fuentes-Fernández, R., Pavón, J., et al.: INGENIAS Development Kit: a visual multi-agent system development environment. In: Proceedings of the 7th International Conference on Autonomous Agents and Multiagent Systems, pp. 1675–1676 (2008)
3. Gutierrez, C., Garcia-Magariño, I.: A Metric Suite for the Communication of Multiagent Systems. J. Phys. Agents 3(2), 7–15 (2009)
4. Helsinger, A., Lazarus, R., Zinky, J.: Tools and techniques for performance measurement of large distributed multiagent systems. In: Proceedings of the 2nd International Joint Conference on Autonomous Agents and Multiagent Systems, pp. 843–850 (2003)
5. Pavon, J., Gomez-Sanz, J.: Agent Oriented Software Engineering with INGENIAS. In: Mařík, V., Müller, J.P., Pěchouček, M. (eds.) CEEMAS 2003. LNCS (LNAI), vol. 2691, pp. 394–403. Springer, Heidelberg (2003)
6. Ruiz, P., Botia, J., Gomez-Skarmeta, A.: Providing QoS through machine-learning-driven adaptive multimedia applications. IEEE Transactions Syst. Man Cybern., Part B 34(3), 1398–1411 (2004)
7. Shakshuki, E., Koo, H.H., Benoit, D., et al.: A distributed multi-agent meeting scheduler. J. Comput. Syst. Sci. 74(2), 279–296 (2008)

Mobile Agents in Vehicular Networks: Taking a First Ride

Oscar Urra, Sergio Ilarri, Thierry Delot, and Eduardo Mena

Abstract. A vehicular ad hoc network (VANET) is a type of mobile network whose nodes are traveling cars which communicate with one another using short-range wireless communications. These cars can exchange and share different information among them, which can lead to the development of interesting applications that require the cooperation of vehicles or using the vehicular network as a distributed computing platform. Along with the opportunities offered by vehicular networks, a number of challenges also arise. To ease the development of applications for vehicular networks, mobile agent technology may be of help.

However, even though previous works have proved the usefulness of mobile agents in wireless environments, the special features of VANETs call for new research efforts. Could mobile agent technology be useful in a VANET? In this paper, we study these issues and illustrate how mobile agents could drive the development of new applications for vehicular networks.

1 Introduction

In the last decade, a number of wireless and small-sized devices (e.g., PDAs or laptops) with increasing computing capabilities have appeared in the market at very affordable costs. These devices have started to be embedded into modern cars in the form of on-board computers, GPS navigators, or even multimedia centers.

This has led to the emergence of vehicular ad hoc networks (VANETs). In this kind of networks, cars traveling along a road can exchange information with other

Oscar Urra · Sergio Ilarri · Eduardo Mena
Department of Computer Science and Systems Engineering, University of Zaragoza, María de Luna 1, 50018, Zaragoza, Spain
e-mail: ourra@ita.es, {silarri,emena}@unizar.es

Thierry Delot
University Lille North of France, LAMIH UMR CNRS/UVHC, Le Mont Houy, 59313 Valenciennes Cedex 9, France
e-mail: Thierry.Delot@univ-valenciennes.fr

Y. Demazeau et al. (Eds.): Advances in PAAMS, AISC 70, pp. 119–124.
springerlink.com

nearby cars. The lack of a fixed communication infrastructure, characteristic of ad hoc networks, implies that vehicles usually can communicate with one another only using short-range wireless communications. Nevertheless, a piece of information can be disseminated and reach a far distance by using many moving cars as intermediates, following multi-hop routing protocols [3].

On the other hand, mobile agents [5] could be defined as computer programs that have the ability to move from one execution place to another, i.e., they can transfer themselves between devices using a communication link. They have a number of features, such as mobility, autonomy, communication capabilities, and flexibility, that make them a good choice for the development of distributed applications.

The question is: Can mobile agents and vehicular networks work together? Apparently, mobile agents may have interesting features that could make of them an interesting approach to build applications for vehicular networks. However, it is not clear how much they can help or whether existing mobile agent technology is ready for this challenge. In this paper, we study these issues and how mobile agents could drive the development of new applications for VANETs. The rest of the paper is structured as follows. In Section 2, we present the technological context of this work. In Section 3, we study the combination of VANETs and mobile agents to identify benefits and challenges. In Section 4 we present a use case scenario, and finally, in Section 5 we draw some conclusions.

2 Technological Context

In this section, we present the technological context of this work. Two central elements are identified: the vehicular ad hoc networks and the mobile agents.

2.1 Vehicular Ad Hoc Networks

A *vehicular ad hoc network* (VANET) [6] is a highly mobile network whose nodes are the vehicles traveling along a road or a highway. The vehicles are equipped with short-range wireless communication devices (e.g. using Wi-fi or UWB [4] technologies) and can establish direct connections with other nearby vehicles. This has a number of advantages over a traditional direct-connection client-server approach, such as: 1) there is no need of a dedicated support infrastructure (expensive to deploy and maintain), 2) the users do not need to pay for the use of these networks, and 3) it allows a very quick and direct exchange of information between vehicles that are within range of each other.

VANETs open up a wide range of opportunities to develop interesting systems for drivers. For example, when a number of vehicles detect that their average speeds are very low for a long time, it probably means that they are in a traffic jam and an information message can be delivered to other vehicles driving towards that area.

However, a number of difficulties also arise: Two vehicles can communicate directly only if their wireless devices are within range of each other (about 100 meters with 802.11 technologies) and, given that the vehicles are constantly moving, the duration of the communication link can be very short (a few seconds). So, to allow the information to reach remote sites, the use of some multi-hop communication protocol is necessary (e.g., see [3]).

The development of applications for vehicular networks requires taking these and other constraints into account.

2.2 Mobile Agents in a Mobile Environment

Mobile agents are software entities that run on an execution environment (traditionally called *place*), and can autonomously travel from place to place and resume their execution at the target node [5]. Thus, they are not bound to the computer where they are initially created and they can move freely among computers. To use mobile agents it is necessary to execute a middleware known as a *mobile agent platform* [8], which provides agents the environment where they can execute.

Mobile agents have been found particularly useful for the development of applications in mobile environments. Such an environment has a number of advantages (e.g., the processing is not tied to a fixed location) but also some drawbacks, such as the limited computational power of mobile devices and the communication constraints imposed by the use of wireless communications (that usually offer a low bandwidth, a high latency, and intermittent/unreliable connectivity). In this scenario, mobile agents can be very useful because they could help to reduce the negative effects of such limitations (e.g., see [7, 10]).

3 Mobile Agents and VANETs

In a VANET there are many vehicles, distributed on a wide geographic area, that exchange data based on certain conditions. The similarity with a situation where software agents move from one computer/device to another makes mobile agents a very suitable option to implement applications for VANETs. In this section, we describe potential benefits and difficulties for the adoption of mobile agents in VANETs.

3.1 Benefits of Using Mobile Agents in VANETs

In VANET applications, data may need to be transported from vehicle to vehicle in order to reach locations that are not directly accessible due to the short range of the wireless communications used. Then, two major problems arise. First, as the propagation of the data can be slow, the information can be outdated when it reaches its destination. Second, it can be difficult to determine the destination itself and how to reach it: It could be a specific car, every vehicle in an area, the vehicles matching a certain condition, etc. To deal with these drawbacks, mobile agents can be very useful because of their adaptability and mobility features. Thus, they can bring a processing task wherever it is needed, and the algorithm or agent's logic can be changed at any time by deploying new versions of the agent code.

Another important advantage of mobile agents in VANETs is that they can move to wherever the data are located, to process and collect only the relevant data (filtering out data which are unnecessary). In other words, they can perform a local data aggregation and filtering and thus reduce the network load.

Finally, we believe that mobile agents can be very useful for data dissemination in vehicular networks. Thus, they can adapt easily to changing environmental conditions in order to improve the dissemination. For example, a basic flooding dissemination protocol such as [2] will fail if the traffic density of the vehicles is low and there are not enough vehicles to re-diffuse the data towards the target. Other dissemination protocols, such as carry-and-forward [12], where the vehicles may hold the data to be transmitted until these data can be relayed to other vehicles, can be used in that case. However, considering the variety of existing dissemination protocols mobile agents seem an ideal technology since the routing decisions lie with the data (encapsulated in the mobile agents) and different dissemination protocols (dynamic and adaptive to the current conditions) can be implemented.

Summing up, we believe that mobile agents could be very useful in vehicular networks and that the applicability of this technology in a VANET calls for new research in the field. As far as we know, [1] is the only work that uses this technology in a vehicular field (besides our initial proposal in [9]). However, it focuses on the case of traffic management with a fixed support infrastructure and does not consider the general case of agents hopping from car to car in a vehicular ad hoc network.

3.2 Difficulties for the Adoption of Mobile Agents in VANETs

While the use of mobile agents in vehicular networks can bring interesting benefits, there are also some difficulties that could hinder their adoption in this context. In particular, it is not clear that current mobile agent platforms are completely ready to be used in an effective way in a vehicular network.

The most important challenges arise from the fact that a VANET is a highly dynamic environment where nodes can come and leave at any time and where it is not easy or convenient to assume the existence of any kind of centralized node that can be relied upon. Therefore, the design of the different services provided by a mobile agent platform should not be based on this assumption.

As an example, existing agent tracking services have been developed with a fixed network in mind and are not appropriate (or even useful) in a mobile environment. This service is used by the platform to track the location of all agents and execution places at every moment, and is also used by agents to communicate with others. In existing mobile agent platforms [8] (e.g., JADE or SPRINGS) this service is usually provided by a centralized component, whose tracking data are updated every time an agent moves to a new place. This approach would be challenging in a VANET, where centralized nodes do not exist and direct communications are not always possible. Instead, the platform could not even need to be aware of every single agent present in the vehicular network, and agents could be able to establish communication only with other nearby agents that are in range of their communication devices.

Other difficulties to use existing mobile agent platforms in a vehicular ad hoc network are related to communication and security issues. Thus, in a traditional distributed fixed network direct communications between nodes is the usual. However, in a vehicular network, when an agent wants to hop to another car it will be broadcasted and any vehicle within communication range could receive it. The agent must be re-started only in one receiving vehicle. Moreover, the agent should be conveniently encrypted and a security mechanism should be applied to allow its execution only if both the agent and the receiving vehicle trust each other. Similarly, other security issues could arise, most of them also existing in fixed networks but heightened by the open character of vehicular networks (e.g., see [11, 10]).

We believe that these are the most important issues to consider for the application of mobile agent technology in vehicular networks. Other potential difficulties, considered for the general case of mobile environments, are cited in [10].

4 Use Case Scenario: Monitoring the Environment

An example application scenario is that of using mobile agents in a VANET to monitor environmental parameters (such as CO_2, temperature, etc.) within a certain geographic area [9]. The vehicles involved in the monitoring tasks can carry one or more environmental sensors so that the interesting parameters can be read from the environment. The agents hop from car to car to reach the area that must be monitored and then try to remain within it boundaries (moving to other vehicles if necessary). So, moving cars are used as both physical carriers of the agents and intermediate points to reach the target area. Using existing vehicles and mobile agents for monitoring greatly increases the flexibility compared to an approach based on the use of static sensors. Thus, it becomes possible to monitor virtually any area that we want.

However, the initial solution proposed in [9] has an important limitation: the monitoring process always ends in a fixed and well-known site, where the monitoring results (carried by mobile agents) are gathered. Thanks to the adaptability and mobility features mentioned in Section 3.1, we are currently improving our approach so that it is also possible to start and end the process on any vehicle in the VANET (i.e., to issue monitoring tasks from a moving vehicle). The main difficulties encountered are: estimating the position of the destination car when the monitoring process ends, and then reaching that position using a multi-hop route.

5 Conclusions

Mobile agent technology can provide interesting benefits thanks to the mobility, autonomy and flexibility of mobile agents. However, while mobile agents have been largely studied in the literature in a variety of application scenarios, including mobile and wireless environments, the use of mobile agents in vehicular networks is a completely unexplored area. This paper represents a first step to cover this gap in the research of mobile agents and vehicular networks. Thus, we have presented the benefits that mobile agents can provide to develop applications for vehicular networks and the difficulties involved.

Acknowledgements. The authors thank the support of the CICYT project TIN2007-68091-C02-02, the ITA Institute of Aragón, and the OPTIMACS project supported by the French ANR agency.

References

1. Chen, B., Cheng, H., Palen, J.: Integrating mobile agent technology with multi-agent systems for distributed traffic detection and management systems. Transportation Research Part C 17(1), 1–10 (2009)
2. Heinzelman, W.R., Kulik, J., Balakrishnan, H.: Adaptive protocols for information dissemination in wireless sensor networks. In: 5th Annual ACM/IEEE International Conference on Mobile Computing and Networking (MobiCom 1999), pp. 174–185 (1999)
3. Lochert, C., Hartenstein, H., Tian, J., Fler, H., Hermann, D., Mauve, M.: A routing strategy for vehicular ad hoc networks in city environments. In: Intelligent Vehicles Symposium (IV 2003), pp. 156–161. IEEE Computer Society, Los Alamitos (2003)
4. Luo, J., Hubaux, J.-P.: A survey of research in inter-vehicle communications. In: Embedded Security in Cars - Securing Current and Future Automotive IT Applications, pp. 111–122. Springer, Heidelberg (2005)
5. Milojicic, D., Douglis, F., Wheeler, R.: Mobility: processes, computers, and agents. ACM, New York (1999)
6. Olariu, S., Weigle, M.C. (eds.): Vehicular Networks: From Theory to Practice. Chapman & Hall/CRC, Boca Raton (2009)
7. Spyrou, C., Samaras, G., Pitoura, E., Evripidou, P.: Mobile agents for wireless computing: the convergence of wireless computational models with mobile-agent technologies. Mobile Networks and Applications 9(5), 517–528 (2004)
8. Trillo, R., Ilarri, S., Mena, E.: Comparison and performance evaluation of mobile agent platforms. In: 3rd Intl. Conf. on Autonomic and Autonomous Systems (ICAS 2007). IEEE Computer Society, Los Alamitos (2007)
9. Urra, O., Ilarri, S., Delot, T., Mena, E.: Using hitchhiker mobile agents for environment monitoring. In: 7th International Conference on Practical Applications of Agents and Multi-Agent Systems (PAAMS 2009), pp. 557–566. Springer, Heidelberg (2009)
10. Urra, O., Ilarri, S., Trillo, R., Mena, E.: Mobile agents and mobile devices: Friendship or difficult relationship? Journal of Physical Agents. Special Issue: Special Session on Practical Applications of Agents and Multiagent Systems 3(2), 27–37 (2009)
11. Vigna, G. (ed.): Mobile Agents and Security. LNCS, vol. 1419. Springer, Heidelberg (1999)
12. Zhao, J., Cao, G.: VADD: Vehicle-assisted data delivery in vehicular ad hoc networks. In: 25th IEEE International Conference on Computer Communications (INFOCOM 2006). IEEE Computer Society, Los Alamitos (2006)

A Multi-Agent System Approach for Interactive Table Using RFID

Yoann Lebrun, Emmanuel Adam, Sébastien Kubicki, and René Mandiau

Abstract. This paper presents a model of Multi-Agent System dedicated to the management of a tabletop that detects traceable objects using RFID technology, which are moved on the table by a set of users during application. In addition, virtual objects can interact with tangible objects. We propose a new model (MAM4IT) to manage this kind of object with situated agents (that we call tangible agents) based on the notion of dynamic roles. The roles will be able to evolve during the application and adapt their behaviors according to the environment and the unpredictable actions of users. In this situation the agents are able to compose new roles. A first concrete case study based on road traffic shows the interaction between tangible and virtual agents.

Keywords: Multi-Agent System, Roles, situated agent, RFID, tabletop, tangible objects.

1 Introduction

The rise of the RFID technology leads up to what is called the *Internet of Things* characterized globally by the fact that the objects of the physical world have an identifier or even a memory allowing them, for example, to connect to Internet and to interact with virtual applications.

This allows a new kind of distributed and multi-user interactions curbing slightly the actions on these new applications through handled objects. The specification of applications dedicated to these new supports of communication, because of the

Yoann Lebrun · Emmanuel Adam · Sébastien Kubicki · René Mandiau
Univ Lille Nord de France, F-59000 Lille, France
UVHC, LAMIH, F-59313 Valenciennes, France
CNRS, FRE 3304, F-59313 Valenciennes, France
e-mail: {yoann.lebrun,emmanuel.adam,
sebastien.kubicki,rene.mandiau}@univ-valenciennes.fr

Y. Demazeau et al. (Eds.): Advances in PAAMS, AISC 70, pp. 125–134.
springerlink.com

dynamic, distributed and partially unforeseeable nature of user's actions, requires adapted methods and models. For that, we think that it is particularly relevant to turn us to the Multi-Agents Systems domain to bring a solution to the modeling, the specification and the design of the applications related to the *Internet of Things*.

The application field of our work is based on an interactive table, within the framework of TTT (interactive Table with Tangible and Traceable objects) project. This table contains 25 tiles (5×5). Each tile contains 64 RFID antennas (8×8) on which users can deposit, to move or to withdraw objects provided with one or several RFID tags.

Our objective in this project is to propose a generic application layer allowing the instantiation of any type of applications using such objects. As these objects comprising roles evolve according to the context of use, we propose an architecture of a system made up of mainly situated agents and based on the concept of dynamic role.

Indeed, the users can: modify, add roles to the tangible objects; associate several objects to define a new object with the combination of the various roles. One should do pay special attention to the concepts of role composition and decomposition by guaranteeing their coherences. This is the main part of our research in the TTT project.

This article describes in a first part our positioning compared to existing work. We describe then our proposal, the model MAM4IT (Multi-Agent Model for Interactive Table), which takes into account specificity of the field of application and detailing the concept of role dynamic. Finally, in the third part, an application of our proposal is presented; this one makes it possible to set up and to validate the various concepts met in our method.

2 Related Works

We think that role composition is a major feature of tangible objects manipulation. Indeed these objects play one or several roles according to the application. The objects can modify their behaviors and sometimes their roles dynamically at run-time according to the context of the application. The existence of dependencies between roles must also be underlined. Indeed, the more dependencies, the more difficult it is to manage relationships between these roles (aggregate/disaggregate).

To allow the use of such an approach, we have studied different models may responding to the project needs. We took inspiration from the "Vowels"[4] formalism in order to structure our criteria: the set of Agents (A), the Environment (E) where agents evolve, under an Organization (O), Interacting (I) between them with their roles and can be User (U) centered. The table 1 presents a sum-up the properties that must own a multi-agent system dedicated to the management of tangible objects:

- Agents, tangible objects have to be managed by 'tangible' agents connected to the physical world; in the opposite, 'virtual' agents represent the virtual part of the application.

- Environment, it represents the Cartesian plan of the application displayed on the interactive table. It must be dynamic, because the user's actions are unpredictable and modifying constantly the environment (for example: when tangible objects are moving).

  Each object of the environment is visible by the users; and according the characteristic of user that uses it or observes it, the object could have to display specific information and adapt the information displaying relatively to the position of the others objects (virtual and tangible). That is why, environment objects have to be managed by virtual agents.

  They update regularly their environment which is partially observable, contrarily to a system agent having a global vision of the environment. Physical constraints are added to the environment, caused by the instability of the system, like the detection of the TAG on the table surface. For example, a unique TAG could be detected at several position simultaneously (i.e on several antennas).
- Interaction, we distinguish two types of interactions: interactions between agents and interactions between roles. The agent can add/delete a role from its list but also transfer roles to another agent. Interactions between roles of an agent represents dialogues between them during step of adding and/or retracting of a role in/from the agent. This role interact with the other in order to check the feasibility of such an action.
- Organization, there is no particular constraint on the organization of a MAS dedicated to the tangible table management, except the fact that the MAS is composed both of 'tangible' agents and 'virtual' agents. The organization mainly come from the application needs, and the action of the users on the agents.
- User, they induce the relatively unforeseen and dynamic part of the global system. Their actions have to be recorded/learned in order to be able to foresee their future ones.

Table 1 Properties description of the application layer

Vowels formalism	Properties
Agent	virtual, tangible, situated, reactive
Environment[13]	dynamic, (partially) observable, discrete
Interaction	between agents through roles; between roles for composition, aggregation, disaggregation, transfer
Organization	spatial localization
User	multiple

To study in detail the role composition, our choice will focus on the detailed description of the role. Many methods exist in the literature, we can cite for example INGENIAS [12] which considers the different views of MAS (agent, organization, goals/tasks, environment, interactions), ADELFE [3] based on the AMAS theory and the concept of emergence, DIAMOND [7], which unify the development of the hardware and software part. It performs a decomposition of the various steps which the main are: needs definition, analysis phase, the generic conception, and implementation), AMOMCASYS [2] which can identify and describe the various roles and model the cooperative actions. Each model describes in a more or less abstract level, the set of criteria and proposes a definition of the roles more or less complete. Some models do not incorporate the notions of pre-requirements and post-condition in the role description. We propose to use the AMOMCASYS method because in this method, rules are distinguished between hard and flexible (not required for the role but an added value).

In the next section, we will present a new model. This model improve the AMOMCASYS model. It possesses new concepts on the role management such as the notion of pre-requirements or post-conditions.

3 A Multi-Agent System for Interactive Table

A preliminary study allowed us to set up and specify the composition of our MAS. In this study, we define the relationships between the different types of agents and their roles based on a class diagram. The MAM4IT (A Multi-Agent Model for Interactive Tables) allows us to instantiate both the agents and their corresponding roles by combining the AMOMCASYS method [1] with the task decomposition [6].

3.1 System Organization

To define the relationships between agents and their respective roles, we propose to follow the UML class diagram given in Fig. 1, inspired by work exposed in [10, 11], which establishes links between agents and role groups. We use an initial set described by the *TTTAgent*. An TTTAgent agent possesses a roles group. This group, initially empty or not (defined by the operator in this case) can evolve during the application: whether the agent adapts by itself its roles group to fulfill its needs; or it receive roles from other agents.

Initially, for an application, roles are stocked and managed by an agent who have the *RoleManager* role. This agent play the directory services role (yellow pages), like the Directory Facilitator recommended by the FIPA. It contains the couple $< agent, roles >$ and $< RFIDtag, roles >$.

AgentTTT and situated agents (that make up the visible instantiated applications) are completely autonomous. A *SituatedAgent* can be extended by either a *TangibleAgent* associated to a real tangible object or a *VirtualAgent* projected on the interaction area. These agents are located in a Cartesian plane defined by the interactive table environment.

For simulation applications, a *TimerAgent* interface allows to define an agent that send signals to specific target agents.

A *Genius* agent plays the scribe role and contains information on the position and address of various system agents (like the AMS agent: Agent Management Service proposed by FIPA).

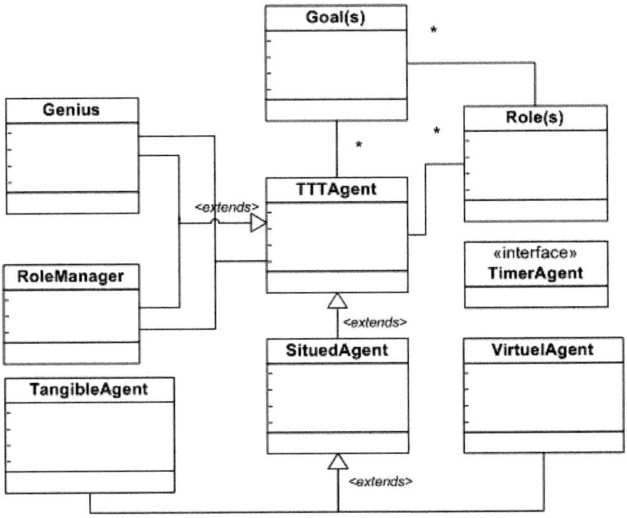

Fig. 1 Main classes of MAS for the TTT project

3.2 System Model: MAM4IT

To model the agents and roles of our system, we use the formalisms defined in [1] which are a continuity of the roles description in the AMOMCASYS method.

Each agent of the MAS is composed of: states, knowledge (social KS, environmental KE, personal KP: containing the goals and properties of the agent), messages, a perception function ($KP \times KE \times KS \times messages_a \Rightarrow A_{action}$) producing an action which modifies the agent state according to knowledge and received messages in order to achieve personal goals and/or collective goal. Knowledge expressed through this agent (KP, KE, KS) are mainly used to make the distinction between virtual and tangible agents. In this case, agent's properties are different. An elementary principle between these agents relies on their interactions: for example a tangible agent can act (in terms of moves) on a virtual and/or tangible agent (i.e the move of a tangible agent can push a virtual agent or another tangible agent), the opposite being impossible.

For a given $agent_a$, each rule is composed of a set of tasks to be executed as well as a priority used to evaluate the agent's preference for this rule.

The role concept characterizes the ability of an agent (see for instance [15]). In our case, we consider a role to be defined by both a set of knowledge (environmental and social) and a set of actions to achieve its goals. The definition of a role R (Cf. eq. 1) is then very similar to the definition of an agent. Indeed, a role gives to an agent the following: data (environmental knowledge) required to complete a role, contacts (KS) related to the role, the role's objective, its conditions and consequences (in personal knowledge), hard rules in which the agent must choose a rule to be considered as associated to this role, flexible rules able to complete this role. Personal knowledge KP^{role} for a given role contains a charge for the agent that will be associated to this role, as well as its objective, its pre-requirements that must validate by the agent in charge of this role and its post-conditions. When the role is done, a set of post-condition can modify the agent behavior. A given agent possesses: permanent knowledge (personal, environmental and social) which enable it to evolve in society. This agent had a temporary knowledge related to the roles that it plays. Depending on the application, temporary knowledge can integrate permanent knowledge .

$$R = \begin{pmatrix} name^{role}, priority^{role}, KP^{role}, KE^{role}, KS^{role}, \\ hardRules^{role}, flexibleRules^{role} \end{pmatrix} \tag{1}$$

We define the rules (see eq. 2) as a set of behavior associated to the role. A rule is composed of: a name, priority and a set of actions plan. The priority order the tasks. The action plan decompose tasks [6] in a specific order of execution. To change direction, the driver must do in order actions "put indicator", "slow down" and "turning wheel".

$$\begin{aligned} rule &= \left(name^{rule}, priority^{rule}, plan^{rule} \right) \\ plan^{rule} &= (t_0; \ldots ; t_{nt})_{nt\ =\ number\ of\ tasks} \end{aligned} \tag{2}$$

Therefore, an agent has a dynamic list of roles. Indeed, an agent may receive a role, losing a role, delegated a role of its acquaintances (our principle of delegation is based on the work [14]).

The roles describe the tasks that agents must perform. The role must be defined with precaution to avoid any error of conception.

We propose a control system for roles management based on three major concepts:

1. the pre-requirements to obtaining a role by an agent (based on agent properties in order to receive a role). From role classification point of view [9], an agent can not obtain a higher role (an agent can not add the role of bus driver (driver's license D) before being driver of vehicle (driver's license B);
2. the implications that define the consequences of a role on an agent (the addition or modification of personal agent property);
3. the management of coherence of the roles aggregation (the combination of roles) by the testimony of agents such as [5].

We have defined and represented a system that can be adapted to our interactive table. To validate our model, we propose in the next section of this paper an application to road traffic simulation relying on the previously described concepts.

4 Model Implementation by an Application

4.1 A Structure in Layers

The total architecture of the system defined by TTT project requirement, is made up of a structure in three layers (quoted in bottom-up order):

- The *Capture and Interface* layer handles tangible objects provided with one or more tags per object and creates a java object by associating it to a form.
- The *Traceability* layer handles events associated to the objects and communicates the modifications of object positions to the applicative layer.
- The *Application* manages the specificity of the application associated to the table. It is made up of the under-layer *Multi-Agents System (MAS)*, able to create an instance of any type of tangible objects in an application; and of the under-layer *Human-Computer Interaction (HCI)* which is given the responsibility of communicating with the users and which makes it possible to transmit virtual information [8].

In this structure, the communication is key point. Many information will be exchanged between layers. In the MAS, it is the *Genius* agent that will transmit information to the IHM layer and to the agents who manage a tangible object. This agent is the only one to have the environmental knowledge of the integrity of the table.

4.2 Road Traffic Management

Our objective relates to the proposal for a multi-agent architecture. It is suitable to communicating objects management which are dynamically associated with roles. We do not plan a particular application, but a set of applications in the Internet of Things context. In order to validate our approach, we have choose a road traffic application, which is currently under development. This application use the different concepts encountered at both the technological level (interactive table, tangible objects, RFID technology) and the proposed model level (agent's knowledge, role composition and transfer).

The application involves a set of road facilities used by security, architecture or transportation experts to manage road traffic. Its main objective is to optimize road traffic by forecasting some vehicle's behaviors to make the traffic flow lighter. In addition, the study of the concept of vehicle line allows us to apply the concepts of role transfer. The lines of vehicles are distinguished by a leader role and a follower role. The leader is on the head of the line and goes to a specific point on the map. When the vehicle is arrived or goes out of the line to go in another direction, it can transmit its leader role to the next agent of the line. The role of follower concerns vehicles which are added to the line. They are going totally or partially to the same direction as the leader. They must respect a safe distance from the vehicle that followed. This network is represented by a graph on which both facility tangible objects (roads, buildings) and signaling tangible objects (traffic lights) are moved by the user. Other tangible objects can also be used as stamps to create new vehicles on the table each time the users put them on the table. Vehicles are displayed together with the road network on the table with a video projector and adapt theirs behaviors according to the situation. The two kind of agents (tangible and virtual) are defined like follow:

- Virtual agents, such as pedestrians and drivers shown on the table, and service agents: the *Genius* agent which allocates the position of the corresponding agent when the environment is modified and transfer information to the upper layer (IHM), and the role manager which stores and/or transmits the different roles used during the course of the application.

- Tangible agents, which are a representations of tangible objects fitted with one or several RFID tags. They correspond to road signs (roundabout, traffic lights), roads or behaviors to be associated with the agent. For instance, with a tangible object provided with the "driver" behavior, we can modify the behavior of a pedestrian who has the "driver's license" pre-requirements. To do so, we just have to place this object at the pedestrian position given by the video projector.

During the initialization step, personal agents will possess the role of driver. For this role to be validated, the agent must be provided with some pre-requirements (such as having a driver's license). Environmental knowledge of this role are used to represent roads and road signs in the vicinity. Social knowledge allow the management of critical situations regarding other agents (such as giving way to vehicle that have priority). The role is composed of a large set of hard rules that follow a course of action. For instance, the RespectTrafficLaw rule relies on the following action plan: ReadRoadSigns → AnalyseRoadSigns → ProceedAccordingly. The rules implementation is distributed according the priority. At a given time, a rule that we call "bend" follows the plan of actions: CalculateBendDegree → "AdjustSpeed", may be priority to an another rule to avoid an accident for example. The actions plan of a rule is ordered and managed with the implementation of a finite state machine where each state corresponds to the execution of an action. These road signs are processed by agents which communicate by sending ACL messages to the agents that are close enough to receive them: traffic light's color, type of road sign or facility, etc.

In the figure 2, roads and vehicles are visualized and initialized when the application is launched. The intersections are managed by tangible road sign (traffic light, stop sign, traffic circle) more precisely by the roles associated to this tangible objects. Roles are transmitted by the manager transmits by ACL message.

When a vehicle is on an intersection, he receives messages from a tangible objects on the intersection that represent road signs. The user can move the road sign at various intersections to see the impact on road traffic.

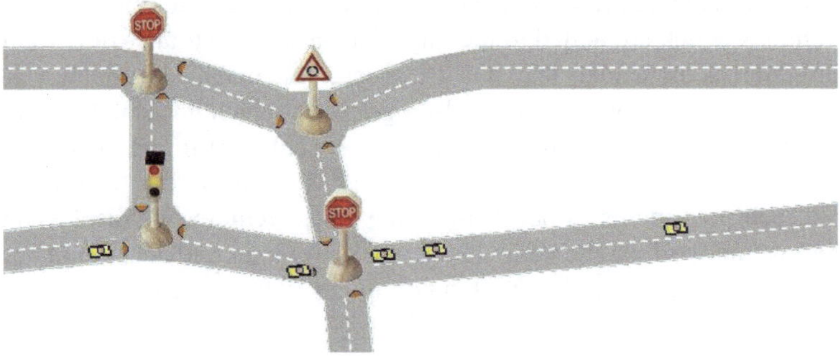

Fig. 2 Representation of the road traffic management

The proposed application is in the prototype stage. It has validated the organization of agents with the MAM4IT model proposed and our concept of roles transfers.

5 Conclusions

The applicative layer based on MAS proposed in this work allows the users to interact with an interactive table by means of tangible objects carrying RFID tags. The main interest of this technology relies on the possibility to store information directly within the RFID chips. This way, tangible objects can be initialized and keep in memory their IP address, and the roles that defined them. They can also be used to keep track of the performed actions so that the "life" of the object can be instantiated. This kind of feature gives some insights regarding novel perspectives to be implemented in future applications in the context of the *Internet of Things*.

To support future applications based on this technology, we propose a new model called MAM4IT. This model presents the advantage of making the distinction between the different types of agents (virtual and tangible) whether or not they are materialized. The dynamic management of their possible interactions is performed both globally and locally on the table. Indeed, according to the interactions with the users, agents (tangible or virtual) having the same role can modify their definition (by changing the priority of the rules that compose them for instance).

Although initial results showed a good appropriateness for our system with different layers of the project and helped to instantiate our model. It is necessary to continue the development to bring new features, for example, when 2 objects are side by side, they will be able to compose their roles to obtain a new role. This composition of role will be the key point in our future developments.

In future works, we will focus on the role formalism based on ontology. This formalisation will help us to develop composition of role facilities when two tangible objects are put together by the users, or when users decide to try several combination of roles on a same virtual agents for example.

Acknowledgment

The present research work has been supported by the French National Research Agency (ANR). We would also like to thank our two partner companies in the TTT project: RFIdes and the CEA. The authors grate-fully acknowledge the support of these institutions.

References

1. Adam, E., Grislin-Le Strugeon, E., Mandiau, R.: Flexible hierarchical organisation of role based agents. In: IEEE International Conference on Self-Adaptive and Self-Organizing Systems Workshops, pp. 186–191 (2008)
2. Adam, E., Mandiau, R., Kolski, C.: Application of a holonic multi-agent system for cooperative work to administrative processes. JASS 2(1), 100–115 (2001)

3. Bernon, C., Camps, V., Gleizes, M.P., Picard, G.: Engineering Self-Adaptive Multi-Agent Systems: the ADELFE Methodology. Idea Group Publishing, USA (2005)
4. Da Silva, J.L.T., Demazeau, Y.: Vowels co-ordination model. In: Gini, M., Ishida, T., Castelfranchi, C., Johnson, W. (eds.) AAMAS 2002: Proceedings of the first international joint conference on Autonomous agents and multiagent systems, pp. 1129–1136. ACM Press, bologna (2002)
5. Duran, F., da Silva, V.T., de Lucena, C.J.P.: Using testimonies to enforce the behavior of agents. In: Ossowski, S., Sichman, J.S. (eds.) AAMAS 2007: Workshop on Coordination, Organization, Institutions and Norms in agent systems (COIN), Honolulu, Hawaii, USA, pp. 25–36 (2007)
6. Hannoun, M., Sichman, J.S., Boissier, O., Sayettat, C.: Dependence relations between roles in a multi-agent system: Towards the detection of inconsistencies in organization. In: Sichman, J.S., Conte, R., Gilbert, N. (eds.) MABS 1998. LNCS (LNAI), vol. 1534, pp. 169–182. Springer, Heidelberg (1998)
7. Jamont, J.P., Occello, M.: Designing embedded collective systems: The diamond multi-agent method. In: ICTAI 2007 -19th IEEE International Conference on Tools with Artificial Intelligence, vol. 2, pp. 91–94. IEEE Computer Society, Patras (2007)
8. Kubicki, S., Lepreux, S., Lebrun, Y., Dos Santos, P., Kolski, C., Caelen, J.: New human-computer interactions using tangible objects: Application on a digital tabletop with rfid technology. In: Jacko, J.A. (ed.) HCI (3), vol. 5612, pp. 446–455. Springer, Heidelberg (2009)
9. Odell, J.J., Parunak, H.V.D., Brueckner, S., Sauter, J.: Temporal aspects of dynamic role assignment. In: Giorgini, P., Müller, J.P., Odell, J.J. (eds.) AOSE 2003. LNCS, vol. 2935, pp. 201–213. Springer, Heidelberg (2004)
10. Odell, J.J., Parunak, H.V.D., Fleischer, M.: Modeling agent organizations using roles. In: Heidelberg, S.B. (ed.) Software and Systems Modeling. Computer Science, vol. 2, pp. 76–81.(2003)
11. Odell, J.J., Parunak, H.V.D., Fleischer, M.: The role of roles in designing effective agent organizations. In: Garcia, A.F., de Lucena, C.J.P., Zambonelli, F., Omicini, A., Castro, J. (eds.) Software Engineering for Large-Scale Multi-Agent Systems. LNCS, vol. 2603, pp. 27–38. Springer, Heidelberg (2003)
12. Pavón, J., Gómez-Sanz, J.: Agent oriented software engineering with ingenias. In: Mařík, V., Müller, J.P., Pěchouček, M. (eds.) CEEMAS 2003. LNCS (LNAI), vol. 2691, p. 1069. Springer, Heidelberg (2003)
13. Russell, S., Norvig, P.: Artificial Intelligence: A Modern Approach, 2nd edn. Prentice Hall, Englewood Cliffs (2003)
14. Saidani, O., Nurcan, S.: A role based approach for modeling flexible business processes. In: The 7th Workshop on Business Process Modelling, Development, and Support (BP-MDS 2006), pp. 111–120. Springer, Heidelberg (2006)
15. Wooldridge, M., Jennings, N., Kinny, D.: The gaia methodology for agent oriented analysis and design. Journal of Autonomous Agents and Multi-Agent Systems 3(3), 285–312 (2000)

Forest Fires Prediction by an Organization Based System

Aitor Mata, Belén Pérez, and Juan M. Corchado

Abstract. In this study, a new organization based system for forest fires prediction is presented. It is an Organization Based System for Forest Fires Forecasting (OBSFFF). The core of the system is based on the Case-Based Reasoning methodology, and it is able to generate a prediction about the evolution of the forest fires in certain areas. CBR uses historical data to create new solutions to current problems. The system employs a distributed multi-agent architecture so that the main components of the system can be remotely accessed. All the elements building the final system, communicate in a distributed way, from different type of interfaces and devices. OBSFFF has been applied to generate predictions in real forest fire situations, using historical data both to train the system and to check the results. Results have demonstrated that the system accurately predicts the evolution of the fires. It has been demonstrated that using a distributed architecture enhances the overall performance of the system.

Keywords: Forest fires, Organization Based Systems, Services Oriented Architectures.

1 Introduction

Prediction and simulation of forest fires propagation is an important problem from the computational point of view due to the complexity of the involved models, the necessity of numerical methods and the required resources for calculation.

The phenomenon of forest fires has not only given as a result an important loss of forests and damage in the economy, but it has also seriously affected human health and environment. The fire fighting should have at its disposal the most

---

Aitor Mata · Belén Pérez · Juan M. Corchado
Departamento Informática y Automática
Universidad de Salamanca
Plaza de la Merced s/n, 37008, Salamanca, Spain
University of Salamanca, Spain
email: {aitor,lancho,corchado}@usal.es

Y. Demazeau et al. (Eds.): Advances in PAAMS, AISC 70, pp. 135–144.
springerlink.com

advanced resources and tools to help the use of available resources in the most efficient way do diminish fire effects as much as possible.

This paper presents OBSFFF, a Organization Based System for Forest Fires Forecasting. This system deploys a prediction model which makes use of intelligent agents and Case-Based Reasoning systems to determine the possibility of finding forest fires in a certain area once a fire has nearby started. It also applies a distributed multi-agent architecture based on Service Oriented Architectures (SOA), modeling most of the system's functionalities as independent applications and services. These functionalities are invoked by deliberative agents acting as coordinators.

The system presented in this paper generates as a solution the future situation of an area, considering the present one. Predictions are created using a Case-Based Reasoning system. The cases used by the CBR system contain atmospheric data (wind, temperature and pressure) and may also contain information about an specific situation to be solved (forest fires) OBSFFF combines artificial intelligence techniques in order to improve the efficiency of the CBR system, thus generating better results. OBSFFF has been trained using historical data. The development of agents is an essential piece in the analysis of data from distributed sensors and gives those sensors the ability to work together and analyze complex situations.

In the following section, a brief description of the forest fires problem is shown. Section 3 explains the main concepts of the organizations of agents. The fourth section describes the OBSFFF system and the final two sections show the results and the conclusions.

2 Forest Fires

Forest fires are a very serious hazard that, every year, causes significant damage around the world from the ecological, social, economical and human point of view [1]. These hazards are particularly dangerous when meteorological conditions are extreme with dry and hot seasons or strong wind. For example, fire is a recurrent factor in Mediterranean areas.

Fires represent a complex environment, where multiple parameters are involved. In this sub-section, a series of applications and possible solutions are explained. They are different approaches to the forest fire problems, including all the main phases existing in the evolution of this kind of problem.

The OBSFFF system presented here was applied to generate predictions in a forest fire scenario. Forest fires represent a great environmental risk. The main approaches that have been used to solve this problem begin with the detection of the fires [2]where different techniques have been applied. Once the fire is detected, it is important to generate predictions that should assist in making a decision in those contingency response situations [3]. Finally, there are complex models that tackle the forest fire problem by trying to forecast its evolution and to minimize its associated risks [4].

The data used to check the OBSFFF system were a subset of the available data that had not been previously used in the training phase. The predicted situation was contrasted with the actual future situation as it was known (historical data was used to train the system and also to test its correction). The proposed solution, in most of the variables, had a near 90\% accuracy rate.

Table 1 Variables that define a case, applying OBSFFF to the forest fires problem

Variable	Definition	Unit
Longitude	Geographical longitude	Degree
Latitude	Geographical latitude	Degree
Date	Day, month and year of the analysis	dd/mm/yyyy
Bottom pressure	Atmospheric pressure in the open sea	Newton/m2
Temperature	Celsius temperature in the area	ºC
Area of the fires	Surface covered by the fires present in the analyzed area	Km2
Meridional Wind	Meridional direction of the wind	m/s
Zonal Wind	Zonal direction of the wind	m/s
Wind Strenght	Wind strength	m/s

To create the cases, the geographical area being analyzed was divided into small squares, each of which was considered a case, with its associated parameters shown in *Table 1*. The squares determine the area to be considered in every case. The problem is represented by the current situation of the area (all its parameters and the presence, or lack thereof, of fire). The solution is represented by the situation in that area in a future moment (same location but parameters changed to those for the next day -or the next step, if less than a day is considered in every step-). The data used are part of the SPREAD project, in particular the Gestosa field experiments that took place in 2002 and 2004 [5]. The experiments of the Gestosa field began in 1998 and finished in December 2004. They were aimed at collecting experimental data to support the development of new concepts and models, and to validate existing methods or models in various fields of fire management.

3 Organizations of Agents

In the multiagent field, the term organization has been mainly used to describe a set of agents that, using some kind of roles, interact with each other coordinating themselves to achieve the global objectives of the system.

L. Gasser assumes that organizations are structured systems with activity, knowledge, culture, history, and ability pattern, different of any particular agent [6]. Organizations exist in a completely different level than individual agents that make up the organizations themselves. Individual agents are replaceable. Organizations are established in a space; it either be geographical, temporal, symbolic, etc. So, an organization of agents proportionates a kind of workspace for the activity and interaction of the agents by defining roles, behavioral expectatives and relations.

Ferber indicates that organizations proportionate a way to divide the system, crating groups or units that form the interaction context of the agents [7]. The organization is then based in two main aspects: structural and dynamic. The structure of the organization represents the remaining components when the individual elements enter or leave the organization. The organization is composed by the set of relationships that allow seeing a number of different elements as unique. The structure defines the way the agents are grouped in organizational units and how those units are related with each other. The roles needed to develop the activities of the organization are also defined in the structure, as long as the relationships and restrictions.

The organizational dynamics is centered in the interaction patterns defined for the roles, describing the way to get into or to leave the organization, the parameters of the roles and the way the roles are assigned to the agents.

Finally, V. Dignum, affirms that the organizations of agents assume that there are global objectives, different from the individual agents' objectives [8]. Roles represent organizational positions that help to achieve those global objectives. Agents may have their own objectives and decide if they take any specific system role or not, determining which among the available protocols are more suitable to achieve their chosen objectives.

4 System Description

In this paper, a new Organization Based System for Forest Fire Forecasting (OBSFFF) is presented. It is formed by an organization of agents that connects some different interface agents, dedicated to interact with the users in different platforms, to an inner CBR system, made by different services, implemented by a series of agents that answer to the different requests of the users.

In OBSFFF, the data collected from different satellites is processed and structured as cases. *Table 1* shows the main variables that defines a case. Cases are the key to obtain solutions to future problems through a CBR system. The functionalities of OBSFFF are accessed using different interfaces executed on PCs or PDAs (Personal Digital Assistant). Users interact with the system by introducing data, requesting a prediction or revising a solution generated (i.e. prediction). The interface agents communicate with the services through the agents' platform and vice versa.

The interface agents perform all the different functionalities which users make use for interacting with OBSFFF. The different phases of the CBR system have been modeled as services, so each phase can be independently requested. For example, one user may only introduce information in the system (e.g. a new case), while another user could request a new prediction.

All information is stored in the case base and OBSFFF is ready to predict future situations. A problem situation must be introduced in the system for generating a prediction. Then, the most similar cases to the current situation are retrieved from the case base. Once a collection of cases are chosen from the case base, they must be used for generating a new solution to the current problem. Growing Radial Basis Functions Networks [9] are used in OBSFFF for combining the chosen cases in order to obtain the new solution.

OBSFFF determines future probabilities in a certain area. OBSFFF divides the area to be analyzed in squares of approximately half a degree side for generating a new prediction. Then, the system calculates the demanded parameters in each square. The squares are colored with different gradation depending on the values of the requested parameters.

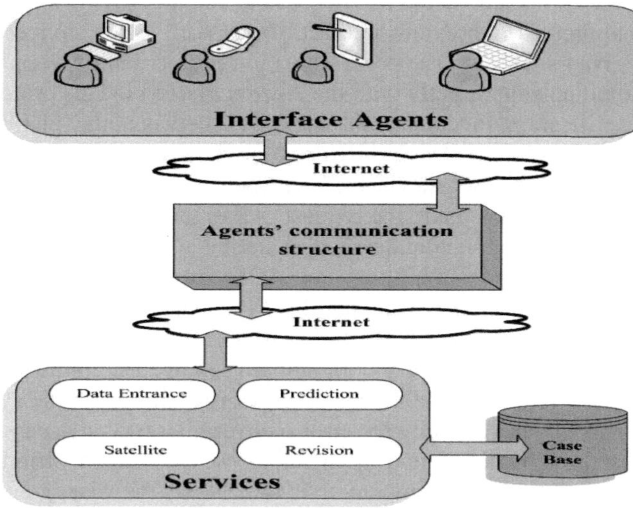

Fig. 1 OBSFFF structure

Figure 1 shows the structure of OBSFFF. There are three basic blocks in OBSFFF: Interface Agents, Services and Agents' Communication Structure. The system covers from the users interfaces, solved by the interface agents, to the data management, solved by the services that can access the case base. These blocks provide all the system functionalities and are described next:

Interfaces Organization represent all the programs that users can use to exploit the system functionalities. Applications are dynamic, reacting differently according to the particular situations and the services invoked. They can be executed locally or remotely, even on mobile devices with limited processing capabilities, because computing tasks are largely delegated to the agents and services.

The *CBR Services Organization* represent the activities that the architecture offers. They are the bulk of the functionalities of the system at the processing, delivery and information acquisition levels. Services are designed to be invoked locally or remotely. Services can be organized as local services, web services, GRID services, or even as individual stand alone services. OBSFFF has a flexible and scalable directory of services, so they can be invoked, modified, added, or eliminated dynamically and on demand. It is absolutely necessary that all services follow a communication protocol to interact with the rest of the components. CBR systems have proved to be quite efficient in prediction tasks in environmental situations[10].

The *Communication Organization* integrates a set of agents, each one with special characteristics and behavior. An important feature in this architecture is that the agents act as controllers and administrators for all applications and services, managing the adequate functioning of the system, from services, applications, communication and performance to reasoning and decision-making. In OBSFFF, services are managed and coordinated by deliberative BDI agents. The agents modify their behavior according to the users' preferences, the knowledge acquired from previous interactions, as well as the choices available to respond to a given situation. The communication protocol allows applications and services to communicate directly with the Agents Platform. This protocol is based on SOAP specification to capture all messages between the platform and the services and applications [11]. Services and applications communicate with the Agents Platform via SOAP messages. A response is sent back to the specific service or application that made the request. All external communications follow the same protocol, while the communication among agents in the platform follows the FIPA Agent Communication Language (ACL) specification.

Agents, applications and services in OBSFFF can communicate in a distributed way, even from mobile devices. This makes it possible to use resources no matter its location. It also allows the starting or stopping of agents, applications, services or devices separately, without affecting the rest of resources, so the system has an elevated adaptability and capacity for error recovery. Users can access to OBSFFF functionalities through distributed applications which run on different types of devices and interfaces (e.g. computers, PDA). Next, the disaggregation of the CBR methodology into services is explained. The four main phases of the CBR cycle are divided into different services covering the entrance of data, the recovery of information from the case base and the solution generation and validation.

The data used to check the OBSFFF system were a subset of the available data that had not been previously used in the training phase. The predicted situation was contrasted with the actual future situation as it was known (historical data was used to train the system and also to test its correction). The proposed solution, in most of the variables, had a near 90% accuracy rate.

4.1 Data Introduction and Case Base Creation

When data about a forest fire is introduced, OBSFFF must complete the information about the area including atmospheric and weather information. OBSFFF uses Fast Iterative Kernel PCA (FIKPCA) which is an evolution of PCA [12]. This technique reduces the number of variables in a set by eliminating those that are linearly dependent, and it is quite faster than the traditional PCA. To improve the convergence of the Kernel Hebbian Algorithm used by Kernel PCA, FIK-PCA set ηt proportional to the reciprocal of the estimated values. Let $\lambda t \in \Re r+$ denote the vector of values associated with the current estimate of the first r eigenvectors. The new KHA algorithm sets de i^{th} component of η_t to the files.

$$[\eta_t]_i = \frac{1}{[\lambda_t]_i} \frac{\tau}{t+\tau} \eta_0 \ , \tag{1}$$

When introducing the data into the case base, Growing Cell Structures (GCS) [13] are used. GCS can create a model from a situation organizing the different cases by their similarity. If a 2D representation is chosen to explain this technique, the most similar cells (i.e. cases) are near one of the other. If there is a relationship between the cells, they are grouped together, and this grouping characteristic helps the CBR system to recover the similar cases in the next phase. When a new cell is introduced in the structure, the closest cells move towards the new one, changing the overall structure of the system. The weights of the winning cell ω_c, and its neighbours ω_n, are changed. The terms ε_c and ε_n represent the learning rates for the winner and its neighbors, respectively. x represents the value of the input vector.

$$\omega_c(t+1) = \omega_c(t) + \varepsilon_c(x - \omega_c) \quad (2)$$

$$\omega_n(t+1) = \omega_n(t) + \varepsilon_n(x - \omega_n) \quad (3)$$

Once the case base has stored the historical data, and the GCS has learned from the original distribution of the variables, the system is ready to receive a new problem. When a new problem comes to the system, GCS are used once again. The stored GCS behaves as if the new problem would be stored in the structure and finds the most similar cells (cases in the CBR system) to the problem introduced in the system. In this case, the GCS does not change its structure because it has being used to obtain the most similar cases to the introduced problem. Only in the retain phase the GCS changes again, introducing the proposed solution if it is correct.

4.2 Prediction Generation after a Request

When a prediction is requested by a user, the system starts recovering from the case base the most similar cases to the problem proposed. Then, it creates a prediction using artificial neural networks. Once the most similar cases are recovered from the case base, they are used to generate the solution. Growing RBF networks [14] are used to obtain the predicted future values corresponding to the proposed problem. This adaptation of the RBF networks allows the system to grow during training gradually increasing the number of elements (prototypes) which play the role of the centers of the radial basis functions. The creation of the Growing RBF must be made automatically which implies an adaptation of the original GRBF system. The error for every pattern is defined by (4).

$$e_i = {}^1\!/_{p} * \sum_{k=1}^{p} ||t_{ik} - y_{ik}||\,, \quad (4)$$

Where t_{ik} is the desired value of the kth output unit of the ith training pattern, y_{ik} the actual values of the kth output unit of the ith training pattern.

Once the GRBF network is created, it is used to generate the solution to the proposed problem. The solution proposed is the output of the GRBF network created with the retrieved cases. The input to the GRBF network, in order to generate the solution, is the data related with the problem to be solved, the values of the variables stored in the case base.

5 Evaluation of the Results

OBSFFF uses different artificial intelligence techniques to cover and solve all the phases of the CBR cycle. Fast Iterative Kernel Principal Component Analysis is used to reduce the number of variables stored in the system, getting about a 60% of reduction in the size of the case base. This adaptation of the PCA also implies a faster recovery of cases from the case base (more than 7% faster than storing the original variables).

Table 2 Percentage of good predictions obtained with different techniques

Number of cases	RBF	CBR	RBF + CBR	OBSFFF
100	43 %	38 %	43 %	45 %
500	48 %	43 %	46 %	54 %
1000	50 %	48 %	58 %	68 %
2000	55 %	53 %	63 %	77 %
3000	59 %	56 %	68 %	84 %
4000	60 %	63 %	73 %	88 %
5000	62 %	68 %	79 %	93 %

The predicted situation was contrasted with the actual future situation. The future situation was known, as long as historical data was used to develop the system and also to test the correction of it. The proposed solution was, in most of the variables, more than a 90% accurate. For every problem defined by an area and its variables, the system offers 9 solutions (i.e. the same area with its proposed variables and the eight closest neighbors). This way of prediction is used in order to clearly observe the direction of the fires which is useful in order to determine the areas that will be affected by the fires.

Table 2 shows a summary of the results obtained after comparing different techniques with the results obtained using OBSFFF. The table shows the evolution of the results along with the increase of the number of cases stored in the case base. All the techniques analyzed improve its results while increasing the number of cases stored. Having more cases in the case base, makes easier to find similar cases to the proposed problem and then, the solution is more accurate. The "RBF" column represents a simple Radial Basis Function Network that is trained with all the data available. The network gives an output that is considered a solution to the problem. The "CBR" column represents a pure CBR system, with no other techniques included; the cases are stored in the case base and recovered considering the Euclidean distance. The most similar cases are selected and after applying a weighted mean depending on the similarity of the selected cases with the inserted problem, a solution s proposed. The "RBF + CBR" column corresponds to the possibility of using a RBF system combined with CBR. The recovery from the CBR is done by the Manhattan distance and the RBF network works in the reuse phase, adapting the selected cases to obtain the new solution. The results of the "RBF+CBR" column are, normally, better than those of the "CBR", mainly because of the elimination of useless data to generate the solution.

Finally, the "OBSFFF" column shows the results obtained by OBSFFF, obtaining better results that the three previous analyzed solutions.

Several tests have been done to compare the overall performance of OBSFFF. The tests consisted of a set of requests delivered to the Prediction Generation Service (PGS) which in turn had to generate solutions for each problem. There were 50 different data sets, each one with 10 different parameters. The data sets were introduced into the PGS through a remote PC running multiple instances of the Prediction Agent. The data sets were divided in five test groups with 1, 5, 10, 20 and 50 data sets respectively. There was one Prediction Agent for each test group. 30 runs for each test group were performed. First, all tests were performed with only one Prediction Service running in the same workstation on which the system was running. Then, five Prediction Services were replicated also in the same workstation. For every new test, the case base of the PGS was deleted in order to avoid a learning capability, thus requiring the service to accomplish the entire prediction process.

6 Conclusions and Future Work

OBSFFF is a new solution for predicting the future situation of the forest fires, analyzing different parameters.. This system presents a distributed multi-agent architecture which allows the interaction of multiple users at the same time. Distributing resources also allows users to interact with the system in different ways depending on their specific needs for each situation (e.g. introducing data or requesting a prediction). This architecture becomes an improvement with previous tools where the information must be centralized and where local interfaces where used. With the vision introduced by OBSFFF, all the different people that may interact with a contingency response system collaborate in a distributed way, being physically located in different places but interchanging information in a collaborative mode.

OBSFFF makes use of the CBR methodology to create new solutions and predictions using past solutions given to past problems. The CBT structure been divided into services in order to optimize the overall performance of OBSFFF.

Generalization must be done in order to improve the system. Applying the methodology explained before to diverse geographical areas will make the results even better, being able to generate good solutions in more different situations. The current system has been mainly developed using data from the Gestosa experiments in central Portugal. With that information, OBSFFF has been able to generate solutions to new situations, based on the available cases. If the amount and variety of cases stored in the case base is increased, the quality of the results will also be boosted.

Although the performed tests have provided us very useful data, it is necessary to continue developing and enhancing OBSFFF. The number of possible interfaces can be augmented, including independent sensors that may send information to the system in real-time. The data received by the system must be analyzed in order to detect new useful information and to generate fast and accurate solutions to existing problems without the direct intervention of the users.

Then, the system will not only be a contingency response but also a kind of supervising system especially in dangerous geographical areas.

References

[1] Long, D.G.: Mapping fire regimes across time and space: Understanding coarse and fine-scale fire patterns. International Journal of Wildland Fire 10, 329–342 (2001)

[2] Mazzeo, G., Marchese, F., Filizzola, C., Pergola, N., et al.: A Multi-temporal Robust Satellite Technique (RST) for Forest Fire Detection. Analysis of Multi-temporal Remote Sensing Images, 1–6 (2007)

[3] Iliadis, L.S.: A decision support system applying an integrated fuzzy model for long-term forest fire risk estimation. Environmental Modelling and Software 20(5), 613–621 (2005)

[4] Serón, F.J., Gutiérrez, D., Magallón, J., Ferragut, L., et al.: The Evolution of a Wildland Forest Fire Front. The Visual Computer 21(3), 152–169 (2005)

[5] GESTOSA, ADAI-CEIF(Center of Forest Fire Studies) (2005), `http://www.adai.pt/ceif/Gestosa/`

[6] Gasser, L.: Perspectives on organizations in multi-agent systems. In: Luck, M., Mařík, V., Štěpánková, O., Trappl, R. (eds.) ACAI 2001 and EASSS 2001. LNCS (LNAI), vol. 2086, pp. 1–16. Springer, Heidelberg (2001)

[7] Ferber, J., Gutknecht, O., Michel, F.: From agents to organizations: an organizational view of multi-agent systems. LNCS, pp. 214–230. Springer, Heidelberg (2004)

[8] Dignum, V., Dignum, F.: A logic for agent organizations. FAMAS@ Agents, 3–7 (2007)

[9] Karayiannis, N.B., Mi, G.W.: Growing radial basis neural networks: merging supervised andunsupervised learning with network growth techniques. IEEE Transactions on Neural Networks 8(6), 1492–1506 (1997)

[10] Mata, A., Corchado, J.M.: Forecasting the probability of finding oil slicks using a CBR system. Expert Systems With Applications 36(4), 8239–8246 (2009)

[11] Cerami, E.: Web Services Essentials Distributed Applications with XML-RPC, SOAP, UDDI & WSDL. O'Reilly & Associates, Inc., Sebastopol (2002)

[12] Gunter, S., Schraudolph, N.N., Vishwanathan, S.V.N.: Fast Iterative Kernel Principal Component Analysis. Journal of Machine Learning Research 8, 1893–1918 (2007)

[13] Fritzke, B.: Growing cell structures—a self-organizing network for unsupervised and supervised learning. Neural Networks 7(9), 1441–1460 (1994)

[14] Ros, F., Pintore, M., Chrétien, J.R.: Automatic design of growing radial basis function neural networks based on neighboorhood concepts. Chemometrics and Intelligent Laboratory Systems 87(2), 231–240 (2007)

Self-adaptive Coordination for Robot Teams Accomplishing Critical Activities

Jean-Pierre Georgé, Marie-Pierre Gleizes, Francisco J. Garijo, Victor Noël, and Jean-Paul Arcangeli

Abstract. This paper presents a self-adaptive cooperation model for autonomous mobile devices, to achieve collaborative goals in crisis management scenarios. The model, which is based on the AMAS theory, allows dynamic team formation, task allocation and reconfiguration. The global behaviour emerges from interactions among individual agents. Task responsibility allocation is done by individual estimations of the degree of difficulty and priority to achieve the task. Then each peer exchanges its evaluation records with the others in order to find out the best suited peer to take the responsibility. Research work has been done in the framework of the ROSACE project. The experimental setting based on forest fire crisis management, and a working example are also described in the paper.

Keywords: self-adaptation, coordination, crisis management.

1 Introduction

Unmanned Aerial Vehicles (UAV), Unmanned Ground Vehicles (UGV) and mobile robots are extensively used in crisis management scenarios where they are responsible for achieving dangerous tasks under close human supervision. However, tight control becomes a serious shortcoming in emergency setting such as fire, where fast evolution of environmental conditions may jeopardize the safety of all actors. New generations of mobile entities helping effectively in crisis management should incorporate Multi-Agent features such as: a) full autonomy to achieve individual and collective goals; b) social abilities for working as a team of mobile cooperating agents; c) self-adaptation to adjust agent's behaviour and team organization to the mission objectives by taking into account unexpected changes in the environment, internal failure and availability of mission resources.

The ROSACE[1] project faces the challenge of producing technology and tools for transforming UAVs and UGVs into Autonomous Adaptive Aerial Vehicles (AAVs) and Autonomous Adaptive Ground Vehicles (AGVs) that are capable of cooperation to achieve collective missions in highly dynamic environments.

Jean-Pierre Georgé · Marie-Pierre Gleizes · Francisco J. Garijo ·
Victor Noël · Jean-Paul Arcangeli
Institut de Recherche en Informatique de Toulouse (IRIT) - Université Paul Sabatier
118 route de Narbonne, 31062 Toulouse Cedex 9, France
email: `{george,gleizes,garijo,vnoel,arcangeli}@irit.fr`

[1] `www.irit.fr/Rosace`

Y. Demazeau et al. (Eds.): Advances in PAAMS, AISC 70, pp. 145–150.
springerlink.com

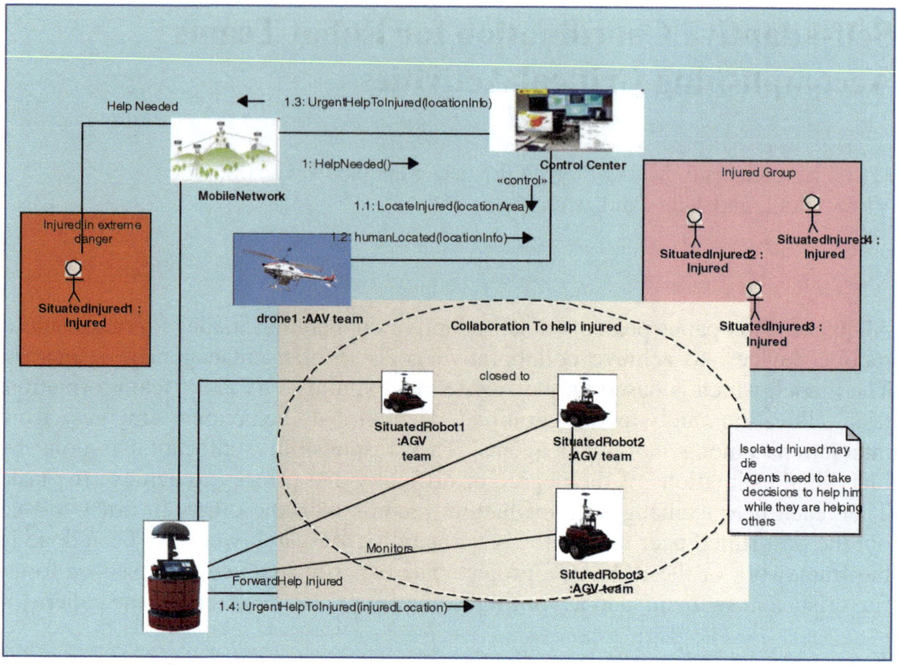

Fig. 1 An example of a critical decision choice case

This paper aims to present the Self-Adaptive Cooperation model that is being developed in the framework of ROSACE, which aims at managing AAV's/AGVs behaviours in order to achieve collective mission goals.

Ongoing work in ROSACE joins research efforts on MAS coordination in other domains (Tate 2006). Works like the FP7 ALIVE[2] project and its application on crisis situations for instance (Quillinan et al. 2009) focus on distributed software architectures and organizational models, and dynamics and adaptation are important compared to more classical approaches.

The Self-Adaptive Coordination approach described in this paper is based on the AMAS theory (Gleizes et al. 2008) which has been applied in numerous application domains: mechanical design (Capera et al. 2004), manufacturing control (Clair et al. 2008), flood forecast (Georgé et al. 2009), ontology creation and maintenance (Sellami et al. 2009). Experimental results have confirmed the benefits of Self-Adaptation in open changing environments where agents have to perform quick reactions and possibly adjust their organization model in order to minimize undesirable effects and maximize system performance. However, incorporating the AMAS self-adapting model into complex physical architectures such as robots and AAVs/AGVs sets up a number of scientific and engineering challenges that are being addressed in the ROSACE project: i) building a robust, flexible and efficient architecture integrating the robotics software layers and considering the specific constraints imposed to the middleware layer corresponding to the real-time

[2] www.ist-alive.eu

embedded systems as well as the management of network resources and communication services; ii) Developing and validating decision models taking into account internal failures and hard external constraints – e.g. lack of communication, immobility, lack of actuation capabilities; iii) Assessment of achievability of mission goals and distributed task allocation; iv) Guarantee of the availability of communication resources for a permanent connectivity preserving the quality of communications.

The organization of the rest of the paper is the following. Section 2 gives a brief outline of the ROSACE experimental setting which is based on mixed teams of humans, robots and Aerial Autonomous Vehicles, cooperating for forest fire-fighting. The AMAS paradigm and its application to critical decisions are presented in section 3 including the processing model cooperation approach and working examples. Finally the conclusions and open issues are summarized in section 4.

2 The ROSACE Experimental Setting

The experimental setting assumes the utilization of AAV Teams and AGV Teams by a Public Organization responsible for territory supervision and forest fire crisis management. The public organization is equipped with an Emergency Management System (EMS) which provides information management and monitoring tools to achieve information fusion, and situation assessment. The EMS also provides mission management tools to help the persons responsible for the mission to assess risks, prepare the mission by recovering intervention plans, mission execution, and resource monitoring and control during mission execution.

AAVs and AGVs are situated at the intervention area to collaborate with humans for: i) Localization tasks, such as localization of people in potential danger; ii) Supervision tasks such as fire progress monitoring; iii) Guidance to safe areas; iv) Provision of first aid to injured; v) Logistics and telecommunication support.

In order to validate the robot control models, a collection of operational scenarios and use cases have been defined primarily focused on agents coordination confronted to critical decision choices. Here is an example of scenario. An AAV has located a group of people jeopardized by fire. Location information is sent to a team of robots that decides to go to help them. While they move to people's location a call is received in the control centre (CC) asking for urgent help. The CC orders an AAV to proceed to the injured location. Once the location has been successfully reached the CC broadcast a message asking teams close to this location to provide urgent help to this injured. The AGVs team receives this message and starts deciding whether to ignore the message and continue their original task, or delegating one member of the team to help this injured. In this last case they should decide who will take the responsibility to go.

3 Applying the AMAS Paradigm Approach

The AMAS provides self-adaptation and self-organization mechanisms for multi-agent systems in open dynamic environments. The adaptation corresponding to a change of the global system behaviour is realized by agent self-organization. The right behaviour is reached by the right organization of the agents. It can also be considered as the right agents acting at the right location at the right moment. Cooperation means not only that agents have to work together in order to share resources and/or competencies, but also that they should try to anticipate and avoid non cooperative situations (cooperation failures), and when non cooperation occurs, they should try to act in order to reach a cooperative situation. Agents are benevolent and not altruistic in the sense of Castelfranchi (Hassas et al. 2006). They only try to help an agent which has more difficulties than themselves if their help does not definitively prevent them to reach their individual goal.

The global behaviour at the system level emerges from interactions resulting from the agent's cooperation model. In ROSACE, the cooperation model is embedded at the decision layer. Agents are supposed to have a cooperative attitude that enables them to take decisions in a given current context, confronted to new and unpredictable events. The context is defined as the agent's knowledge about itself, and about the perceived environment.

AAVs and AGVs as agents. AAVs and AGVs are considered as physical agents with sensing, actuation, communication, and decision capabilities. They share a multi-layered component based architecture where the overall behaviour of the entity is managed by a Robot Global Control Component (RGC) which is in charge of orchestrating the internal components behaviours to achieve a coherent global behaviour. The RGC gathers elaborated information, make choices, orders execution of actions, and monitor results. RGC's control model is based on a declarative goal processor (Garijo et al. 2004) that manages a goal space, and a working memory. Strategic and tactics criteria for generating goals and for selecting task to try to achieve goals are declarative. They are defined by situation-action rules, where the situation part specifies a partial state of the working memory including the objective and its internal state, and the action part contains statements for executing tasks. The processing cycle is driven by incoming information, which is stored in the working memory. Then control rules are used to decide either to generate new goals, to focus on a new goal, to verify the resolution of pending goals, or to proceed to the resolution of pending goals by executing new tasks and actions.

The environment in ROSACE plays a critical role since it may jeopardize the normal functioning of the whole entity. Internal components are dependent on environmental parameters such as topography or distance among networking nodes. For example the communication component is needed for coordination among cooperating peers. Decisions should then be made by taking into account both internal and external constraints.

Decision process implementation. Decision making in AGVs Global Control is modelled as a concurrent process where the AMAS principles are applied to generate new goals in the goal space of the agent, and/or to select a goal to be achieved. The generic process for adaptive cooperation is the following. Each agent: i) Evaluates their own capabilities to achieve the new goal; ii) Sends its own evaluation record to the team members; iii) Receives evaluation records from team members; iv) Takes a decision to get the responsibility for the goal based on the best evaluation record.

Team consensus is reached when the best evaluation record exists. In this case the agent that generates this optimal record should take the responsibility to achieve the goal. When there are two or more records satisfying optimality conditions the agents which generate the records should update their evaluations to allow one of them to take the responsibility of the goal.

Working Example. While going to assist the group of jeopardized people, the goal space of each AGV in the team is focused on the same goal, which is to *provide assistance to a group of people*. The message broadcasted by the control centre is received by all the members of the team. Interpretation and evaluation of the message lead to the generation of a new goal which is *to provide assistance to an injured person at location z*. Achievement of this goal has higher priority that the goal being achieved by the team. Then a decision should be taken by the team members on who will take the responsibility of achieving this goal. Individually this means that each agent should find out evidence either to continue achieving its current goal, or to change its focus to the new goal.

Evaluating the change of focus first consists in gathering relevant information (knowledge about the current context of the agent: energy, distance...), then in analysing if the new goal could be achieved or not by the agent (impact, risk...). This analysis is based on the cooperative attitude of the agent and on the local point of view it has about the situation. The agent must check its own constraints such as: have I enough energy? Has my current goal more priority? Have I all competencies?... This analysis provides a degree of difficulty to adopt this new goal or to participate to this new goal. So, the assessment task generates a goal achievability report which is used to determine whether or not the goal could be achieved by the agent and if it could the degree of difficulty for the agent to do it. If it concludes that the goal could be achieved it generates a goal achievement report which summarizes the cost estimation for achieving the goal.

Then, the agent has to get its neighbours to analyse its perceptions. Its neighbours can be: all agents in the perception area of its camera, all agents with which it can communicate to using its networking resources. The neighbours at a given time can be considered as the temporary team. This temporary team can change dynamically over time (for example between two perception steps). All agents have to exchange their achievability report concerning a given goal, and given their cooperative attitude they will choose the right agent for the right task by comparing the difficulty and criticality of each agent.

4 Conclusions and Future Challenges

The self-adaptive cooperation model has been implemented in a simulated environment based on Blender[3] with a limited number of robots. In comparison to more sophisticated cooperative models based on agent's teams where agents have

[3] `www.blender.org`

fixed roles, and have complex decision making mechanisms, the proposed solution is simple, easy to implement and efficient.

Self-adapting agents are able to assimilate the changes in the environment for improving the achievement of its tasks, and also for making decisions to assume mission tasks taking into account the point of view of its cooperating peers. As agents have common mechanisms to avoid non cooperative situations, possible conflicts which could block task allocation are minimized or deleted. Tasks are assumed by the best situated agent, so at the global level cooperation succeeds.

While initial results seems promising, research work should continue to assess model performance taking into account scalability issues as well as internal components failures and external constraints such as lack of communications, week energy, immobility, uncertainty of perceived data, and others. At the theoretical level formal demonstration of the effectiveness of evaluation functions for task allocation and decision making is also necessary.

Acknowledgments. This project is supported by the RTRA STAE[4] foundation.

References

Capera, D., Gleizes, M.-P., Glize, P.: Mechanism Type Synthesis based on Self-Assembling Agents. Journal of Applied Artificial Intelligence 18(9-10), 921–936 (2004)

Clair, G., Kaddoum, E., Gleizes, M.-P., Picard, G.: Self-Regulation in Self-Organising Multi-Agent Systems for Adaptive and Intelligent Manufacturing Control. In: IEEE Int. Conf. on Self-Adaptive and Self-Organizing Systems. IEEE CS Press, Los Alamitos (2008)

Garijo, F., Bravo, S., Gonzalez, J., Bobadilla, E.: BOGAR LN: An agent based component framework for developing multi-modal services using natural language. In: Conejo, R., Urretavizcaya, M., Pérez-de-la-Cruz, J.-L. (eds.) CAEPIA/TTIA 2003. LNCS (LNAI), vol. 3040, pp. 207–220. Springer, Heidelberg (2004)

Georgé, J.-P., Peyruqueou, S., Régis, C., Glize, P.: Experiencing Self-Adaptive MAS for Real-Time Decision Support Systems. In: Int. Conf. on Practical Applications of Agents and Multiagent Systems (PAAMS 2009), pp. 302–309. Springer, Heidelberg (2009)

Gleizes, M.-P., Camps, V., Georgé, J.-P., Capera, D.: Engineering Systems which Generate Emergent Functionalities. In: Weyns, D., Brueckner, S.A., Demazeau, Y. (eds.) EEMMAS 2007. LNCS (LNAI), vol. 5049, pp. 58–75. Springer, Heidelberg (2008)

Hassas, S., Castelfranchi, C., Di Marzo Serugendo, G.A.K.: Self-Organising Mechanisms from Social and Business/Economics Approaches. Informatica 30(1) (2006)

Quillinan, T.B., Brazier, F., Aldewereld, H., Dignum, F., Dignum, V., Penserini, L., Wijngaards, N.: Developing Agent-based Organizational Models for Crisis Management. In: Proc. of the Industry Track of the 8th Int. Joint Conf. on Autonomous Agents and Multi-Agent Systems (AAMAS 2009), Budapest, Hungary (2009)

Sellami, Z., Gleizes, M.-P., Aussenac-Gilles, N., Rougemaille, S.: Dynamic ontology co-construction based on adaptive multi-agent technology. In: Int. Conf. on Knowledge Engineering and Ontology Development (KEOD 2009). Springer, Heidelberg (2009)

Tate, A.: The Helpful Environment: Geographically Dispersed Intelligent Agents That Collaborate. IEEE Intelligent System 21(3) (2006)

[4] `www.fondation-stae.net`

A Cooperative Communications Platform for Safety Critical Robotics: An Experimental Evaluation

Frederico M. Cunha, Rodrigo A.M. Braga, and Luis P. Reis

Abstract. As the number of handicapped people increases worldwide, Intelligent Wheelchairs (IW) are becoming the solution to enable a higher degree of independence for wheelchair users. In addition, IW Projects relevance is increasing, mainly in the fields of robotics and safety-related systems due to their inherent and still unresolved problems related with environment uncertainty, safe communications and collaboration methodologies. This paper describes the development of new communication system, based on multi-agent systems (MAS) and motivated by Intelligent Wheelchair systems. It uses a message-oriented paradigm as a mean for fault-tolerant communications in open transmission systems as well as a facilitator for entity collaboration. It provides an overview of the related work in the area, the background and the main constraints to system development, contextualized an IW development project. The results achieved enable us to conclude on the effectiveness and adequacy of the proposed communication model to the field of mobile robots, as well as to conclude that it may outperform JADE in several test scenarios.

1 Introduction

The World Health Organization (WHO), estimates that around 2% of world population (130 million people) live with physical handicaps. The most common aid for this kind of mobility problem is the wheelchair, specifically the electric wheelchair. Numerous Intelligent Wheelchair (IW) related projects have been announced, and are under development in the last years. The increase study of this field, led to a globally accepted view of the main functional requirements for such systems. According to [10] and further developed in [2], the Communication systems is one of

Frederico M. Cunha · Rodrigo A.M. Braga · Luis P. Reis
Artificial Intelligence and Computer Science Laboratory - LIACC,
Faculty of Eng. of University of Porto - FEUP, Rua Dr. Roberto Frias s/n, Porto, Portugal
e-mail: fredericom.cunha@gmail.com, rodrigo.braga@fe.up.pt,
lpreis@fe.up.pt

Y. Demazeau et al. (Eds.): Advances in PAAMS, AISC 70, pp. 151–156.
springerlink.com

the main functions of an IW, as it enables interaction with other devices like other IWs, intelligent robots and remotely operated control software for medical staff. Although many IW projects exist, the communication system used is rarely described and scarcely treated as an important and vital piece of an Intelligent Wheelchair. A common solution seen for the communication system is the use of CORBA based systems, or other technologies that enable communication through object sharing techniques, as seen in [11]. The SENA robot, presented in [7], is one of the few IW projects that approaches the communication system. Currently, it uses a custom Multi-agent System (MAS) to implement communications [6], thus taking advantage of MAS maturity, robustness, scalability, easy management of information and agents.

This article will focus in the field of mobile robotics, specifically IW communications and will: identify the main requirements to a safe and secure communication system, propose solutions that can address the found constraints, as well as, test the proposed solutions and compare the results with commonly used platforms. The rest of this article is divided into 3 additional sections. Section 2 provides a description of the applicable constrains, linking them to the described project and to the field of mobile robotics. A detailed description of the proposed solutions is also given in this section. Section 3 describes the test environment, the applied test methodology and presents the test results. Section 4 discusses the relevance of the test results and the applicability of the proposed methods to the field of mobile robots and robots in dynamic and safety-critical environments.

2 System Description

The IntellWheels' main objective is to create a complete intelligent system, hardware and software, which can be easily integrated into any commercially available electric wheelchair with minor modifications to its structure. Currently it implements a multi-layer control architecture, described in [1] and [2], distributed through 4 agents. The user interacts with the system through a multi-modal interface, that allows different input sources to be used [2]. Also a simulator is available, enabling a mix of different realities to be used when testing the platform. The following communication system's requirements were identified: multiple agent support, compatibility with other communication languages and communication systems, system's reconfiguration according to changes in the physical and networked environment, applicable to open transmission systems, use of defensive methodologies to protect system communication.

Normally, a multi-agent platform, such as Jade, would be used to enable communications, organize and manage the different agents. However with common multi-agent frameworks it is not possible to easily customize or enhance system functionality in order to adapt the system to a specific reality. Moreover, to implement fault-tolerant methods, it is required to use federated communication models with support on centralized databases, as resolve the issues related with firewall traversal and dependability on proprietary extensions. Thus, in Intellwhells case, the solution was to develop a new communication system, based on MAS methodologies that could address the system's requirements.

Taking the requirements into account, the new system was based on the FIPA standard [5], and uses the FIPA-ACL communicative acts as well as FIPA-SL. It was developed to work in two separate environments. The first, a global network applied to all agents connected to a network, and the second applied to the agents present in the same pc. Central in this architecture is the election of a Container entity, similar to Jade [9], and the distribution of a Local Agents List (LAL), as well as a Global Agent List (GAL), using a message-oriented paradigm. These lists contain the applications' configurations that enable communications and distribution of the public encryption key between agents. The Container was designed to be responsible for the lists maintenance operations that include creation, update and deletion. However and contrary to other systems, the Container was not designed as a separate entity or as the base for agents' creation and their activity. The idea behind this is that it is admissible and probable for a wheelchair to lose network connectivity or to change its network configuration but it is no acceptable for these changes to cause a system's malfunction.

2.1 Architecture

As above mentioned, the local platform is organized and maintained by a Container entity. It is not however, a separate software. The system was designed for the container algorithms to remain inside the communication's structure, thus making it a part of all agents. This way it is also possible to start an agent and for it not to depend directly on the communication system's configuration, state machine and resources to perform its function. The system's architecture was designed as five separate layers with their respective receiving and sending handling methods and interfaces, as seen in Fig. 1 a), running in parallel. This way, it becomes possible for the user to choose which layers should be applied to the application, without compromising the agent's functionality while following the OSI Reference Model and implementing fault tolerant methods described in [3] and recommended by [8].

The **Communications layer** is responsible for receiving and sending messages from and to the message transport layer. It allows the user to choose between TCP/IP, UDP or even HTTP messages. This layer also prevents the interpretation of repeated messages, present in the physical media, and enables the retransmission of messages thus preventing packet loss at network level. It also prevents the application from receiving messages with a size larger than the one specified by the user during agent implementation. When not specified the default messages' maximum size is 16kBytes for TCP and 8kBytes for UDP.

The **Decryption and Encryption layer** is responsible for the message's security, preventing the interception and modification of messages. The Encryption method is chosen accordingly to the message's destination and the knowledge at the moment. The possible encryption methods involve the use of a private and public key pair

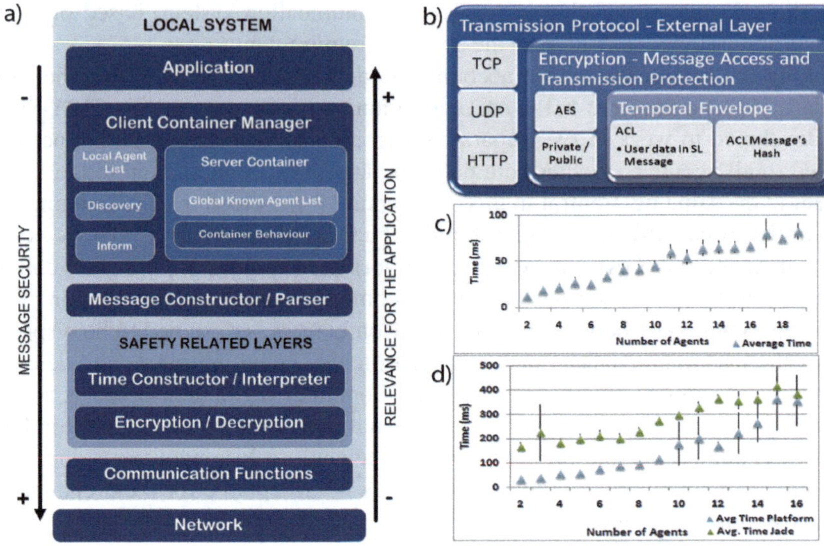

Fig. 1 a) Platform's layered structure; b) Resulting message envelope; c) Results in the second test; d) Results in the third test

or an AES pre-shared key. It also performs message integrity checking by cross-referencing the message with the transmitted message's hash.

The **Time layer** is responsible for adding a time stamp to the message to be sent, and for organizing the received messages according to the time stamp. It also performs the detection and elimination of injected packets. Another function that it performs is the configuration and synchronization of the local system's clock with a networked NTP clock, if available. This configuration is done automatically and only if the application is the local Container.

The **Message Constructor and Parser layer** is responsible for the construction of the message according to the FIPA-ACL standard and represented using FIPA-SL. It also selects the messages that are accepted by the application according to their correct structure configuration and to the sender's presence in the platform, thus stopping any communication from an unauthenticated application.

The **Client Container Manager** manages the application's organization and integration in the local and networked environments, implementing methods like replication, fault detection, recovery and discovery. It also implements the user interface with the communication system by enabling direct access to the LAL and GAL.

3 Tests: Scenario, Protocol and Results

For validation purposes, the above described architecture, was implemented in Pascal, using Borland Delphi Professional v7.2 IDE. Jade v3.7 and the Eclipse Platform v3.4.2 were used to develop and test the Jade agents described. For all tests the application Process Explorer from Sysinternals, was used as a resource monitor. The tests were made in a PC with an Intel Core 2 Duo 2,53GHz cpu, 4GB of RAM, with Windows Vista and using a 802.11n wireless network card. All tests were repeated 20 times with the same conditions and all data was analysed and represented with a confidence interval of 95%. To validate the platform, three type of tests categories were established and implemented.

The first tests' objectives were to evaluate the effectiveness, performance and to determine the platform's limits, when submitted to stress conditions. It used messages sent in short intervals with a increasing size. A two agent interaction was implemented, one serving as a initiator and the other a message repeater. The initiator would compare the sent message with the one received and then increase the message's size with 100 new bytes. The test results, seen in Fig. 2, prove that the platform was able to transmit all messages until the maximum size limits, for 4h and with small time deviations between rounds.

The second test's objective was to evaluate the performance, effectiveness and scalibity of the communication platform. It consisted on a agent sending a message, with a fixed size of 500 Bytes, that would be redistributed to all agents in a serialized manner. The time that the message took to be passed between all agents and return to the initiator, was measured. The test was repeated using the maximum number of 19 agent instantiations. The results, seen in Fig. 1c), show that the platform was able to deliver all messages in the correct order without errors, and that it scales linearly with the amount of agents present.

The third test's objective was to compare the platform's performance and scalability with Jade's, using a Book Seller Book Buyer agent protocol, described in [4]. In the test, only one buyer agent was instantiated, while the number of seller agents increased in each test round to the maximum of 15 agents. The time needed for a successful interaction was measured and can be seen in Fig. 1d). These show that Jade is slower than the proposed platform with a small number of agents and both

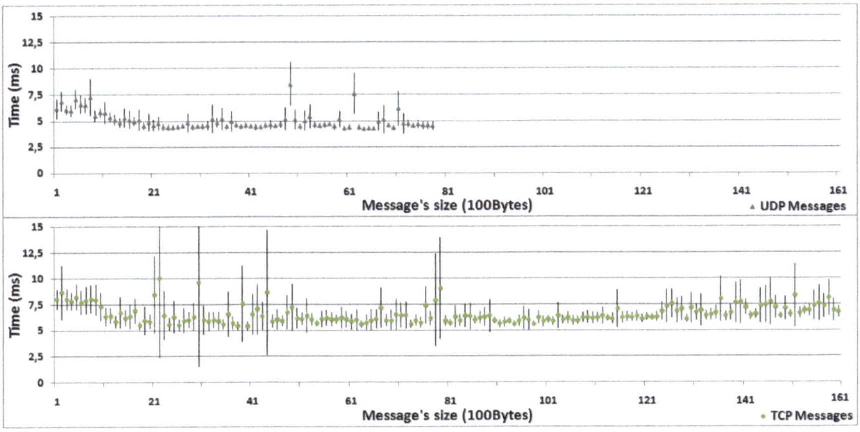

Fig. 2 Results achieved in the first test's scenario

with small variations. However when the number of agents increased the proposed platform's results intervals became wider.

4 Conclusions

In the field of IW, object interaction is absolutely needed, to enable cooperation between distinct robots. The results achieved using the proposed architecture, allow us to conclude that the platform can in fact provide safe communications between a large number of agents. Additionally, the implemented defensive procedures, proved to be effective when applied to this field. Moreover, the collected data shows that the framework is able to outperform JADE in several test scenarios.

References

1. Braga, R.A.M., Petry, M., Moreira, A.P., Reis, L.P.: Intellwheels: A Development Plat-Form for Intelligent Wheelchairs for Disabled People. In: Proc. of the 5th Int. Conf. on Informatics in Control, Aut. and Robotics., May 2008, vol. I, pp. 115–121 (2008)
2. Braga, R.A.M., Petry, M., Moreira, A.P., Reis, L.P.: Concept and Design of the Intellwheels Platform for Developing Intelligent Wheelchairs. In: Informatics in Control, Aut. and Robotics, vol. 37, pp. 191–203 (April 2009)
3. EN 50159-2, Railway applications – Communication, signaling and processing systems, Part 2: Safety related communication in open transmission systems. In: Vol. Europ. Commitee for Electr. Standardization (March 2001)
4. Bellifemine, F.L., Caire, G., Greenwood, D.: Developing Multi-Agent Systems with JADE. Wiley Series in Agent Technology. Wiley, Chichester (2007)
5. FIPA, Found. for Intelligent Physical Agents (October 2009), `http://www.fipa.org`
6. Galindo, C., Cruz-Martin, A., Blanco, J.L., Fernãndez-Madrigal, J.A., Gonzalez, J.: A Multi-Agent Control Architecture for a Robotic Wheelchair. In: App. Bionics and Biomechanics III, pp. 179–189 (2006)
7. Galindo, C., Gonzalez, J., Fernandez-Madrigal, J.A.: A Control Architecture for Human-Robot Integration. Application to a Robotic Wheelchair. IEEE Trans. Systems, Man Cybernet-Part B 36 (2006)
8. IEC 61508. Functional safety of electrical/electronic/programmable electronic safety-related systems. In: Vol. European Committee for Electrotechnical Standardization
9. JADE, Java Agent Dev. Framework (October 2009), `http://jade.tilab.com`
10. Jia, P., Hu, H.: Head Gesture based Control of an Intelligent Wheelchair. In: CACSUK-11th Ann. Conf. Chinese Aut. Comp. Society in the UK, Sheffield, UK (2005)
11. Prenzel, O., Feuser, J., Graser, A.: Rehabilitation robot in intelligent home environment - software architecture and implementation of a distributed system. In: 9th Int. Conf. Rehab. Robotics, ICORR 2005, June 28-July 1, pp. 530–535 (2005)

A Real Time Approach for Task Allocation in a Disaster Scenario

Silvia A. Suárez B., Christian G. Quintero M., and Josep Lluis de la Rosa

Abstract. A disaster scenario is a very dynamic environment where agents have to face surrounding environmental changes constantly. In this kind of scenarios, the quality of the tasks allocation is highly relevant because every minute spent reduces the chance of successfully rescuing the victims. In this approach, a real time sequencing technique is presented in order to schedule victims taking into account his/her death time. To do that, our rescue agents search for the victims and predict their death time to establish a victim priority order in our decentralized scheduling algorithm. Moreover, the approach in this paper is compared with a centralized task allocation approach. The RoboCup Rescue simulator has been used for experimentation.

1 Introduction

In recent years, task allocation in MAS has been the focus of considerable research, particularly in the rescue and crisis management area. The RoboCup rescue project [1] is a simulated scenario of an earthquake in a city, where rescue agents must minimize the damage caused by an earthquake. There are three kinds of rescue agents: fire-fighters, policemen and ambulance teams. As in real situations, agents have a limited scope. Agent brigades can see visual information within a radius of 10 meters. In addition, one agent is capable to say or listen a fixed maximum of messages each simulation cycle. Regarding task allocation and coordination, agents have to make decisions about task distribution. These decisions are related to, for instance, which victims the agents have to rescue first. This rescue problem can be

Silvia A. Suárez B. · Josep Lluis de la Rosa
University of Girona
e-mail: sasuarez@eia.udg.edu, peplluis@eia.udg.edu

Christian G. Quintero M.
Universidad del Norte
e-mail: christianq@uninorte.edu.co

Y. Demazeau et al. (Eds.): Advances in PAAMS, AISC 70, pp. 157–162.
springerlink.com

modeled as a task scheduling problem. We have focused, on victims' allocation for each ambulance team. The rescue agents arrange in sequence the victims according to a due-date-based technique. In addition, the rescue agents are coordinated by a central agent, who has information about the victims to be rescued. Decision-making in agents is based on the victims' damage and their death time estimations. The rest of this paper is as follows: First, the real time approach for task allocation is presented in section 2. Then, experimental results are provided in section 3. Finally, some conclusions are outlined in section 4.

2 Real Time Approach for Task Allocation

The rescue allocation problem can be modeled as a task scheduling problem in which each task consists of a single operation [2] [3]. Our approach is concerned with the special case of finite sequencing for a single machine and, particularly, with the sequencing theory according to due-date $d_{[i]}$. In this sense, due-dates are used to sequence the tasks so that: $d_{[1]} \leq d_{[2]} \leq ... \leq d_{[n]}$. According to the scheduling theorem by [4], when scheduling a *n/1//Lmax* scheduling problem, the maximum task lateness and maximum task tardiness are minimized by sequencing the tasks in non-decreasing due-date order.

2.1 Scheduling Process When Rescuing Victims

In our rescue problem, we have *n* victims considered as *n* tasks and each ambulance agent is considered as one resource (machine). At the beginning of the simulation victims are distributed throughout the scenario and ambulance agents are exploring the surrounding area. Once ambulance agents know about a victim or group of victims, they have to make the decision about how to rescue them. In this sense, our victims' scheduling algorithm, which is shown in Figure 1 (Left), has two relevant issues. First, victims' scheduling that refers to organizing the set of victims to be rescued: victims are scheduled according to death time. Secondly, the algorithm selects highest priority victims: it is a key issue in this scenario to first rescue civilians who will die without our help meanwhile lowest priority victims are reported to the ambulance center. In the scheduling algorithm in Figure 1, the V=V_1,V_2,...,V_n set are the victims. The *m* parameter is the number of ambulance teams rescuing the civilian. The δ value is fixed depending of the simulation results and *id* corresponds to the identification number of the victim. *VictToRescue* is the list of victims allocated for rescue. The *deathTime(V)* parameter is a computation of the remaining time before the victim dies. The death time estimation method is presented in section 2.2.

The rescue time for one victim *rescueTime(v)* is computed by using (1)

$$rescueTime(v) = \frac{buriedness(v)}{m} + carry\_unload \qquad (1)$$

```
For V1,.... Vn
    deathTime(Vi)= deathTime_estimation();
    m=Number of Ambulances;
    RescueTime(Vi)=Buriedness(Vi)/m +
    carry_and_unload_time();
    earlyComplete=earlyComplete+RescueTime(Vi);
    if (deathTime(Vi)>earlyComplete) &&
    (earlyComplete/deathTime(Vi)>δ)
            VictToRescue.put(deathTime(Vi),Vi)
    else
            Tell(id(Vi),deathTime(Vi)) To Central
End For
  Sort(victToRescue) in nondecreasing deathTime order

If (VictToRescue isNotEmpty)
    Rescue(VictToRescue)

else
    Continue exploring the surrounding.
```

```
For each Vi
Hear (deathTime(Vi)) from Ambulance Agent;
    deathTime_Rec(Vi)= deathTime(Vi);
    VictToRescue_Rec.put(deathTime_Rec(Vi),Vi)

Sort(VictToRescue_Rec) in nondecreasing deathTime order.

If (VictToRescue_Rec isNotEmpty)
    Send(First victim in VictToRescue_Rec) to Ambulance
```

Fig. 1 Left: Scheduling algorithm in ambulance team agents. Right: Scheduling algorithm in ambulance center agent

Regarding the m parameter (m=number of ambulance teams), a greater number of ambulances can rescue a victim in less time. The *buriedness(v)* parameter shows how much the civilian v is buried in the collapse of buildings: a value of more than one means that it cannot move by him/herself. Each ambulance can decrease the *buriedness* of the buried victim by a value of 1 each time and can dig up and carry only a single victim. The *carry_unload* value is the needed time to carry the victim to a safety shelter and unloading him or her. An estimated mean time has been used for this action.

An ambulance center's scheduling algorithm is presented in Figure 1 (Right). This algorithm is in charge of the lowest priority victims discarded by the previous figure 1 (left) algorithm. The center's role is to communicate these victims to free ambulance agents. The algorithm in Figure 1 (Right) also schedules victims according to a non-decreasing death time order.

Before introducing the victim into the scheduling, agents have to decide whether the victim can be rescued alive and also if he or she has high or low rescue priority.

$$PossibilityOfRsc(V) = \left\{ \frac{earlyComplete}{deathTime} < 1 \right. \tag{2}$$

$$earlyComplete(V) = RscTime(V) + \sum_{i=0}^{n-1} RscTime(i) \tag{3}$$

In equation 2 if *PossibilityOfRsc(V)<1*, then the victim can be rescued alive. The *earlyComplete(V)* parameter is the earliest time at which the agent can efficiently rescue the set of victims inside the scheduling and in this case n is the number of victims. The V parameter represents one victim or a group of victims, depending on the number of detected victims on the rescue site. In addition, agents calculate the victims' emergency or priority using equation 4.

$$priority(V) = \left\{ \frac{earlyComplete}{deathTime} > \delta \right. \tag{4}$$

Then, if *PossibilityOfRsc(V)<1* and *priority(V)>*δ, victim *V* is introduced into the ambulance agent's scheduling.

2.2 Victim's Death Time Estimation

In order to compute accurate enough death-time values, we have used a case based approach to model behavior by logging its output for various inputs in the miscellaneous sub-simulator of the RoboCupRescue simulator. In this sense, we retrieved previous experiences in our simulations, in order to create a complete case base of civilians' death times in each time. For that purpose, the outputs of three different maps of the simulator have been used (Kobe, Foligno and Random maps). Figures 2 (left) show the hp-time curve in the Kobe map. The health point *(hp)* parameter indicates the life quantity of the victim. The highest *hp* value is 10000, the lowest one is 0, which means dead. The *hp* value decreases according to victim's damage in each cycle of simulation. Figure 2 (right) depicts the damage-time evolution for each victim during 300 cycles on all the maps. The time-hp and time-damage curve can be estimated by an exponential function with a reasonable error. The time-hp and time-damage curves are similar on each map. Based on outputs from these maps we have created a case base of *(hp/damage − > death time)* pairs which is used for death time estimation in future simulations. Death time values obtained from the case base which match with the current hp and damage input values are used to compute three different kinds of death times: *median, mean and minimum.*

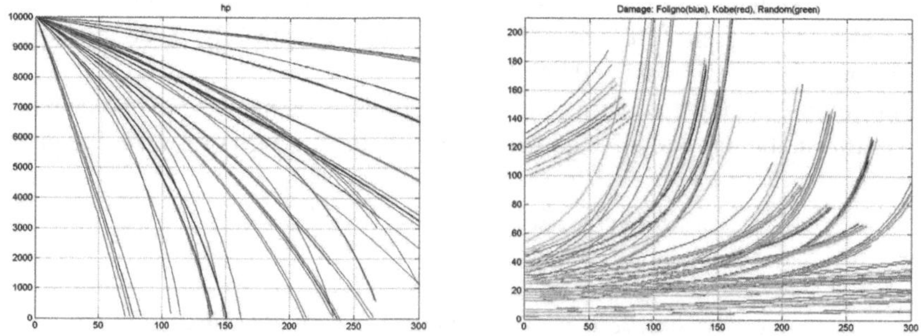

Fig. 2 Left: Health points (hp) of victims on the Kobe Map. Right: Damage to victim on all three maps

3 Experimental Results

The experiments have been done in the RoboCup Rescue simulator. Figure 3 shows the system performance with the three different death time estimation measures. Outputs of 15 simulations have been used for these estimations. The mean death time estimation obtains the best results as a higher percentage of victims were rescued. Furthermore, the approach in this paper has been compared with the approach of task allocation using combinatorial auctions proposed in [5] (see figure 4 Left). The decentralized task allocation mechanism in this paper is in real time; so, agents avoid the waste of time due the solution wait. On the other hand, with this procedure, the number of exchanged messages is considerably reduced as shown in Figure 4 (Right), where the averages results of exchanged messages number during a single time step taken every 30 cycles are represented. In this sense, there is a 40 percent of reduction of the quantity of information sent with respect to a

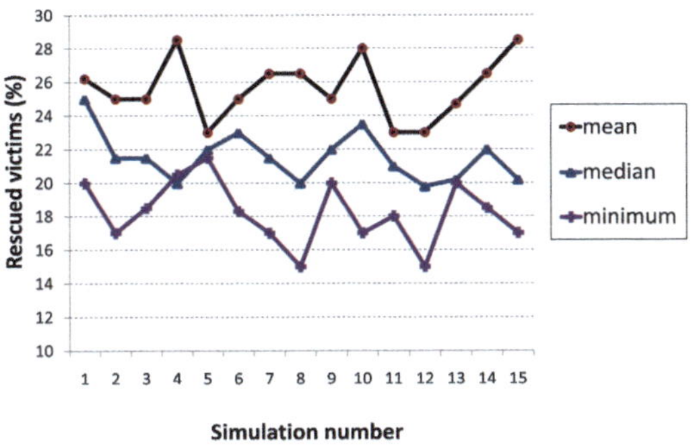

Fig. 3 Simulation results for three different methods of victims' death time estimation: simulation results using mean, median, and minimum death time estimation

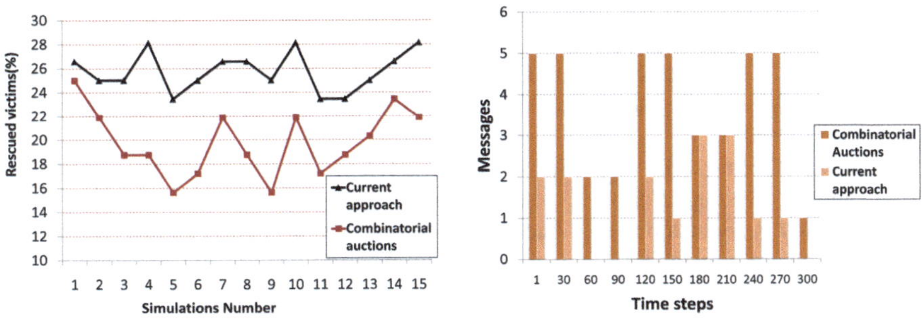

Fig. 4 Left: Comparison of performance among two methods for task allocation. Results using the real time algorithm (current approach) and simulation results using task allocation based on combinatorial auctions. Right: Comparison of exchanged messages number in both approaches

centralized one. This is mainly due to the fact that the agents do not have to send bids or other information they know, but only information about unattended victims to the central. With these results we demonstrate that the proposed approach presented in this paper has improved the rescue operation over the approach using combinatorial auctions mechanism.

4 Conclusion

In this paper an approach for allocating civilians to ambulance agents has been developed. Our scheduling algorithm sequences victims in non-decreasing death time order. The advantage of modeling the problem using sequencing is that it becomes solvable in polynomial time. For death time's estimation, a new method has been presented. The proposed method avoids agent's perception errors, which may lead to erroneous death-time predictions. In addition, a comparison between two different proposed mechanisms for task allocation has been made. The algorithm presented here performs better due to two aspects: first, the number of exchanged messages is drastically reduced; and secondly, it is a real time algorithm (in combinatorial auctions, agents have to wait some time steps for the central decision). The experimentation results have proved the feasibility of the approach presented in this work and the RoboCup Rescue simulator has been used for experimentation.

Acknowledgements. This research is funded by the Spanish MCYT project DPI2007-66872-C02-01, Arquitecturas Multiagente de Control e Interacción de Robots Móviles en su Aplicación al Rescate de Supervivientes de Catástrofes con Agua, the CSI-ref.2009SGR-1202 for the group CSI and 2009 BE-1-00229 of the AGAUR awarded to Josep Lluis de la Rosa.

References

1. Kitano, H., Tadokoro, S.: RoboCup rescue: a grand challenge for multiagent and intelligent systems. AI Magazine 22(1), 39–52 (2001)
2. Paquet, S., Bernier, N., Chaib-draa, B.: Multiagent Systems viewed as Distributed Scheduling Systems: Methodology and Experiments. Advances in Artificial Intelligence, 43–47 (2005)
3. Conway, R., Maxwell, W., Miller, L.: Theory of Scheduling. Addison-Wesley Publishing Company, Reading (1967)
4. Jackson, J.R.: Scheduling a production line to minimize maximum tardiness. In: Management science research project. Univ. of California (1955)
5. Suárez, S., Collins, J., López, B.: Improving Rescue Operation in Disasters. In: Approaches about Task Allocation and Re-scheduling, PLANSIG 2005 (2005) ISSN: 1368-5708

ASGARD – A Graphical Monitoring Tool for Distributed Agent Infrastructures

Jakob Tonn and Silvan Kaiser

Abstract. Monitoring the runtime behaviour of a distributed agent system for debugging or demonstration purposes provides a challenge to agent system developers. As a lot of the components in such a system are possibly executed on several different physical systems, maintaining an overview over the complete environment is extremely difficult. Methods taken from the development of monolithic software projects such as log files, debug outputs or step-by-step execution of a program do not easily translate to these scenarios due to the distributed nature of the system. In this paper we describe our concept for a graphical monitoring and management tool "ASGARD" (**A**dvanced **S**tructured **G**raphical **A**gent **R**ealm **D**isplay), which provides an easy-to-use and intuitively understandable method for monitoring and demonstrating Multi Agent System Infrastructures. ASGARD provides a graphical representation of the connected systems using a 3D visualization. The very promising results from empirical evaluation show that an administrator's overview over such MAS at runtime is vastly improved.

1 Introduction

Applications based on Multi Agent Systems (MAS) have gained more and more importance in software development during the recent years. Agent Frameworks such as JIAC V [4] provide a standardized approach for the development and implementation of such systems. Developing distributed MAS is a complex process [6] and requires more effort than developing a non-distributed application. Generally, the

Jakob Tonn
DAI-Labor, Technische Universität Berlin, Ernst-Reuter-Platz 7, 10587 Berlin, Germany
e-mail: `jakob.tonn@dai-labor.de`

Silvan Kaiser
DAI-Labor, Technische Universität Berlin, Ernst-Reuter-Platz 7, 10587 Berlin, Germany
e-mail: `silvan.kaiser@dai-labor.de`

Y. Demazeau et al. (Eds.): Advances in PAAMS, AISC 70, pp. 163–173.
springerlink.com

development process can be separated into the analysis and design phase, the implementation and the testing stage. Whereas the design process has been made a lot easier with standardized modeling languages and tools, and the process of implementation became partially automated using those models and readily-available libraries and frameworks, the process of monitoring the runtime behaviour of MAS has had far less attention by research projects and industrial solutions. Developers and testers are thus required to either use monitoring methods taken from conventional software development, which are partially suited for the distributed MAS at most, or create their own project-specific methods to indicate the runtime behaviour of their implementation. Another aspect is the demonstration of a distributed MAS in technical discussions. Just showing design documentation and the result of a working implementation usually raises the question of how far the design patterns are connected to the actual implementation. Due to the complex nature of agent systems, this step requires a lot of abstraction in the spectator.

These issues can be approached with the concept of software visualization. The idea behind the ASGARD approach is to overcome the "invisible" nature of software by visualizing programs, program artifacts and program behaviour with the goal of making a system easier to understand [1]. With the application of software visualization technologies, autonomous parts of a software system (like agents) can be easily identified, and animated entity representations create a runtime overview on processes and interaction. The ASGARD concept creates a generic solution for administration and demonstration of distributed MAS infrastructures. With the base of an intuitive visualization of MAS applications at runtime, understanding the internal structures, states and interactions takes less effort for both developers and spectators. The concept relies on the basic agent metaphor as used in Agent Oriented Software Engineering and continues this metaphor into the runtime system. Normally a MAS is monitored like a classical object oriented/imperative/petri net/etc. application. With ASGARD agents and their environment can be seen as entities of their metaphors, not only at design time but also inside the running system. Furthermore, developers have access to management functions which help them debug their applications. This is an important utility in the process of finding solutions for runtime problems. The focus of the project lies on visualizing large-scale distributed MAS on one hand and the ability to visualize behaviour in a single entity in detail on the other. This requires a great deal of scalability in the graphical representation, which is provided by the use of 3D technology in ASGARD. Another advantage of a 3D graphical environment over 2-dimensional raster images or scalable vector graphics is the enhanced freedom of placing objects, which allows for better layout of entities and larger systems to be visualized. Users are common to 3D visuals both because of their real-world experiences as well as the increased use of it in recent technology. We believe using 3D technology is the best way to achieve our goal of creating an easy-to-use and intuitive tool.

This paper continues by describing the concept of the ASGARD visualization, including some implementation information about the prototype. The underlying technology follows, a short evaluation, state of the art comparison and finally the conclusion containing future work.

2 The ASGARD Concept

The central feature of ASGARD is the visualization of agents and nodes in a distributed MAS Infrastructure employing 3D graphics technologies (see Fig. 1). We chose a 3 dimensional approach for a number of reasons. We live in a 3D environment which gives us a familiar understanding of relations between entities in such an environment. This creates possibilities to use these common relations for displaying system information. Furthermore 3D environments allow to display large numbers of entities that cannot be displayed in 2D without scaling issues. This allows us to communicate more clearly laid out information to a user than other classical 2D concepts in the same display size. The 3D environment in ASGARD both visualizes the hierarchy and structure of the incorporated entities by their placement in space, as well as interaction and processes by using animation.

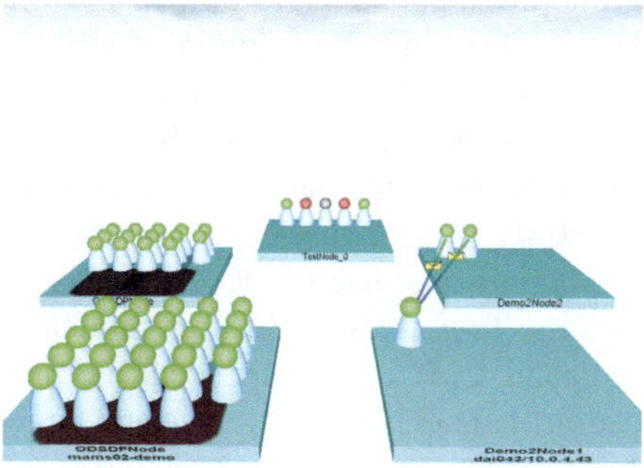

Fig. 1 ASGARD prototype visualization of several JIAC agentnodes

2.1 Metaphors and Visualization Details

A critical point in creating an understandable visualization is the choice of metaphors that represent the different entities. A good representation should enable the spectator to intuitively recognize and classify the entity according to his knowledge [3]. In our case of distributed agent systems, we have to make sure that metaphors make it easy to recognize the concepts known from theory about agent systems. The software agent as a generic actor in an environment already constitutes a strong metaphor in Agent Oriented Technologies (AOT) system design. It is an approach that uses a uniform concept better suited for visualization metaphors than other concepts, e.g. like object oriented or imperative programming entities.

Agents are the key entities in an agent framework, as the name already states. In a MAS, agents are the entities where all application-specific functionality is happening. Agents provide the lifecycle for service implementation and are able to perceive information and communicate. As all this defines a kind of human-like behaviour, a good metaphor would be an abstraction of a human figure. A simple and common metaphor that basically is an abstraction of a human body shape is the one of a play figure from common board games like 'Chinese Checkers" or "Ludo". As this metaphor is commonly used in modeling notations for multi agent systems, the use of it furthermore extends the connection between design documents and runtime system structures.

Agent Nodes in a MAS are the environments for agents on a single system. They provide the infrastructure required to execute functionality in agents, so that agents can cooperate to realize complex applications. A simple metaphor to represent this function as a base provider to agents is a platform on which the contained agents stand (which can be easily recognized in Fig. 1). The platform can be used to display further information about the node, like the network address as textual representation or security support by applying a texture on the platform surface.

Communication is one of the key features of interaction in distributed agent systems. A common metaphor for message communication is a letter envelope. The use of this metaphor raises the problem that a real-time display of the message transport would be way too fast for a human. The solution of artificially slowing down the visualization of a message transfer has proven to be problematic in the use of ASGARD's predecessor, the JIAC NodeMonitor (see Sect. 5). It created the impression to users less familiar with the implementation of communication that the message transfer feature was painfully slow. Thus, our approach is the use of a combination of two metaphors. The message itself is visualized as a letter moving between the communication partners. To make sure the communication event is visible long enough for the user to catch, a partially translucent line is used as a metaphor to display the communication channel between source and target. This line will slowly dissolve to invisibility if no new messages are transferred, which creates the effect that frequently-used communication channels will appear far stronger than seldom-used ones.

States of entities can be visualized in an understandable way by colouring parts of the entity representation's surface. As state is a detail information about a certain entity, visualizing the state as colour is preferable over a shape change [3]. This is made especially well-understandable by picking colours users are common to from their real-life experience. ASGARD uses colours to visualize states is the agent "heads", which are coloured according to the lifecycle state of the agent, so that a red head indicates a stopped and a green head a running agent.

Layout and level of detail: As the ASGARD concept is especially targeted at large distributed agent systems, the level of detail displayed in the current view has to be adjusted according to the amount of visible objects. As we are using a first-person perspective view, basing the level of detail displayed on the distance from an object to the camera is a convenient method. It protects ASGARD from overflowing the user with a massive amount of small objects. Further displaying of detail information can be done using tooltip effects and selection of entities.

The layout of the visualized entities (which means their placement in the visual space) can be used to visualize structures and hierarchies in the MAS. An already-mentioned employment is the placement of agents on Agent Nodes to provide information about the location and hierarchic position of agents.

Human-Machine Interaction: Human-Machine-Interaction in ASGARD is provided by employing Graphical User Interface (GUI) standards like mouse interaction, drag and drop, popup menus and standard operating system GUI elements. These technologies are used to enhance the level of information provided by the visualization and to create a comfortable and intuitive management interface. Mouse over effects are used to display more information. An object property table GUI component is used to display additional information about entities which are not visualized as metaphoric representations due to either their textual nature or where an intuitively understandable metaphor is not available. Entity selection can be performed by mouse clicks to display an entity's properties permanently in the property table. Furthermore, object selection is used to control the viewpoint. ASGARD is able to focus the viewpoint on selected objects to overcome the sometimes unintuitive camera controls in a 3D environment. Commonly used camera controls like zoom and rotation are offered as well. A context menu is used to access management functionality for the selected entity.

2.2 Implementation

The implementation of ASGARD applies the generic concept of metaphoric visualizations to JIAC V as an existing MAS which is used in various projects.

Program Structure: To maintain a clean and modular program structure which supports future extensions of the ASGARD concept, a well-thought program structure is required. Model-View-Controller (MVC) patterns have proven to be a good way of modularizing between functionality, graphical rendering and data in GUI developments [10]. Having those aspects in separate units of code helps maintaining a clean implementation and makes adjustments and implementations easy. As the structure of entities in the JIAC framework is a tree hierarchy, it is suitable to use this kind of structure to organize entity objects in ASGARD as well. This is even more convenient as the JME graphics engine (see section 3) organizes visual elements in a scene graph, which is essentially a tree structure as well.

Extensions and PlugIns: To make ASGARD a versatile, flexible and useful tool, it requires methods to extend it's functionality. An interface for plugins provides the possibility to adapt the program to the user's needs. Plugins can access the data from connected JIAC V entities as well as the JMX connectors and thus create additional monitoring and management functions. They can be used to display additional or application-specific data or increase the level of detail in which information are displayed.

The **Load Measurement Plugin** (see Fig. 2) is an exemplaric concept of an extension plugin for ASGARD. An optional JIAC V agent bean is able to provide

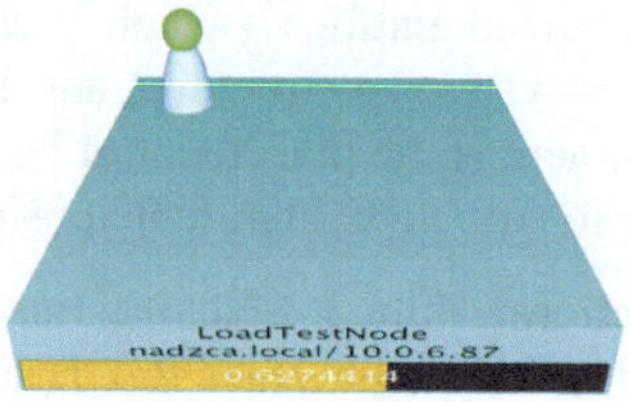

Fig. 2 ASGARD Load Measurement Plugin

information about the current system load. The load measurement plugin is able to read those values over the management interface and use them to create a visualization that shows the current system load of an Agent Node.

3 ASGARD Prototype Technology

The functionality of the ASGARD prototype employs several already-existing technologies.

JIAC V (**J**ava-based **I**ntelligent **A**gent **C**omponentware) is a java-based agent framework with particular emphasis on industrial requirements [4]. JIAC combines agent technology with a service-oriented approach, and furthermore provides transparent distribution, security and management features and dynamic reconfiguration in distributed environments. As the latest version of JIAC, JIAC V is used in a broad spectrum of projects, ranging from service delivery platforms to simulation environments, a generic monitoring and debugging tool greatly improves the workflow for JIAC V application developers. For the creation of such a tool for JIAC V the already integrated management interface based on the JMX technology comes at great advance.

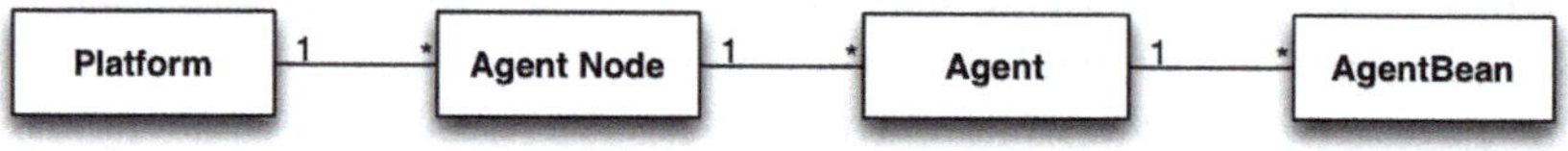

Fig. 3 JIAC V basic structure

In the JIAC V basic structure (see Fig. 3), *Agents* simply provide a lifecycle, any additional features are implemented in *Agent Beans*. Each Agent Bean provides a special capability in a modular way, and by "plugging" it into an Agent, the Agent gets equipped with the bean's feature. *Agent Nodes* provide an execution environment for multiple agents and consist of a Java Virtual Machine (JVM) running the JIAC V agent infrastructure. A *Platform* consists of multiple distributed Agent Nodes, connected via a communication infrastructure, and provides the functionality for JIAC V applications.

JMX (**J**ava **M**anagement E**x**tensions) [13] provide runtime access to Java objects from outside of a Java Virtual Machine for monitoring and management purposes.

It has the benefits of requiring only very small changes to an exisiting Java implementation and providing a great level of abstraction to network technologies. These features make JMX an useful technology for all kinds of remote monitoring and management problems. As JIAC V already has a JMX-based management interface implemented, accessing JIAC applications this way is the most convenient method for ASGARD to provide runtime monitoring and management.

JME (**J**ava **M**onkey **E**ngine) [7] is a 3D engine implemented in Java. Although its main purpose is providing a base for java-based games, the provided capabilities make it useful for all kinds of projects that require a three-dimensional graphical visualization.

JME is based on the OpenGL [12] interface, which creates the possibility of implementing applications based on JME for various platforms. The binding to OpenGL is realized using native libraries for all supported operating systems so that JME provides a high performance environment for rendering 3D graphics. JME offers functionality like pre-made geometric primitives, import methods for loading common geometric mesh file formats, a scene graph implementation and animation controllers. Thus, the use of this engine greatly reduces the amount of work needed to implement the high-quality graphical visualization ASGARD offers.

4 Evaluation

In order to evaluate ASGARD, a first prototype of the concept was implemented. The prototype is able to show a quick perspective on LAN-wide Agent Nodes currently in use, as well as detail information about agents. As the main motivation for the ASGARD concept was to create an easy-to-understand and use visual representation of a MAS, user tests were chosen as the best evaluation method. The prototype was furthermore used in the debugging process of the MAMSplus [9, 14] project. By using ASGARD to visualize the central MAS, developers were immediately able to identify agents that had been stopped without prior notice because of communication issues with a third party API. Those tests showed a demand for introspection and generated requests for more features that should be implemented (see Future Work). This proved that the path proposed in the initial concept is worth extending. The tests also showed the limitation of ASGARD not being able to run over a remote desktop connection due to the use of hardware 3D acceleration, this issue should be taken care of in future. Future evaluation will be done by tests with a larger and more diverse group of users, and the use of ASGARD as monitoring and demonstration tool in other projects.

5 Related Work

As the problem of monitoring and managing a complex software infrastructure like a distributed MAS is inherent in the structure of such systems, there are several other approaches, projects and implementations that focus on offering a solution to it by employing software visualization technologies.

Shell based application monitoring is probably the most senior and still commonly applied technique for monitoring a running MAS. Generally direct program output and/or log files are written to the console. For distributed MAS this generally means several shell windows, concurrently showing the output of different entities of the MAS.

This approach allows a low level insight into running programs but has a range of serious drawbacks. The most prominent being insufficient overview when concurrent events take place on different machines, as well as the lack of support for representing connections between the different events. These connections have to be searched for manually. Typically running program outputs are visually scanned for relevant textual patterns, an approach better suited for retrospective analysis but not for live monitoring.

Richter-Peill provides a concept partially similar to ASGARD in his bachelor's thesis [11]. His concept is focused on creating a 3D visualization of the interaction between agents at runtime in a Java Agent DEvelopment Framework (JADE) [2] MAS. Richter-Peill connects to the agent framework by adding an agent that gathers information and sends them to the visualization renderer. This approach has the disadvantage of changing the current MAS structure, which may lead to side effects in the system's processes. As Richter-Peill focuses on visualizing interaction, a visualization of the MAS structure is not provided by his solution. This makes it impossible to use his concept to compare an implemented MAS to a design model.

The **ADAM3D** [5] tool is also based on the motivation of making the debugging process of MAS implementations easier by creating a 3D visualization of the MAS. Ilarri et al. chose to use the 3rd dimension as temporal dimension to show the interaction between agents over time, sort of like a 3-dimensional sequence diagram. This approach makes ADAM3D a versatile tool for analysis, but does not provide insights into the structure of the MAS or a management interface. A similar time-based visualization could be added to ASGARD by means of a plugin in future.

Ilarri et al. employ log output analysis as the base for their visualization. Whereas this approach has the advantage of working both during runtime as well as on saved output after a system's run, it may require large-scale adaptions in an already existing MAS application.

The **JIAC NodeMonitor** [8, p. 126] is the predecessor to the ASGARD concept. It provides a visual monitoring and management tool for a single Agent Node. The NodeMonitor provides a 2D visualization of the Agents on the connected Agent Node and features state visualization using icons and animated interaction representation. Furthermore, management of the visualized entities is provided using standard GUI technology like popup menus. Connection to the distributed MAS is established via JIAC V's JMX management interface.

The NodeMonitor's limitations are the focus on a single Agent Node and the lack of scalable visualization. When using it to monitor large and especially widely distributed applications, a lot of the entity relations in the system are not shown to the user. The lack of scalable visualization, that is caused by using two-dimensional raster image technology (in this case, Swing and Java2D), further enhances the difficulty of visualizing large-scale systems using the JIAC NodeMonitor.

6 Conclusion

Contribution: The main achievement of ASGARD is the consequent use of metaphors known from AOT design notations throughout the whole monitoring process. This makes the visualizations intuitively understandable. Thus, MAS have a clear and consistent visualization throughout the whole development lifecycle. The current implementation and structure of a MAS can be directly visualized at runtime using ASGARD and easily be matched and compared with the design concepts.

Another important achievement of ASGARD is the ability to provide a quick overview over large-scale systems, and especially over distributed entities in a single application window. ASGARD generates the possibility to identify problems visually at runtime, instead of requiring the developer to analyze large log files after the application has been shutdown.

A third aspect is ASGARDs extensibility. The plugin interface allows for extensions that can be goal-specific to an extremely high degree. This makes ASGARD a versatile tool for the debugging process of MAS and provides the possibility to create specific demonstrations of running systems.

Summary: Our motivation for the ASGARD concept is that a visual monitoring and management tool makes the process of administrating MAS easier and enables to verify implementations against design concepts. The JIAC V management interface forms an ideal base for an implementation of ASGARD. The use of visual metaphors known from the design process ensures easy understandability. As the ASGARD implementation follows a strict MVC pattern, modularity and extensibility through plugins are achieved. Compared to classical monitoring using log output on shells, connections between entities are much better visualized, and problems are much more easy to identify at runtime. Other concepts or tools try to visualize MAS as well, but either do not take the step back to design metaphors (Richter-Peill's visualization) or are limited in their graphical abilities (JIAC V NodeMonitor), which ASGARD avoids by using highly scalable 3D visuals.

Future Work: The ASGARD concept with it's modular structure can easily be extended to a greater range of functionality. In order to visualize running JIAC V systems in depth, visual representations of more structural entities are planned.

Identifying Agents: Currently all Agents and Agent Nodes have a uniform appearance showing their individual basic entity types and states. For better orientation in larger platforms ways to distinguish between entities of the same type are needed.

Different approaches, e.g. distinction based on roles, components or owner identities have to be researched.

Visualization Plugins: In order to create an in-depth visual representation of a MAS infrastructure, plugins that visualize special aspects in detail, e.g. the entities inside of an Agent, are needed.

Extended layouting: The modular concept for layout managers in ASGARD allows for new algorithms for entity layout to be implemented. While our current approach is effective when displaying approximately up to 30 Nodes with a few hundred Agents each, larger platforms require new layout scaling mechanisms for the ASGARD.

Entity Selection: Last but not least search & group selection mechanisms are needed when dealing with large numbers of entities in the ASGARD. This will allow focussing and manipulating filtered subsets of the platforms entities.

References

1. Ball, T., Eick, S.G.: Software Visualization in the Large. IEEE Computer 29(4), 33–43 (1996)
2. Bellifemine, F., Poggi, A., Rimassa, G.: JADE - a FIPA-compliant agent framework. In: Proceedings of the Practical Applications of Intelligent Agents (1999)
3. Diehl, S.: Software Visualization - Visualizing the Strucutre, Behaviour and Evolution of Software. Springer, Heidelberg (2007), ISBN 978-3-540-46504-1
4. Hirsch, B., Konnerth, T., Heßler, A.: Merging Agents and Services — the JIAC Agent Platform. In: Bordini, R.H., Dastani, M., Dix, J., El Fallah Seghrouchni, A. (eds.) Multi-Agent Programming: Languages, Tools and Applications, pp. 159–185. Springer, Heidelberg (2009)
5. Ilarri, S., Serrano, J.L., Mena, E., Trillo, R.: 3D Monitoring of Distributed Multiagent Systems. In: WEBIST 2007 - International Conference on Web Information Systems and Technologies, pp. 439–442 (2007)
6. Jennings, N.R., Sycara, K., Wooldridge, M.J.: A roadmap of agent research and development. Autonomous Agents and Multi-Agent Systems 1, 275–306 (1998)
7. JME Development Team. Java Monkey Engine (2009), `http://www.jmonkeyengine.com`
8. Keiser, J.: MIAS: Management Infrastruktur für agentenbasierte Systeme. PhD thesis, Technische Universität Berlin (September 2008)
9. Konnerth, T., Kaiser, S., Thiele, A., Keiser, J.: MAMS Service Framework. In: Decker, K.S., Sichman, J.S., Sierra, C., Castelfranchi, C. (eds.) AAMAS 2009: Proceedings of The 8th International Conference on Autonomous Agents and Multiagent Systems, Richland, SC, May 10– 15, pp. 1351–1352. International Foundation for Autonomous Agents and Multiagent Systems (2009)
10. Reenskaug, T.: Models-Views-Controllers. Technical report, Xerox-Parc, 12 (1979)

11. Richter-Peill, J.: Visualisierung der Interaktion in Multiagentensystemen. Bachelor's thesis, University of Hamburg, Germany (October 2005)
12. Segal, M., Akeley, K.: The Design of the OpenGL Graphics Interface. Technical report, Sun Microsystems, Inc. (1994)
13. Sun Microsystems, Inc. Java Management Extensions (JMX) Specification, version 1.4 (November 2006)
14. Thiele, A., Konnerth, T., Kaiser, S., Keiser, J., Hirsch, B.: Applying jiac v to real world problems: The mams case. In: Braubach, L., van der Hoek, W., Petta, P., Pokahr, A. (eds.) MATES 2009. LNCS, vol. 5774, pp. 268–277. Springer, Heidelberg (2009)

Comparing Three Computational Models of Affect

Tibor Bosse, Jonathan Gratch, Johan F. Hoorn, Matthijs Portier,
and Ghazanfar F. Siddiqui

Abstract. In aiming for behavioral fidelity, artificial intelligence cannot and no longer ignores the formalization of human affect. Affect modeling plays a vital role in faithfully simulating human emotion and in emotionally-evocative technology that aims at being real. This paper offers a short expose about three models concerning the generation and regulation of affect: CoMERG, EMA and I-PEFiC^{ADM}, which each in their own right are successfully applied in the agent and robot domain. We argue that the three models partly overlap and where distinct, they complement one another. We provide an analysis of the theoretical concepts, and provide a blueprint of an integration, which should result in a more precise representation of affect simulation in virtual humans.

Keywords: Affect modeling, Cognitive modeling, Virtual agents.

1 Introduction

Over the last decade, a virtual explosion can be observed in the amount of novel computational models of affect. Nevertheless, current affect models in software agents are still simplifications compared to human affective complexity. Although many agents currently have the ability to show different emotions by means of facial expressions, it is quite difficult for them to show the right emotion at the right moment. In anticipation of richer interactions between user and agent, this

Tibor Bosse · Matthijs Portier · Ghazanfar F. Siddiqui
VU University, Department of Artificial Intelligence
De Boelelaan 1081, 1081 HV Amsterdam, The Netherlands
email: `{tbosse, mpr210, ghazanfa}@few.vu.nl`

Johan F. Hoorn · Matthijs Portier · Ghazanfar F. Siddiqui
VU University, Center for Advanced Media Research Amsterdam
Buitenveldertselaan 3, 1082 VA Amsterdam, The Netherlands
email: `jf.hoorn@camera.vu.nl`

Jonathan Gratch
University of Southern California, Institute for Creative Technologies
email: `gratch@ict.usc.edu`

Ghazanfar F. Siddiqui
Quaid-i-Azam University Islamabad, Department of Computer Science,
45320, Pakistan

Y. Demazeau et al. (Eds.): Advances in PAAMS, AISC 70, pp. 175–184.
springerlink.com

paper explores the possibility to integrate a number of models that are sufficiently similar, while preserving their individual qualities. As a first step into that direction, we compared three models (CoMERG, EMA, and I-PEFiC^{ADM}) of agent affect-generation and affect-regulation (or coping).

We selected three models inspired by some of the most influential theories in the emotion domain to achieve more realistic affective behavior in agents. The theory of Emotion and Adaptation of Smith and Lazarus [12] was formalized by Gratch and Marsella [10] into *EMA*, a model to create agents that demonstrate and cope with (negative) affect. The emotion regulation theory of Gross [5] was used as inspiration by Bosse, Pontier, and Treur [2] to develop *CoMERG* (the Cognitive Model for Emotion Regulation based on Gross), which simulates the various emotion regulation strategies described by Gross. The concern-driven theory of Frijda [4] was used by Hoorn, Pontier, and Siddiqui [6] to design *I-PEFiC^{ADM}*, a model for building robots that can trade rational for affective choices. We consider these theories because of their adequate mechanisms, simplicity and coherence. Together, they (Frijda, Smith & Lazarus, Gross) cover a large part of emotion theory. All three were inspired by an appraisal model of emotion, which makes them well suited for integration. In addition, these models are already implemented as computational models, which makes it easier to integrate them. All three approaches point at important aspects of human affective behavior, but also miss out on something. CoMERG [2] and EMA [10] address the regulation of affective states, but EMA does not regulate positive affect. CoMERG, on the other hand, has no provisions for generating affect, and does not explicitly account for a causal interpretation of the world state. I-PEFiC^{ADM} [6] generates and balances affect but is mute about the different regulation mechanisms. Because the models are complementary to each other, it makes sense to integrate them.

As a first step, the present contribution attempts to align and contrast different affect models as they were derived from the original emotion theories[1]. We will point out what deficiencies should be overcome to build a better artifact for human-agent interaction and to gain more insight into human affective processes.

2 CoMERG, EMA, and I-PEFiCADM

2.1 CoMERG

According to Gross [5], humans use strategies to influence the level of emotional response to a given type of emotion; for instance, to prevent a person from having a too high or low response level.

In [2], Gross' theory was taken as a basis to develop the emotion regulation model CoMERG. This model, which consists of a set of difference equations

[1] Note that the presented models embody a particular variant of an affect theory in that they have some unique properties that distinguish them from their original source. Many design choices underlying such models arise from the need to create a working computational system, a challenge the original theorists have never confronted.

combined with logical rules, can be used to simulate the dynamics of the various emotion regulation strategies described by Gross. CoMERG was incorporated into agents in a virtual storytelling application [1]. Following Gross' theory, CoMERG distinguishes five different emotion regulation strategies, which can be applied at different points in the process of emotion generation: *situation selection, situation modification, attentional deployment, cognitive change,* and *response modulation.*

2.2 Emotion and Adaption (EMA) Model

EMA is a computational model of the cognitive antecedents and consequences of emotions posited by appraisal theory, particularly as conceptualized by Smith and Lazarus [12]. A central tenet in cognitive appraisal theories is that appraisal and coping center around a person's *interpretation* of their relationship with the environment. This interpretation is constructed by cognitive processes, summarized by appraisal variables and altered by coping responses. To capture this process in computational terms, EMA maintains an explicit symbolic representation of the relationship between events and an agent's internal beliefs, desires and intentions, by building on AI planning to represent the physical relationship between events and their consequences, and BDI frameworks to represent the epistemic factors that underlie human (particularly social) activities.

Appraisal processes characterize this representation in terms of individual appraisal judgments, extending traditional AI concerns with utility and probability:

- Desirability: what is the utility (positive or negative) of the event
- Likelihood: how probable is the outcome of the event.
- Causal attribution: who deserves credit/blame.
- Controllability: can the outcome be altered by actions of the agent.
- Changeability: can the outcome change on its own.

Patterns of appraisal elicit emotional displays, but they also initiate coping processes to regulate the agent's cognitive response to the generated emotion. Coping strategies work in the reverse direction of appraisal, identifying plans, beliefs, desires or intentions to maintain or alter in order to reduce negative emotional appraisals:

- Planning: form an intention to perform some act
- Seek instrumental support: ask someone that controls outcome for help.
- Procrastination: wait for an external event to change the current circumstances.
- Denial: lower the perceived likelihood of an undesirable outcome.
- Mental disengagement: lower utility of desired state.
- Shift blame: shift responsibility for an action toward some other agent.

Strategies give input to the cognitive processes that actually execute these directives. For example, planful coping generates an intention to act, leading a planning system associated with EMA to generate and execute a valid plan to accomplish this act. Alternatively, coping strategies might abandon the goal, lower the goal's importance, or re-assess who is to blame.

EMA is a fully implemented model and has been applied to a number of systems that must simulate realistic human emotional responses. Several empirical studies have demonstrated EMA's effectiveness in modeling emotion [9].

2.3 I-PEFiC^{ADM}

Originally, the empirically validated framework for Perceiving and Experiencing Fictional Characters (PEFiC) described the receiver's reception of literature, theater, and movie characters [7]. Later versions were applied to the embodied-agent domain and supplemented with user interaction possibilities, resulting into the Interactive PEFiC model. I-PEFiC was then used to model affective behavior of robots as a module for Affective Decision Making was added to simulate irrational robot behavior, hence I-PEFiC^{ADM} [7].

The groundwork of I-PEFiC^{ADM} is formed by the cognitive process triplet of an encoding, a comparison, and a response phase. During *encoding*, the robot perceives the user and the situation the user is in. The features of the 'user in a situation' are indexed on four dimensions as a description of what someone is like or does. The robot attributes a level of *ethics* to the user, that is, the robot tries to figure whether the user's character is good or bad. *Aesthetics* is a level of beauty or ugliness that the robot perceives in the user. *Epistemics* is a measure for the realistic or unrealistic representations that the user conveys about him or herself. During the encoding, moreover, the robot looks at the user in terms of *affordances*. Certain aspects of the user may count as helpful or as an obstacle.

In the *comparison* phase, the user's features are appraised for *relevance* to robot goals (relevant or irrelevant) and *valence* to goals (positive or negative outcome expectancies). User features (e.g., intelligence) encoded as positive (e.g., 'helpful') may afford the facilitation of a desired robot goal. This instigates positive outcome expectancy. The comparison between the features of robot and user establishes a level of *similarity* (similar or dissimilar). The measures in the encode phase - mediated by relevance and valence in the comparison phase and moderated by similarity - determine the robot's responses.

In the *response* phase, the robot establishes the levels of *involvement* with and *distance* towards the user. Involvement and distance are two tendencies that occur in parallel and compensate one another [14]. In addition, the robot calculates a value for the so called *use intentions*, the willingness to employ the user again as a tool to achieve robot goals. Together with involvement and distance, the use intentions determine the overall satisfaction of the robot with its user.

Based on this level of *satisfaction*, the robot may decide to continue or stop the interaction and turn to another user. In the Affective Decision Making module [7], the robot makes a decision on the more rationally generated use intentions in unison with the more affectively generated involvement-distance trade-off. The action that promises the highest expected satisfaction during interaction is selected.

3 Triple Comparison

Fig. 1 depicts the similarities and differences between CoMERG, EMA, and I-PEFiC^{ADM}. Although explicitly mentioned in I-PEFiC^{ADM} alone, it is not hard to apply the encode-compare-respond phases to CoMERG and EMA. In the next sections, we offer a comparison of models, using Fig. 1 as our reference point.

3.1 Encode

According to CoMERG, people can select different *situations*, or modify the situation they are currently in, to regulate their emotions. In CoMERG the evaluation of how 'good' a certain situation is, is assumed to be given. In EMA, situations are appraised using the utility of state predicates about the situation and a causal interpretation of this situation. The agents can cope with these situations to change the person-environment-relationship, either by motivating changes to the interpretation of this relationship or by motivating actions that change the environment. I-PEFiC^{ADM} regards the user, agent, or character as part of a situation and that situation primes the features that are selected and how they are perceived.

Features in CoMERG are called aspects. According to CoMERG, a person can focus on one or another aspect (feature) of the world to regulate his or her emotions. In EMA, "current state predicates" in effect relate to those features considered in the subjectively construed situation. State predicates are statements about features of the environment which can be true or false. In I-PEFiC^{ADM}, features receive a certain weight according to frequency of occurrence, salience, or familiarity. Weights can change because of attentional shifts or situation changes.

The *appraisal domains* of I-PEFiC focus on characters. There is a host of evidence that for the judgment of fictional characters [7] and embodied agents [14], users classify features as good or bad, beautiful or ugly, realistic or unrealistic, and as aids or obstacles. According to EMA, on the other hand, agents perceive the world according to a causal interpretation of past and ongoing world events, including plans and intentions of self and others and past actions.

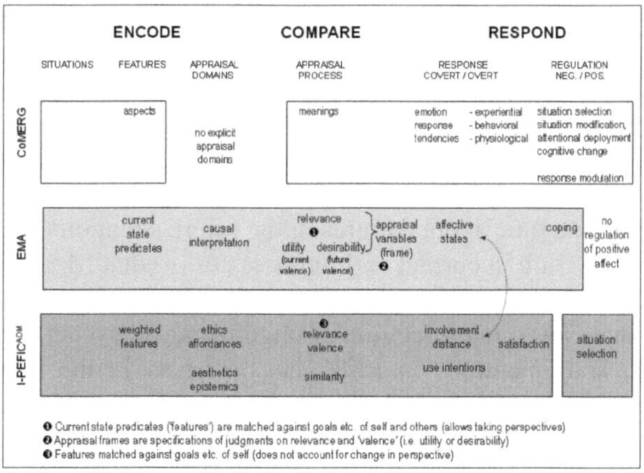

Fig. 1 Overview of CoMERG, EMA, and I-PEFiC^{ADM}

3.2 Compare

CoMERG refers to the *appraisal process* by "cognitive meanings." According to CoMERG, a person can perform the emotion regulation strategy 'cognitive change,' by attaching a different cognitive meaning to a situation. One type of cognitive change, *reappraisal*, means that an individual cognitively re-evaluates a potentially emotion-eliciting situation in terms that decrease its emotional impact [5]. In the view advocated by I-PEFiC^{ADM}, personal meaning is attached to a feature through relevance and valence. In a particular (imagined) situation, an object or feature may potentially benefit or harm someone's goals, beliefs, or concerns and as such, acquires 'meaning' ([4] cf. primary appraisal in [8]).

In EMA, this meaning is acquired through an appraisal process which is modeled in much detail. In this process, multiple appraisal frames are generated to allow for taking different perspectives. These appraisal frames are generated using many appraisal variables, which are taken from the theory of Smith & Lazarus, who call them the appraisal components, and Roseman, who calls them appraisal dimensions. Most of these appraisal variables could be mapped to relevance and valence used in I-PEFiC^{ADM}. According to EMA, relevance measures the significance of an event for the agent. Unlike Frijda, however, EMA equates significance with utility, which in Frijda's terms would be 'valence.' An event outcome is only deemed significant in EMA if it facilitates or inhibits a state predicate with non-zero utility. Valence is not explicitly mentioned in EMA although "utility" and "desirability" can be regarded as two instantiations of it.

Utility is a measure of the relative satisfaction from (or desirability of) environmental features. EMA represents preferences over environmental features as numeric utility over the truth-value of state predicates. Utilities may be either intrinsic (meaning that the agent assigns intrinsic worth to this environmental feature) or extrinsic (meaning that they inherit worth through their probabilistic contribution to an intrinsically valuable state feature). Utility, then, may be viewed as positive or negative outcome expectations about features in the current situation and is expressed in current state predicates (hence, 'current valence').

Desirability covers both a notion of intrinsic pleasantness and goal congruence (in Scherer's typology), as well as a measure of importance or relevance. It captures the appraised valence of an event with regard to an agent's preferences. An event is desirable, from some agent's perspective, if it facilitates a state to which the agent attributes positive utility or if it inhibits a state with negative utility. Like utility, desirability may be viewed as positive or negative outcome expectations but this time about features in the future situation ('future valence').

The explicit division in current and future states is what I-PEFiC^{ADM} is missing, as well as the possibility to change perspectives. EMA and I-PEFiC^{ADM} resemble each other in that causal interpretation of ongoing world events in terms of beliefs, desires, plans, and intentions in EMA is comprised in the beliefs, goals, and concerns that are checked for relevance and valence in I-PEFiC^{ADM}. However, EMA uses a number of variables, called appraisal frames, to cover the appraisal process, whereas in I-PEFiC^{ADM}, these appraisal frames appear to pertain to the more general concepts of relevance and valence. For example, urgency would be a clear-cut specification of relevance (cf. [4]) and ego involvement could be seen as a part of valence. However, EMA also uses some variables (such as causal attribution and coping potential) which are more related to the environment and less to the character, and which are somewhat broader than relevance and valence.

3.3 Respond

Fig. 1 exemplifies that in EMA, relevance of an event as well as utility and desirability (current / future valence) of features are mapped via an appraisal frame onto emotion instances of a particular category and intensity. These are called affective states. This may be seen as a covert response to the situation – an internal affective state that does not yet nor necessarily translate into overt actions. In I-PEFiC^{ADM}, affective states as such are not the focus but rather the involvement-distance trade-off, which is seen as the central process of engagement.

What comes closest to EMA's affective states are involvement and distance (Fig. 1, curved arrows). On this view, emotions emerge during the trade-off. For example, if a girl is asked for a date by a boy she loves, her involvement with him may be accompanied by happiness. When the boy looks at other girls on this date, the girl may still be involved with the boy but this time she feels challenged.

The involvement-distance trade-off could also count as the concretization of the emotion response tendencies that CoMERG hinges on. In CoMERG, these tendencies result in several responses: experiential, behavioral, and physiological. EMA and I-PEFiC^{ADM} are restricted to the experiential and behavioral domain. In EMA, affective states lead to coping behavior. For example, if your car makes strange noises, you might adopt emotion-focused coping (e.g., wishful thinking: tell yourself it is not that important and will probably stop by itself) which will inform the next decision; or you might adopt problem-focused coping to take a specific overt action to address the threat (e.g., have your car checked at the garage). In I-PEFiC^{ADM}, the combination of involvement, distance, and use intentions predicate the level of satisfaction (experiential), which feeds into affective decision making. This results into overt responses (behavior) such as kissing, kicking, or walking away.

CoMERG describes five *emotion regulation strategies* (see Sec. 2.1). Following Gross, CoMERG predicts that strategies that are performed earlier in the process of emotion generation are more effective to regulate one's emotions. EMA provides a more specific model which focuses (in much detail) on coping. Situation selection and situation modification are implemented in EMA via problem-focused coping strategies (i.e., take-action) and avoidance. Attentional deployment corresponds to EMA's strategies of seek/suppress information. Cognitive change corresponds to EMA's various emotion-directed strategies. EMA does not model suppression. I-PEFiC^{ADM} focuses on situation selection. Another difference is that CoMERG and I-PEFiC^{ADM} allow the regulation of affect by increasing, maintaining, or decreasing the positive or negative response, whereas

EMA focuses on decreasing negative affect alone. In EMA, being overenthusiastic is left uncontrolled, whereas in CoMERG and I-PEFiC^{ADM}, positive affect can be down-regulated or compensated for. As a result, one can state that coping in EMA is one of the instantiations of emotion regulation in CoMERG.

For EMA, there must be an explicit causal connection between coping strategies and the emotions they are regulating whereas for CoMERG that is not a prerequisite. In CoMERG, people perform strategies to change their level of emotion, which are simply modeled via difference equations. EMA gives a more detailed and formal description of how emotion regulation works. For example, reappraisal as a general emotion regulation strategy in CoMERG is in EMA described in terms of a change in causal interpretation.

4 Integration

In our attempt to integrate the above models, we will adhere to the naming convention of 'features' instead of 'aspects' that the agent can detect in a situation because both EMA and I-PEFiC^{ADM} use that concept and it is interchangeable with 'aspects' in CoMERG. Only I-PEFiC^{ADM} explicitly mentions the appraisal domains that are important in perceiving features. Therefore, the agent will use ethics, affordances, aesthetics, and epistemics as the main domains through which features are funneled into the appraisal process.

CoMERG, EMA, and I-PEFiC^{ADM} all assume or elaborate an appraisal process. CoMERG is least explicit and the concept of 'meaning' can easily be attached to 'personal significance' and 'personal relevance' in both EMA and I-PEFiC^{ADM}. In EMA and I-PEFiC^{ADM}, relevance and valence play an active role, but EMA models the different manifestations rather than the general concepts. In unison, we will use the term relevance to indicate importance or meaning to (dynamic) personal goals, concerns, beliefs, intentions, plans, etc. and valence as (current) utility or (future) desirability of features in a situation. This may instantiate in the form of, for example, urgency as an aspect of relevance and likelihood or unexpectedness as an aspect of valence.

On the response side, EMA focuses on mood and emotions whereas I-PEFiC^{ADM} emphasizes the more general trends of involvement, distance, and use intentions. Yet, they are two sides of the coin that could be called 'affective states.' Emotions and moods may evolve from involvement-distance trade-offs and both the specific (e.g., happy emotions) and general experiential response (e.g., involvement) may be liable to regulation strategies.

CoMERG provides the most profound distinctions with respect to the type of responses (experiential, behavioral, and physiological) and the number of regulation strategies. However, in no way are these distinctions at odds with EMA or I-PEFiC^{ADM}. Coping is best worked out by EMA and situation selection by I-PEFiC^{ADM}, encompassing a module for affective decision making that on the basis of expected satisfaction chooses from several domain actions.

Fig. 2 shows a blueprint for the integration of CoMERG, EMA, and I-PEFiC^{ADM} into a framework for computerized affect generation and regulation. On the far left of the figure, we see a virtual agent. She can perform attentional deployment to weigh the features of her interaction partners. The agent develops state predicates about others in a certain context. Features receive indices for different appraisal domains. The observed other acquires personal meaning or significance for the agent because she compares their features with her personal goals, beliefs, and concerns. This establishes the relevance and valence of others to her goals and concerns. While relevance determines the intensity of affect, valence governs its direction. The agent can also look at others through the eyes of another agent.

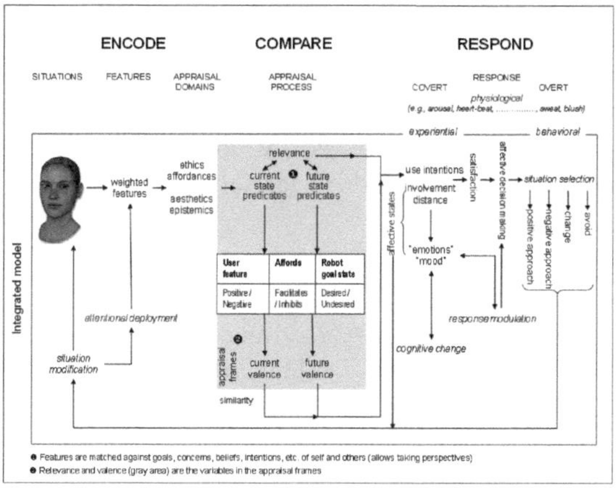

Fig. 2 Proposed integration of the three models

When the (initial) appraisal process is completed, the agent is ready to affectively respond. Relevance, current and future valence form an appraisal frame that feeds into her (un)willingness to 'use' someone for her purposes (e.g., having a conversation) and that helps her to trade friendship (involvement) for keeping her cool (distance). Inside, the agent will 'experience' several (perhaps ambiguous) emotions. On a physiological level, she may be aroused (e.g., increased heart-rate). All this is not visible to others yet; they are covert responses.

During affective decision making, the agent selects the option that promises the highest expected satisfaction. This may be accompanied by physiological reactions such as blushing and trembling. Response modulation may influence the affective decision making. The performed action leads to a new situation.

5 Conclusion

Various researchers from different fields have proposed formal models that describe the processes related to emotion elicitation and regulation (e.g., [2, 3, 7, 10]). For this reason, it is impossible to provide a complete comparison of existing models within one paper. Instead, the approach taken in this article was to select three of the more influential models, which share that they can be used to enhance believability of virtual characters: CoMERG, EMA, and I-PEFiC^{ADM}. The theories by which they were inspired cover most psychological literature in affect-related processes, including the works of Frijda [4], Lazarus [8], and Gross [5].

In this article, we have argued that each of the three approaches has its specific focus. For example, CoMERG covers a wide variety of emotion regulation strategies, whereas I-PEFiC^{ADM} provides an elaborated mechanism for encoding of different appraisal domains, which have empirically shown to be crucial in human-robot interaction. EMA on its turn contains very sophisticated mechanisms for both appraisal and coping, which have already proved their value in various applications. Because several of these features are complementary to each other, this paper explores the possibilities to integrate them into one combined model of affect for virtual humans. For a first attempt to implement this integrated model, see [11]. As a next step, it is planned to perform systematic user-tests in order to assess whether our integration indeed results in more human-like affective behavior than the three sub-models do separately.

References

1. Bosse, T., Pontier, M.A., Siddiqui, G.F., Treur, J.: Incorporating Emotion Regulation into Virtual Stories. In: Pelachaud, C., Martin, J.-C., André, E., Chollet, G., Karpouzis, K., Pelé, D. (eds.) IVA 2007. LNCS (LNAI), vol. 4722, pp. 339–347. Springer, Heidelberg (2007)
2. Bosse, T., Pontier, M.A., Treur, J.: A Dynamical System Modelling Approach to Gross Model of Emotion Regulation. In: Lewis, R.L., Polk, T.A., Laird, J.E. (eds.) Proc. of the 8th Int. Cf. on Cognitive Modeling, ICCM 2007, pp. 187–192 (2007)
3. Breazeal, C.: Emotion and sociable humanoid robots. In: Hudlika, E. (ed.) International Journal of Human Computer Interaction, vol. 59, pp. 119–155 (2003)
4. Frijda, N.H.: The Emotions. Cambridge University, New York (1986)
5. Gross, J.J.: Emotion Regulation in Adulthood: Timing is Everything. Current directions in psychological science 10(6), 214–219 (2001)
6. Hoorn, J.F., Pontier, M.A., Siddiqui, G.F.: When the user is instrumental to robot goals. First try: Agent uses agent. In: Proceedings of Web Intelligence and Intelligent Agent Technology 2008 (WI-IAT 2008), vol. 2, pp. 296–301 (2008)
7. Konijn, E.A., Hoorn, J.F.: Some like it bad. Testing a model for perceiving and experiencing fictional characters. Media Psychology 7(2), 107–144 (2005)
8. Lazarus, R.S.: Emotion and Adaptation. Oxford University, New York (1991)
9. Mao, W., Gratch, J.: Evaluating a computational model of social causality and responsibility. In: 5th Int. Joint Conference on Autonomous Agents and Multiagent Systems, Hakodate, Japan (2006)
10. Marsella, S., Gratch, J.: EMA: A Model of Emotional Dynamics. Cognitive Systems Research 10(1), 70–90 (2009)
11. Pontier, M.A., Siddiqui, G.F.: Silicon Coppélia: Integrating Three Affect-Related Models for Establishing Richer Agent Interaction. In: Proc. of Web Intelligence and Intelligent Agent Technology 2009 (WI-IAT 2009), vol. 2, pp. 279–284 (2009)
12. Smith, C.A., Lazarus, R.S.: Emotion and Adaptation. In: Pervin, L.A. (ed.) Handbook of Personality: theory & research, pp. 609–637. Guilford Press, NY (1990)

A Generic Architecture for Realistic Simulations of Complex Financial Dynamics

Philippe Mathieu and Olivier Brandouy

Artificial Stock Markets (here after ASM) have received an increasing amount of academic interest since the seminal works of [20] or [17]. Such platforms have benefitted from advances and new methods developed in the field of multi agent-systems (see for example [10], [14] and [27]). These Agent-Based virtual environments are particularly useful to study various aspects of the financial world in an entirely controlled environment, opening new perspectives for policy makers, regulatory institutions and firms developing business solutions in the financial industry (for example asset management or trading). There is little doubt ASM could become a key system in the post financial crisis risk-management toolbox to overcome the weaknesses of traditional approaches. Agents-based modeling and simulation offer frameworks to study the impact of a Tobin's tax for example, or to develop new stress tests for assessing financial resilience to economic shocks or to develop new automatic trading techniques.

In this research paper, we introduce a new, highly flexible Agent-Based model of financial markets in an API form. This application offers a solution for implementing realistic simulations of complex financial dynamics using both artificial intelligence, distributed agents and realistic market algorithms. We consider that if various questions in Finance can be solved with Agent-Based Modeling solutions, Multi-Agent Systems directly benefit from Financial questions as well : ASMs, with driving simulators (see [9]), offer one of the richest environments to evolve software engineering for multi-agent systems . Therefore, the punchline of this paper is that development of artificial markets offers the whole variety of issues one can

Philippe Mathieu
Université des Sciences et Technologies, Computer Science Dept. &
LIFL (UMR CNRS-USTL 8022)
e-mail: philippe.mathieu@lifl.fr

Olivier Brandouy
Sorbonne Graduate School of Business, Dept. of Finance &
GREGOR (EA MESR-U.Paris1 2474)
e-mail: olivier.brandouy@univ-paris1.fr

Y. Demazeau et al. (Eds.): Advances in PAAMS, AISC 70, pp. 185–197.
springerlink.com

face in agent based modeling. Among others, ASM are grounded on an individual-based approach with local interaction, distributed knowledge and resources, heterogeneous environments, agent autonomy, artificial intelligence, speech acts, discrete scheduling and simulation. We defend this punchline with an illustration derived from the building of the ArTificial Open Market API (here-after ATOM, see `http://atom.univ-lille1.fr`).

This paper is organized as follows. In a first section we briefly present the relevant literature around artificial markets and Agent-Based modeling in Finance. In a second section, we extensively introduce the ArTificial Open Market API and present the mains computing issues linked to its development. In the third section, we present agent's behavior and introduce a series of tests geared at verifying the efficiency of our API. We then conclude and open new research and technological perspectives.

1 Elements of Literature in Agent-Based Finance

There is a long and fruitful research tradition in Economics based on the so-called "methodological individualism" which could be referred as to the "individual-based approach" in computer science (see for example [24]). Methodological individualism proposes a reductionist perspective for the analysis of macroscopic economic phenomena. Along this approach, these latter must ultimately be explained by agents' preferences and actions. The complex system approach initiated by [1] in physics and [2] and [3] in economics has profoundly renovated this philosophy. This approach has also altered the common comprehension of financial markets built by neoclassical economists (see [5]). In this perspective, Agent-Based models of financial markets were developed to renew the analysis of various ill-understood issues in Finance. Acknowledging that financial systems are evenly characterized by order and disorder, that grasping economic complexity could not be done easily with traditional techniques, a few researchers used Agent-Based models and simulations to study financial markets emergent dynamics (for instance, [13] or [20]).

Developing ASM necessarily implies to select components in the model that are ex-ante considered as necessary from a theoretical point of view. On the one hand, a fine selection of these elements is definitely critical and should be conducted keeping in mind the Occam's razor principle. On the other hand, ignoring some essential parameters in the model could strongly weaken its relevance. For example, in Agent-Based Finance, market microstructure, which refers to the way transaction are organized, is often reduced to a mere equation linking price formation to the imbalance between supply and demand (see, among others [16], [7] or [11]). This simplification erases one important feature in real-world markets : trading is generally not synchronous.Other approaches have overcome this issue introducing directly in the Agent-Based model an order-book and a scheduler organizing non-synchronous trading. Generally speaking, these models do not accept the whole variety of orders and only focus on limit and market-orders (for instance [21]) or cannot be used as a replay engine for existing order-flows (for instance [19]). They also suffer from a lack of flexibility and must be viewed as software rather than as APIs : this means that they perform quite well in doing what they are supposed to do, but cannot be used to explore a wide range of financial issues due to some structural choices made by the developer during the coding phase.

In the following section we introduce the ArTificial Open Market API. Among others, we make a specific point on the ability of this new, generic package to overcome the issues mentioned previously.

2 The Artificial Trading Open Market API: General Principles and Distinctive Features

ATOM is a general environment for Agent-Based simulations of stock markets. It is based on an architecture close to the Euronext-NYSE Stock Exchange one. Agent-Based artificial stock markets aim at matching orders sent by virtual traders to fix quotation prices. This might be done using imbalance equations as evoked in section 1. In the case of ATOM, price formation is ruled by a negotiation system between sellers and buyers based on an asynchronous, double auction mechanism structured in an order book. Using this API, one can generate, play or replay order flows (whatever the origin of these order flows, real world or virtual agents population). It also allows distributed simulations with many computers interacting through a network as well as local-host, extremely fast simulations. ATOM can be used to design experiments mixing human beings and artificial traders. One of the main advantages of ATOM consists in its modularity. This means that it can be viewed as a system where three interacting main components interact: i) *Agents*, and their behaviors, ii) *Markets* defined in terms of microstructure and iii) the *Artificial Economic World* (including an information engine and, potentially, several economic institutions such as banks, brokers, dealers...). The two first components can be used independently or together. Depending upon the researcher targets, the *Artificial Economic World* can be plugged or not in the simulations. For example, one can use the system for the evaluation of new regulation policies or market procedures, for assessing potential effects of taxes or new trading strategies in a sophisticated artificial financial environment. Thanks to its high modularity and its ability to mimic real-world environments, it can also serve as a research tool in Portfolio Management, Algorithmic Trading or Risk Management among others. From a pure technological point of view, ATOM can also be viewed as an order-flow replay engine. This means that bankers can test their algorithmic-trading strategies using historical data without modifying the existing price series or backtest the impact of their trading-agents in totally new price motions or market regimes generated by artificial traders.

Six distinctive aspects of ATOM can be highlighted:

1. It can be used without any agent. One can directly send orders written in a text file (for example, a set of orders as it arrived on a given day, for a given real-world stock market) to each order-book implemented in the simulation. In this case, ATOM serves as a "replay-engine" and simulations merely rely on market

microstructure. It therefore runs really fast (an entire day of trading in less than 5 seconds).

2. Agents can be viewed as simple nutshells in certain cases : they only take actions they are urged to by a third party. These agents are called "Hollow Agents". For example, a human trader can act through such agents. By definition, "Hollow Agents" do not have any artificial intelligence and can be assimilated to human-machine wrappers.
3. Beyond "Hollow Agents", ATOM can use various kind of sophisticated agents with their own behaviors and intelligence (see section 3). Thousands of these agents can evolve simultaneously, creating a truly heterogeneous population. Once designed, agents evolve by themselves, learning and adapting to their (financial) environment.
4. In any of the previous cases, ATOM generates two files; the first file records orders emitted by the agents, with the platform time stamp fixed at the very moment they arrive to a given order-book. Notice again this file can be used in a "replay-engine" configuration. The second generated file collects prices resulting from the orders.
5. ATOM can mix human-beings and artificial traders in a single market using its network capabilities. This allows for a wide variety of configurations, from "experimental finance" classrooms with students, to competing strategies run independently and distantly by several banks or research labs. The scheduler can be set so to allow human agents to freeze the market during their decision process or not (see above, section 2.4).
6. Any artificial stock market should be tested rigorously to verify if it matches the following criteria: i) ASM must have the ability to replay perfectly an order flow actually sent to a given market with the same microstructure. The resulting price series (on the one hand, the "real-world" one and on the other hand, the "artificial" one) should overlap perfectly. ii) Given a population of agents, the ASM should generate stylized facts qualitatively similar to the market it is geared at mimicking. As it will be shown later, ATOM succeeds in both cases.

2.1 Artificial Stock Markets as Complex, Adaptive Systems

Artificial Stock Markets are environments allowing to express all classical notions used in multi-agent systems. First of all, the environment in which agents evolve, as well as their behaviors and own dynamics, communication or interactions. ASM, like any other MAS, are suited for the study of various emergents phenomena. Using the so-called "vowels" approach [23], the definition of `AEIO` (`A` Agents, `E` Environment, `I` Interactions, `O` Organization) is straightforward. Nevertheless, if one wants to build an efficient platform, several issues can be identified and must be precisely and strictly regulated.

It is hardly possible to describe the complex algorithmic structures that are necessary for the realization of such multi-agent platforms; therefore we have chosen to introduce three of these structures that appear to be representative of the difficulties one must face while developing an ASM : i) the management of orders' ID, ii) the scheduling system, and iii) the introduction of a human-being in the simulation loop (here-after *"human-in-the-loop"* problem).

2.2 A Unique Identity for Orders

In its simplest form, an order is a triplet constituted by a direction (purchase or sale), a quantity and a price. Usually this type of order is called a *"Limit Order"*. In the Euronext-NYSE system, several other orders are used (see *"EURONEXT" Rule Book* at `http://www.euronext.com`). Once constructed by an agent, the order is sent to the order-book. It is then ranked in the corresponding auction-queue ("Bid" or "Ask" if it is an order to "Buy", respectively to "Sell") where are stacked the other pending orders using a "price-then-time" priority rule. As soon as two pending orders can be matched, they are processed as a "deal", which delivers a new price. Notice that the clearing mechanism implies that cash is transferred from the buyer to the seller and stocks from the seller to the buyer.

For various reasons, financial institutions may need to be able to process again an historical record of orders (for example, for the optimization of algorithmic trading methods). Such historical records collect the expression of human behaviors in specific circumstances. To be able to replay this order-flow, a first difficulty consists in interpreting exactly the order flow as it is expressed in the real-world. If *"Limit Orders"* had been present in this record exclusive of any other order type, the sequentiality of orders would have been sufficient to guarantee a perfect reproducibility. Unfortunately, issued orders can be modified or deleted are in many markets. This implies one must be able to identify clearly which previously issued order an "Update" or a "Delete" order points to. Thus, a generic platform has to use a unique ID for orders. This is particularly important in situations where several possible identification keys for orders potentially coexists : in the replay-engine situation (real-world orders ID and time stamps), in the Agent-Based platform mixing human beings and artificial traders situation (orders ID, platform time-stamp) or in any combination of these states. To our knowledge, this is neither the case for the Genoa Artificial Stock Market (see [22]) or in the Santa-Fe ASM for example (see [15]).

How should an ASM deal with this issue?

A first idea would be to use the time-stamp imprinted on each order. This information is particularly important if one wants to work on the time-distribution of orders. Unfortunately this idea is technically irrelevant. The time-stamp of standard operating systems is given using the third decimal places of seconds. However, it is perfectly possible to process several orders in one out of one thousand second. The time stamp is therefore necessary, but can be under no circumstances used as an ID.

Agents, human-beings or artificial traders, do not have to fix orders' ID. This task is devoted to the order-book itself. The order-book must stamp orders at reception, mainly to avoid fraudulent manipulations from agents. One platform-ID is therefore affected to each order. This latter must be different from any other possible identification number indicated in the order file or corresponding to a time-stamp. This means that the platform can handle three different identifiers, which makes its structure rather complex. Nevertheless, this additional complexity is mandatory if one wants to be able to use the ASM as a replay-engine and with artificial agents as well.

2.3 The Scheduling System

The scheduler is a particularly critical element in all multi-agents systems. This component manages the very moment and situations in which agents have word. The scheduling system aims at avoiding possible biases in the simulation. However, this fundamental component in MAS systems is seldom discussed.

Outside the Computer Science community, it is often believed that using independent processes for each agent is a guarantee of autonomy. This is definitely not the case. Using threads consists in letting the operating system scheduler decide which agent will have the word at the next step in the simulation. Another misunderstanding consists in believing that threads will allow agents to work in parallel : using a single processor, there is necessarily one and only one single process running at each time step. Parallelism is simply simulated by the operating system. Nevertheless, the main disadvantage in this approach is that the system scheduler does neither exclusively consider agents nor even the MAS application itself; it also manages all applications running in the computer. Therefore, except on specialized real-time systems, there is no chance to observe agents solicited exactly at the same (relative) moment when two executions of the same simulation are processed. Results cannot be reproduced perfectly and the developer loses command on the code execution. It is therefore mandatory to code a specific scheduler to avoid these shortcomings.

When Can the Agents Express Their Intentions?

One should not desire performing a loop in the simulation that keeps the word order among agents unchanged. This would introduce biases in the simulation : the first chosen agent would have systematically a priority over other agents; the last one might wait a long time before being allowed to express its intentions. Performing a uniform randomization of agent's word would lead to the same issues as well. In this last case, a few agents can theoretically stay unselected for a long time and even be ignored by the system.

Simulations in ATOM are organized as "round table discussions" and are grounded on an *equitably random* scheduler. Within every "round table discussion", agents are randomly interrogated using a uniformly distributed order. This latter feature ensures that each of them has an equal *possibility* of expressing its intentions. Notice that the API offers a random generator that is shared by all agents. The reproducibility of experiments is therefore guaranteed : one can either use a seed during the initialization of an experiment, or use ATOM in the "replay-engine" configuration since, as mentioned before, any simulation delivers a record of all the orders.

How Do They Proceed?

In real life, investors do not share the same attitudes. Some will be more reactive than others, or will implement more complex strategies leading to a higher rate of activity. In ATOM having the possibility to express an intention does not necessarily imply that a new order is issued. Since agents are autonomous, they always have the possibility to decline this opportunity. Developing an agent that sends twice less orders than any other agent can be made in programming her behavior such as she will decline word on odd turns, while others accept it each time they have the possibility to do so.

Moreover, if an agent had been allowed to send several orders when interrogated, this would have led to an equity problem similar to the one described before. To overcome this issue, agents are just allowed to send at most *one single* order to *a given order book* (*i.e.* one order at most per stock) within the same "round table discussion". However, if an agent plans to issue several orders concerning the same stock (thus, the same order book), she must act as a finite state automata. Each time she is allowed to express herself, she will change her state and send a new order. Developers can use this technique to set up various experiments without sacrificing a fair equity between agents or a perfect reproducibility of their protocols.

However, notice that agents have the possibility to send several orders within the same "round table discussion" : this ability is simply constrained by the "one order to each book" rule. If the ASM is settled such as it runs a multi-stock experiment, an agent can therefore rebalance her portfolio using one order per category of stocks she holds, provided the scheduler has offered the possibility to do so.

2.4 *Human in the Loop*

ATOM can include human-beings in the simulation loop. This is an important feature that is seldom offered in multi-agent artificial stock markets, if simply possible with respect to the algorithmic choices made in other platforms. Following the example of driving or crowd simulators (see for example [25] or [9]), human agents and virtual agents can evolve together. Human agents do not differ from artificial agents in their philosophy : they share the same general characteristics as other agents. The so-called "vowels" approach is respected, even if `U` (for "Users") is subsumed by `A` – Agents –. A human agent is an interface allowing for human-machine interaction. Through this interface one can create and send orders. Notice that human agents do not have any artificial intelligence : they just embed human intelligence in a formalism that is accepted by the system.

To allow the introduction of human in the loop, ATOM has been designed to deal with communications over the network. Human agents can be run on different machines and the system allows client-server configurations. This approach is particularly fruitful for a pedagogic use of the platform during Finance class for example. In this latter case, several students have their own trading interface on their computers. In other terms, each of them runs a human agent linked to the ATOM server through the network. However, the presence of human agents does not alter the way the scheduler operates.

Two kinds of human agents can co-exist in ATOM : Modal Human Agents (MHA) and Non Modal Human agent (NMH).

- MHA can stop the scheduling system while running. As long as her human owner does not express her intentions (to issue a new order or to stay unchanged), the simulation is temporary frozen. In a classroom, this aspect is particularly important and leaves time to students for deciding their actions.
- NMH cannot freeze the simulation, which means that human agents compete in real time with artificial traders. Even if human agents can have a hard time in this situation, it remains realistic in a financial world where algorithmic trading is more and more frequent.

In this section we have presented three major technical points that characterize ATOM and should also concern many ASM. Even if other important technical issues could not be mentioned in this article, we have stressed that the development of artificial stock market platforms put forward a series of complex issues in terms of computer science. In the next section, we introduce some additional elements relative to the artificial intelligence of virtual agents that can be run in our platform. This question is of main concern for computer scientists and for financial researchers alike.

3 Agents Behaviors and Validation Tests

As mentioned previously (see section 2), every ASM should succeed in processing perfectly a given order flow collected from a real-world stock market at a specific date. The result is obtained confronting prices delivered by the market at this date and the prices generated by the ASM using the same set of orders. It should also generate relevant "stylized facts" with regard to their real-world counterpart : these stylized facts are statistical characteristics of financial time series that prove to be systematically observed in various contexts (different assets, periods of time, countries).

This section presents how ATOM fulfill this requirements; it also develops one key element in the system that has not been introduced for the moment : agent's behaviors.

3.1 Artificial Traders: From Basic Reactive Agents to Highly Sophisticated Entities

Many ASM can run large populations of homogeneous, respectively heterogeneous artificial traders. This is also the case for ATOM although the API allows for facilities which are not available in other platforms. Generally speaking, artificial traders are characterized by their initial endowments (financial resources and information), and by their artificial intelligence or behavior. For example, the following types of agents can be implemented :

Zero Intelligence Traders (ZIT): This behavior is merely based on stochastic choices (orders are composed of prices, quantities and directions treated as random variables, *i.e.* uniformly drawn in a bounded interval). This kind of behavior has been popularized in economics by [12]. ZIT constantly provide orders but neither extract information from the market nor from any other component of the ASM. Despite their extreme simplicity, these agents are widely used because more sophisticated forms of rationality appear to be useless to explain the emergence of the main financial stylized facts at the intraday level (see above, section 3.3). Notice that several types of ZIT can be coded, which question the "zero-intelligence" concept itself.

Technical Traders: "Chartists" are a specific population of technical traders. These agents try to identify patterns in past prices (using charts or statistical signals) that could be used to predict future prices and henceforth send appropriate orders. One can find an example of such behavior in [4]. From a software engineering perspective, these agents need to have some feedback from the market and some kind of learning process as well (reinforcement learning for large sets of rules is generally used). This lead to some complex algorithmic issues. For example, if one considers a population of a few thousand Technical Traders, it is highly desirable to avoid that each agent compute the same indicators, or simply store themselves the whole price series.

Sophisticated Intelligence Traders (SIT): Three kind of SIT can evolve in ATOM:

i) *Finite-State Agents* and *Hollow Agents*; by construction one agent can only send one order per time tick ("round of word"). If this agent wants to send a series of orders to the same order book, this must be divided into several orders, one per time-tick. Finite-State Agents can deal with this issue, which implies a minimum level of sophistication in their behaviors. Hollow agents serve as a media to send orders formulated by other agents (human beings or artificial traders). These latter agents are especially useful for the replay-engine instantiation of ATOM.
ii) *Cognitive Agents* generally have a full artificial intelligence, although it can be designed to be rather minimal (usual features to develop such agents are memory, information analysis processes, expectations, strategies and learning capacities). For example, an agent buying at a specific price and sending immediately a "stop order" to short her position if the price drops under $\theta\%$ times the current price, will fall in this category. Agents using strategic order splitting (see for example [26]) or exploiting sophisticated strategies (for instance, [6]) can also be considered as Cognitive Agents.
iii) *Evolutionary Agents* are the ultimate form of SIT; they outperform Cognitive Agents in terms of complexity since they are able to evolve with their environment. These agents can also generate new rules or strategies (this can require genetic algorithms for example). Each and every of these agents can be implemented in ATOM. Notice they all can manage asset portfolios if required. This is one of the flexibilities proposed in the API. Several assets can be traded at each time step by any-kind

of agent, using sophisticated or extremely basic strategies (such as optimal, respectively nave diversification). In these cases, agents should use information about the state of the artificial economy at a given time horizon. All these information are provided by the *Artificial Economic World* component, which adds the possibility of describing the complete set of temporal dimensions in the system (past, present, future). In this latter case, agents could eventually use past prices for the computation of co-moments among assets and future values for expected returns and volatility (see [18]). Notice again that if each kind of agents can be mixed with others, ATOM also allows for human beings to be added into any artificial stock market through a HMI.

3.2 ATOM Reality-Check

In this section, we report a series of tests conducted to check whether ATOM can generate financial dynamics in line with the ones of the Euronext-NYSE stock-exchange or not. The first series of test is devoted to the ability of ATOM at generating unbiased prices when it deals with a real-world order-flow.

Figure 1(a) and Figure 1(b) report results of the first reality-check (top Figures report results produced with the ATOM data, bottom Figures being those based on Euronext-NYSE data). We ran ATOM with a Hollow Agent reading the entire set of 83616 orders concerning the French blue-chip France-Telecom (FTE) recorded on June 26th 2008 between H9.02'.14".813"' and H17.24'.59".917"'. As mentioned previously, handling time in simulations is particularly complex and may lead to unsolvable dilemma. We cannot guarantee an exact matching of waiting times but rather a coherent distribution of these values delivered by the simulator engine with regard to the observed waiting times.

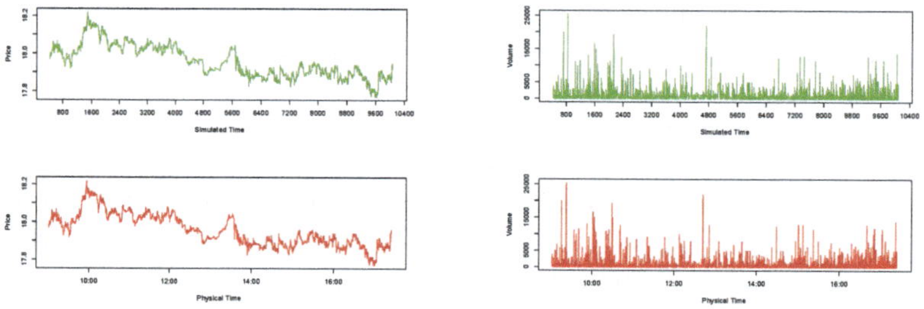

(a) Prices : ATOM vs. Euronext-NYSE

(b) Volumes : ATOM vs. Euronext-NYSE

Fig. 1 Results of the "Reality Check" procedure

Notice that ATOM performs rather decently in satisfying the first reality check procedure.

3.3 Stylized Facts

The second subset of tests focuses on the ability of ATOM at generating realistic artificial prices when populated with artificial agents. We ran a series of simulations to verify if ATOM can generate major stylized facts that are usually reported in the literature (see for example [8]). For the sake of simplicity and space-saving, we only report in a pictorial form of the classical departure from Normality of asset returns at the intraday level (Figures 2(a) and 2(b)). Notice again that these statistics are reported on the left-hand Figures when based on ATOM prices and on the right hand when based on Euronext-Nyse data. Real data are those used previously for the reality check, artificial data were generated using a population of ZIT as described in the ATOM API.

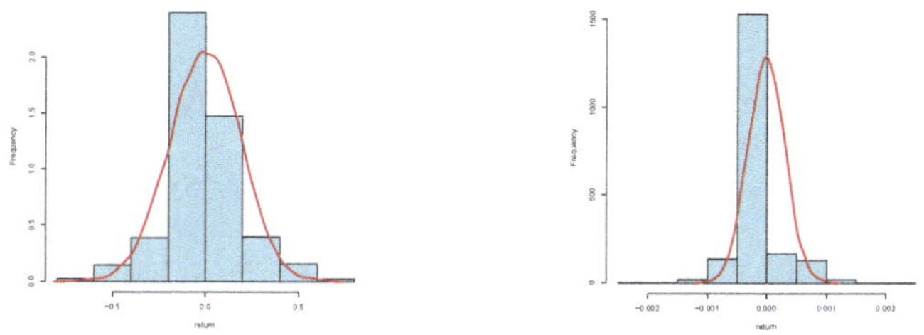

(a) Departure from Normality, ATOM

(b) Departure from Normality, Euronext-NYSE

Fig. 2 Stylized facts, ATOM vs. Euronext-NYSE

In clearly appears that ATOM produces stylized facts in line with those observed for a specific stock listed on Euronext-NYSE : this means that the platform can both generate realistic price series with a population of ZIT and process without bias a real order-flow.

4 Conclusion

The recent financial crisis has stressed the need for new research tools that can deal with the high level of complexity of the economic world. Agent based methods propose a powerful alternative to traditional approaches developed in finance such as econometric or statistical models. Among others, Artificial Stock Markets are particularly interesting in that they offer a completely controlled environment to test new regulations, new exchange structures or new investment strategies.

In this paper we have defended that these models offer one of the richest environments to illustrate multi-agent systems notions. We have particularly highlighted the importance of a polymorphic platform : it therefore can be used for a wide range of experiments, including or not artificial agents, sophisticated behaviors, communication over the network... We also discussed a series of software engineering problems arising when the ultimate goal is to develop a complete API for market simulation. A precise proof mechanism that can be used to validate any artificial market platform has been introduced. This proof is based on two tests : on the one hand, each platform should be able to replay perfectly a real set of orders and deliver exactly the corresponding real price series; on the other hand, while running with an artificial agents population, it should also generate prices that match the so-called financial "stylized facts".

We have presented how these notions have governed the development of the ATOM (ArTifical Open Market) API. This platform can evolve complex AI agents, the only limitation in this domain being researchers' imagination.

References

1. Anderson, P.: More is different. Science 177, 393–396 (1972)
2. Anderson, P., Arrow, K.J., Pines, D. (eds.): The Economy as an Evolving Complex System. Santa Fe Institute Studies in the Sciences of Complexity, vol. 5. Addison Wesley, Redwood City (1988)
3. Arthur, B.: The Economy as a Complex System. In: Complex Systems. Wiley, Chichester (1989)
4. Arthur, B.: Inductive reasoning and bounded rationality: the el-farol problem. American Economic Review 84, 406–417 (1994)
5. Arthur, B.: Complexity in economic and financial markets. Complexity 1(1), 20–25 (1995)
6. Brandouy, O., Mathieu, P., Veryzhenko, I.: Ex-post optimal strategy for the trading of a single financial asset. SSRN eLibrary (2009)
7. Cincotti, S., Ponta, L., Pastore, S.: Information-based multi-assets artificial stock market with heterogeneous agents. In: Workshop on the Economics of Heterogeneous Interacting Agents 2006 WEHIA 2006 (2006)
8. Cont, R.: Empirical properties of asset returns: stylized facts and statistical issues. Quantitative Finance 1, 223–236 (2001)
9. Dresner, K.M., Stone, P.: A multiagent approach to autonomous intersection management. J. Artif. Intell. Res (JAIR) 31, 591–656 (2008)
10. Ferber, J.: Multi-agent system: An introduction to distributed artificial intelligence. J. Artificial Societies and Social Simulation 4(2) (2001)
11. Ghoulmie, F., Cont, R., Nadal, J.P.: Heterogeneity and feedback in an agent-based market model. Journal of Physics: Condensed Matter 17, 1259–1268 (2005)
12. Gode, D.K., Sunder, S.: Allocative efficiency of market with zero-intelligence traders: Market as a partial substitute for individual rationality. Journal of Political Economy 101(1), 119–137 (1993)
13. Kim, G., Markowitz, H.: Investment rules, margin, and market volatility. Journal of Portfolio Management 16(1), 45–52 (1989)
14. Kubera, Y., Mathieu, P., Picault, S.: Interaction-oriented agent simulations: From theory to implementation, pp. 383–387 (2008)
15. Le Baron., B.: Building the santa fe artificial stock market. Working Paper. Brandeis University (June 2002)
16. Le Baron, B., Arthur, W.B., Palmer, R.: Time series properties of an artificial stock markets. Journal of Economic Dynamics and Control 23, 1487–1516 (1999)

17. Levy, H., Levy, M., Solomon, S.: A microscopic model of the stock market: Cycles, booms, and crashes. Economic Letters 45(1), 103–111 (1994)
18. Markowitz, H.: Portfolio selection. Journal of Finance 7(1), 77–91 (1952)
19. Muchnik, L., Solomon, S.: Markov nets and the natlab platform; application to continuous double auction. New Economic Windows (2006)
20. Palmer, R.G., Arthur, W.B., Holland, J.H., LeBaron, B., Tayler, P.: Artificial economic life: A simple model of a stockmarket. Physica D 75, 264–274 (1994)
21. Raberto, M., Cincotti, S., Dose, C., Focardi, S.M.: Price formation in an artificial market: limit order book versus matching of supply and demand. Nonlinear Dynamics and Heterogenous Interacting Agents (2005)
22. Raberto, M., Cincotti, S., Focardi, S., Marchesi, M.: Traders' long-run wealth in an artificial financial market. Computational Economics 22, 255–272 (2003)
23. Ricordel, P.-M., Demazeau, Y.: Volcano, a vowels-oriented multi-agent platform. In: Dunin-Keplicz, B., Nawarecki, E. (eds.) CEEMAS 2001. LNCS (LNAI), vol. 2296, pp. 253–262. Springer, Heidelberg (2002)
24. Schumpeter, J.: On the concept of social value. Quarterly Journal of Economics 23(2), 213–232 (1909), `http://socserv2.socsci.mcmaster.ca/~econ/ugcm/3113/schumpeter/socialval.html`
25. Shao, W., Terzopoulos, D.: Autonomous pedestrians. Graphical Models 69(5-6), 246–274 (2007)
26. Tkatch, I., Alam, Z.S.: Strategic order splitting in automated markets. SSRN eLibrary (2009)
27. Wooldridge, M.: An Introduction to MultiAgent Systems. John Wiley & Sons, Chichester (2009) ISBN 978-0521899437

SinCity 2.0: An Environment for Exploring the Effectiveness of Multi-agent Learning Techniques

A. Peleteiro-Ramallo, J.C. Burguillo-Rial, P.S. Rodríguez-Hernández, and E. Costa-Montenegro

Abstract. In this paper we present an extensive and practical analysis of Multi-agent learning strategies using our open simulator SinCity 2.0. SinCity has been developed in NetLogo and it can be considered as an extension of the simple predator-prey pursuit problem. In our case, the predators are substituted by police cars, the prey by a thief and the chase is performed in an urban grid environment. SinCity allows to model, in a graphical friendly environment, different strategies for both the police and the thief, also implying coordination and communication among the agent set. We build this model, introducing traffic and more agent's behaviors to have a more realistic and complex scenario. We also present the results of multiple experiments performed, comparing some classical learning strategies and their performance.

1 Introduction

The predator-prey pursuit problem is one of the first and well-known testbed for learning strategies in Multi-agent Systems. It consists of a set of agents, named predators, that aim at surrounding another agent, named prey, that must escape from them.

In this paper we present our simulator, SinCity 2.0, which is a city model build using Netlogo [1], where we deploy a more complex version of the predator-prey problem. We perform experiments to show the influence of the changes on the system, as well as the effect of the variation of different parameters over the simulation.

The rest of the paper is organized as follows. First we describe the simulation scenario developed. Next, we introduce our police-thief pursuit model. After that we explain the learning techniques used in our simulations and the results we have obtained. Finally, we present the conclusions and future research work.

A. Peleteiro-Ramallo · J.C. Burguillo-Rial · P.S. Rodríguez-Hernández ·
E. Costa-Montenegro
Telematics Engineering Department, University of Vigo, Spain
e-mail: {apeleteiro,jrial,pedro,kike}@det.uvigo.es

Y. Demazeau et al. (Eds.): Advances in PAAMS, AISC 70, pp. 199–204.
springerlink.com

2 The SinCity 2.0 Pursuit Problem

Our simulator SinCity 2.0 is an extension of the predator-prey pursuit problem, where the prey is substituted by a thief car and the predators by a set of police cars. Our model SinCity 2.0 is an improvement of previous Netlogo city models: the first approach was Traffic Basic [3], then Traffic Grid [2] was presented, after Gershenson presented the Self-Organizing Traffic Lights model (SOTL) [4] and finally, our previous simulator, SinCity [7]. In that simulator we improved the design of the city and the traffic, adding for example bidirectional roads, a ring-road, four possible directions for traffic in intersections, etc. This provided a more realistic scenario.

In SinCity 2.0, on every challenge, the thief robs the bank, and after that, it escapes to its hideout. The police patrols the city until the bank alarm is triggered, and then it tries to identify the thief car (police has to see it). When this happens, the thief position is notified to all polices and the chase starts. If the thief car gets out of sight, the notified polices keep the point where it was last seen, but the police who actually saw the thief marks the direction in which it escapes, and tries to follow it. If any police car arrives to the point where the thief was first seen but it is not in sight, then the chase stops and prevents the other polices to go to that place and the patrol continues. We consider that the thief is captured if it is surrounded by police or when the police is at a predefined distance to the thief. We consider that the thief wins when it reaches its hideout.

The thief has two different strategies to learn (Table 1a). The strategy *Hideout* is used to go from its current location to the hideout when there is no police car in sight. *Runaway* is used during the chase to escape from the police. The thief considers all the polices that he can see, not only the closest one.

The police also has two different strategies to learn (Table 1b). The strategy *Chase* is used when a police sees a thief, or when a police is told about the position of the thief, but have not seen it. The strategy *Saw&lost* is used when a police sees a thief, but then it loses it (the thief is not in its sight anymore). In this case the police uses the information of the direction in which it saw the thief escaping.

Table 1 Number of states for each algorithm and each strategy to learn

(a) Thief states

	Runaway	Hideout
QL	1215	120
LA	60	120
SOM	3840	-

(b) Police states

	Chase	Saw&lost
QL	120	60
LA	120	60

We include traffic in our model to evaluate its influence in the model behavior and to have a more realistic view of the city. We define two types of 'normal' cars. Ones repetitively go from a fixed initial point to a fixed destination, park there for a random time and come back to their routines. The others repetitively choose a random destination and do their trip. Point out that we can change the number of 'normal' cars, police cars, thief cars and the size of the city.

3 Algorithms and Learning Techniques

Korf's approach [5] is a non-learning algorithm, which uses a fitness function that makes that each predator is 'attracted' by the prey and repelled from the closest predators. The idea was to chase the prey arranging predators in an stretching circle.

A Self-Organizing Map (SOM) [6] is a type of artificial neural network that is trained using unsupervised learning. We use the formula:

$$W_{n(t+1)} = W_{n(t)} + \Theta(t) \cdot \alpha(t) \cdot (I(t) - W_{n(t)})) \quad (1)$$

where $\alpha(t)$ is a monotonically decreasing learning coefficient, $I(t)$ is the data input vector and $\Theta(t)$ is the neighborhood.

The Learning Automata (LA) [8] is a type of reinforcement learning that uses two very simple rules:

$$P(s,a) = P(s,a) + \alpha \cdot (1 - P(s,a)) \quad (2)$$

$$P(s,b) = (1 - \alpha) \cdot P(s,b) \;\; for\, b \neq a \quad (3)$$

where α is a small learning factor. In case of success, Eq. 2 is used to increase the probability of action a in state s, while Eq. 3 is used to decrease the probability of the rest of the actions.

The Q-Learning Algorithm (QL) [9] is another type of reinforcement learning, which uses the formula:

$$Q(s,a) = Q(s,a) + \alpha \cdot (R(s) + \gamma \cdot max_a Q(s',a') - Q(s,a)) \quad (4)$$

In Eq. 4, s is the current state, s' is the next state, α is the learning rate, $Q(s,a)$ is the value of Q for the present state and action chosen and $Q(s',a')$ is the value of Q for the next state (s') and action (a').

4 Simulation Results

In this section we present our simulation conditions and results. Note that the thief and police cars take decisions about what road to take only at every intersection, which is where they can change direction.

We compare the results using the learning techniques in a run. A run (set of challenges) stops when the average standard deviation of the thief wins in last 500 challenges is lower than a 3%, provided than at least 1000 challenges have taken place.

We present in Table 2 a comparison among the three learning strategies. The number of 'normal' cars is 200 and the learning parameters we use are: α=0.1 for LA, QL and the probability distribution of every neuron in the SOM. Besides, in QL we set $R(s)$=0.25 and γ=0.2. Finally, for the SOM case we have $\alpha(t)$= $1/(1+0.01 \cdot t)$ and $\theta(t)$ = $\alpha(t)$=$1/(1+0.01 \cdot t)$, where t is the learning iteration. We point out that

SOM is only used by the thief. We observe that introducing traffic does not have a big influence in the results, since we observe that LA is clearly the winner in all cases, and on the other side, QL obtains the worst results. However, this scenario is more realistic.

Table 2 percentage of thief victories among the learning strategies: LA, QL and SOM (the last used only by the thief) in a scenario with 200 'normal' cars

Thief algorithm	Police algorithm	Police cars	% Thief wins
LA	QL	2	69.8
SOM	QL	2	51.2
SOM	LA	2	40.6
QL	LA	2	26.4
LA	QL	3	57.6
SOM	QL	3	35.4
SOM	LA	3	24.2
QL	LA	3	15.0
LA	QL	4	46.6
SOM	QL	4	20.4
SOM	LA	4	12.2
QL	LA	4	7.2

In Fig. 1, we show the results of the simulation where the police algorithm is set to Korf and we vary the thief algorithm. The number of 'normal' cars in this experiment is 200. Korf's algorithm does not learn and may be used to compare thief success when using the learning techniques. As we may see in Fig. 1, the best results are obtained in average by the LA algorithm, followed by SOM and QL algorithms. This is even more interesting considering that LA uses less states than the others and only considers if there are police cars in a direction, but without determining their precise distance. QL is a more complex and complete algorithm, and obtains worse results than a more 'rudimentary' one. To try to explain this fact, we made some changes in our model, so that QL has the same number of states as LA. The results are shown in Table 3. We see that the QL performance is improved, but obtaining LA better results still. In Table 4, we show the results when QL has more states than LA. We also considered the parameters set as a possible explanation for this behavior. For this reason, we run the simulation with a variation of parameters, using γ=[0.5-0.9] and α=[0.05-0.3] for QL, and α=[0.05-0.3] for LA. Table 5 shows the results when the thief uses LA. We observe that while we increase α (the learning rate), the success increases, reaching a maximum in α=0.25 and decreasing in α=0.3. Table 6 shows the results when the thief uses QL. We observe that the success percentage increases while α raises, except in α=0.15 and α=0.3. It is also shown how increasing γ (which determines the importance of future rewards) decreases the success percentage. Comparing these tables we see that LA still performs better in every simulation (the maximum value is in bold type in both tables). Therefore we show how in this scenario simple solutions are better than complex ones.

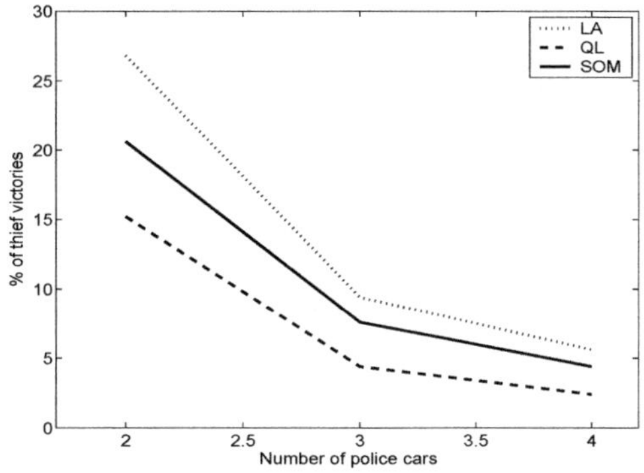

Fig. 1 Results with police using Korf and the thief LA, QL and SOM in a scenario with 200 'normal' cars

Table 3 QL with the same number of states as LA

Thief algorithm	Police algorithm	Police cars	% Thief wins
LA	Korf	2	27.4
QL	Korf	2	17.2
LA	Korf	3	16.2
QL	Korf	3	7.6
LA	Korf	4	6.4
QL	Korf	4	3.2

Table 4 QL with more states than LA

Thief algorithm	Police algorithm	Police cars	% Thief wins
LA	Korf	2	31.6
QL	Korf	2	16.4
LA	Korf	3	13.8
QL	Korf	3	5.8
LA	Korf	4	8.4
QL	Korf	4	3.6

Table 5 Thief algorithm LA, police algorithm Korf, variation of parameter α

α	0.05	0.1	0.15	0.2	0.25	0.3
% Thief wins	17.4	24.6	25.6	28.8	**31**	28.1

Table 6 Thief algorithm QL, police algorithm Korf, variation of parameters α and γ

α	0.05	0.1	0.15	0.2	0.25	0.3
γ=0.5	15.4	17.4	16.8	18.2	16.6	15.8
γ=0.6	13	14.8	18.6	17.2	**21**	17.6
γ=0.7	12	14.8	12.6	14	17.4	16.6
γ=0.8	11.4	14.0	12.4	16.8	18	11.8
γ=0.9	9.8	11.6	8.6	12.9	13	12.2

5 Conclusions and Future Work

In this paper we have presented SinCity 2.0, an open city simulator we have developed to obtain a highly flexible and efficient testbed for MAS learning techniques. We present multiple experiments in this more realistic and complex model.

The main contributions of the paper are the new version of the simulator in NetLogo and the results obtained from the simulations performed. In the practical part, we present the simulation results of different experiments to compare the surprising behavior obtained with QL and LA. As a final contribution, we show how a simpler solution (LA) obtains better results than a more complex algorithm (QL).

As future work, we plan to increase the complexity of the learning techniques, extend their number, increase the number of cooperating agents and also consider evolutionary programming techniques for generating the agents' decision rules.

References

1. Wilensky, U.: NetLogo: Center for connected learning and computer-based modeling. Northwestern University, Evanston (2007), `http://ccl.northwestern.edu/netlogo`
2. Wilensky, U.: NetLogo Traffic Grid model (2004)
3. Wiering, M., Vreeken, J., van Veenen, J., Koopman, A.: Simulation and optimization of traffic in a city. In: IEEE Intelligent Vehicles Symposium, pp. 453–458 (2004)
4. Gershenson, C.: Self-organizing traffic lights. Complex Systems 16(29) (2004)
5. Korf, R.: A simple solution to pursuit games. In: Proceedings of the 11th International Workshop on Distributed Artificial Intelligence, Glen Arbor, MI (1992)
6. Kohonen, T.: Self-Organizing Maps (2001)
7. Peleteiro-Ramallo, A., Burguillo-Rial, J., Rodríguez-Hernández, P., Costa-Montenegro, E.: Sincity: a pedagogical testbed for checking multi-agent learning techniques. In: ECMS 2009: 23rd European Conference on Modelling and Simulation (2009)
8. Narendra, K., Thathachar, M.A.L.: Learning automata: an introduction. Prentice-Hall, Englewood Cliffs (1989)
9. Kaelbling, L.P., Littman, M.L., Moore, A.P.: Reinforcement Learning: A Survey. Journal of Artificial Intelligence Research 4, 237–285 (1996)

A Tracing System Architecture for Self-adaptive Multiagent Systems*

Luis Búrdalo, Andrés Terrasa, Vicente Julián, and Ana García-Fornes

Abstract. This paper proposes an event trace architecture, based in the concept of service, which can be used to increase the amount and quality of the information that agents perceive from both their physical and social environments in order to know when to trigger a reorganisation.

1 Introduction

Complex or dynamic systems require multiagent system approaches to be *self-adaptive* [4] in order to dynamically adapt to changes in the environment. Self-adaptive multiagent systems must not only be able to modify their structure and behaviour by means of adding, removing or substituting some of their components, but they must also be able to evaluate their environment and their own health in order to determine when an adaptation is necessary [2]. However, monitoring the system in order to detect situations that may trigger a reorganization process, usually introduces centralistic solutions or dramatically increases the number of messages exchanged and the time necessary for a service request to be served. As a consequence, it is difficult for running entities to notice when an important change has taken place or to precisely determine what has changed in the system.

This paper proposes an event trace architecture which can be used in multiagent systems in order to increase the amount and quality of the information that agents perceive from both their physical and social environments. Event tracing can be used by entities in the system as a way to share information unattendedly and to model changes in their environment. Therefore, trace information can be used in order

Luis Búrdalo · Andrés Terrasa · Vicente Julián · Ana García-Fornes
Departamento de Sistemas Informáticos y Computación,
Universidad Politécnica de Valencia, cno/ de Vera S/N, 46022, Valencia, Spain
e-mail: {lburdalo,aterrasa,vinglada,agarcia}@dsic.upv.es

* This work is partially supported by projects PROMETEO/2008/051, CSD2007-022, TIN2008-04446 and TIN2009-13839-C03-01

Y. Demazeau et al. (Eds.): Advances in PAAMS, AISC 70, pp. 205–210.
springerlink.com

to trigger a reorganisation process when the situation makes it necessary. All this tracing information is offered as tracing services. Entities in the multiagent system which are interested in receiving tracing information from other entities have to request the corresponding tracing service to them.

2 Tracing System Architecture

This section describes a generic architecture to provide event tracing support to a multiagent system, considering a set of requirements previously described in [1]. This architecture can be implemented in any multiagent platform in order to give support to any type of trace events, while supporting selective event tracing.

From the viewpoint of the architecture, a multiagent system can be considered to be formed by a set of *tracing entities* or components which are susceptible of generating and/or receiving certain information related to their activity as *trace events*. Tracing entities can generate and receive trace events in a non-exclusive way at run time. A *trace event* is a piece of data representing an action which has taken place during the execution of an agent or any other component of the multiagent system. Each trace event has an origin entity, which generated it, and a *timestamp*, which indicates the global time at which the event was generated. In this way, tracing entities are able to chronologically sort trace events produced anywhere in the multiagent system. Also, each trace event has an *event type*, which tracing entities will use to identify the nature of the information represented by the trace event. Depending on the event type, each trace event may have some extra *attached data*.

This architecture is based on the concept of *tracing service*, which is used in order to model the concept of event type. Tracing services are special services which are offered by tracing entities to share their trace events. Each tracing entity may offer a set of tracing services, corresponding to the different event types which it generates.

In this architecture, tracing entities can be considered to be playing two different tracing roles. When they are generating trace events, tracing entities are considered *Event Source* entities (*ES*). When they are receiving trace events, tracing entities are considered *Event Receiver* entities (*ER*). Any tracing entity can start and stop playing any of these two roles, or both, at any time.

The Trace Manager is the main component of the tracing system. The Trace Manager (described in more detail in Section 3) is in charge of coordinating the entire tracing process, recording trace events generated by ES entities and delivering them to ER entities. Despite not being strictly necessary, the Trace Manager is designed to be integrated within the multiagent platform. If it were implemeted outside the multiagent platform, some of its functionalities, like Domain Independent Tracing Services, which will be later explained in Section 2.2 may not be available. Also, implementing the Trace Manager within the multiagent platform is better in order to obtain a good quality of service and to minimize the amount of resources consumed, which is also an important issue mentioned in [1].

2.1 Tracing Entities

This architecture considers three different kinds of tracing entities: *Agents*, *artifacts* and *aggregations*.

On the one hand, *agents* are all those autonomous and proactive entities which define the multiagent system behaviour. On the other hand, *artifacts* are all those passive elements in the multiagent system (databases, physical sensors and actuators, etc.) susceptible of generating events at run time or receiving them as an input [3]. Artifacts can combine in order to perform more complex tasks, generating or receiving trace events as a tracing individual. From the point of view of the tracing system, these combinations of artifacts are also modelled as single artifacts.

If the multiagent system supports aggregations of agents (or agents and artifacts), such as teams or organizations, then such aggregations are considered by the tracing system as *single* tracing entities, in the sense that trace events can be generated from or delivered to these entities as tracing individuals. Agents and artifacts within an aggregation are still tracing entities and thus, they can also generate and receive trace events individually, not only as members of the aggregation.

From the point of view of the architecture, the multiagent platform can be seen as a set of agents and artifacts. Therefore, the components of the platform are also susceptible of generating and receiving trace events.

When a tracing entity is playing the ER tracing role, the tracing system provides it with a *stream*, which can be seen as a special mailbox where the Trace Manager delivers the trace events for this ER entity. These streams can either be pieces of memory or log files. In both cases, the ER entity which owns the stream has to limit its size in order not to overload its resources.

2.2 Tracing Services

Event types are modeled in this architecture as *tracing services*. A tracing service is a special service which is offered by an ES entity to share its trace events, in a similar way to a traditional service. Each ES entity can offer different tracing services, and the same tracing service can be offered by many different ES entities.

Attending to the tracing entity which offers them, tracing services can be classified in: *domain dependent* (*DD*) and *domain independent* (*DI*).

- **Domain independent tracing services (DI):** Tracing services offered by the multiagent platform itself. They can be present in any multiagent system running in the platform.
- **Domain dependent tracing services (DD):** Tracing services designed ad-hoc for a specific multiagent system.

Trace events can be processed or even composed in order to generate compound trace events, which can be used to represent more complex information. Both domain dependent and domain independent tracing services can also be classified in simple and compound and, as a result, tracing services can be classified in four groups: *Simple Domain-Independent* (*SDI*), *Simple Domain-Dependent* (*SDD*),

Compound Domain-Independent (*CDI*) and *Compound Domain-Dependent* (*CDD*) tracing services.

The functionality of tracing services is based on three aspects which are further explained in the rest of the section: Request of tracing services, efficient resource management and security.

As with traditional services, when an ER entity is interested in receiving trace events of a specific event type which are generated by a given ES, it has to request the correponding service. From that moment on, the Trace Manager starts recording the corresponding trace events and delivering them directly to the ER stream until the ER cancels the request. The Trace Manager only records those trace events which have been requested by an ER entity, so that no resources are spent in recording and delivering trace events which have not been requested by any ER entity.

The Trace Manager provides a list of all the available tracing services and the ES entities which offer them. When an ES entity wants to offer any tracing information, it must inform the Trace Manager in order to publish the corresponding tracing service so that other tracing entities can request it if they are interested in its trace events. When a tracing entity does not want to receive certain trace events anymore it has to cancel the request to the corresponding tracing service.

In order to let ES entities decide which ER entities can receive their trace events, when an ES entity publishes a tracing service, it has also to specify which agent roles are authorized to request that service to that ES entity (*direct authorization*). In this way, when an ER entity wants to request a tracing service to an ES, it has to be able to assume one of the authorized agent roles. ER entities which are authorized to request a tracing service to certain ES entity can also authorize other roles to request the same tracing service to that ES entity. This is defined as *authorization by delegation*. In this way, the tracing system mantains an authorization graph for each tracing service which is being offered by each ES. This authorization graph is dynamic, since tracing entities can add and remove authorizations at run time. When an authorization, direct or by delegation, is removed, all those delegated authorizations which depended on the removed one are also removed.

The tracing system does not control which entities can assume each role in order to request or to add authorizations for a tracing service. It is the multiagent platform which has to provide the necessary security mechanisms no prevent agents from assuming inappropiate roles.

3 The Trace Manager

As commented in Section 2, the Trace Manager is the main component of the tracing system, in charge of recording trace events generated in ES entities, delivering them to ER entities and, in general, coordinating the entire tracing process. It is possible to implement the Trace Manager outside the platform; however, in order to provide DI tracing services and to minimize the amount of resources consumed by the tracing process, it is necessary to implement it inside the multiagent platform. Internally, this module has four main components which are necessary to give support to the architecture:

A *Trace Entity Module (TEM)*, in charge of registering and managing all the tracing entities, a *Tracing Services Module (TSM)*, in charge of registering and managing tracing services offered by ES entities, a *Subscription Module (SUBM)*, which stores and manages subscriptions to each tracing service and ES entity, and an *Authorization Module (AM)*, which stores and manages the authorizartion graph for each tracing service and ES entity.

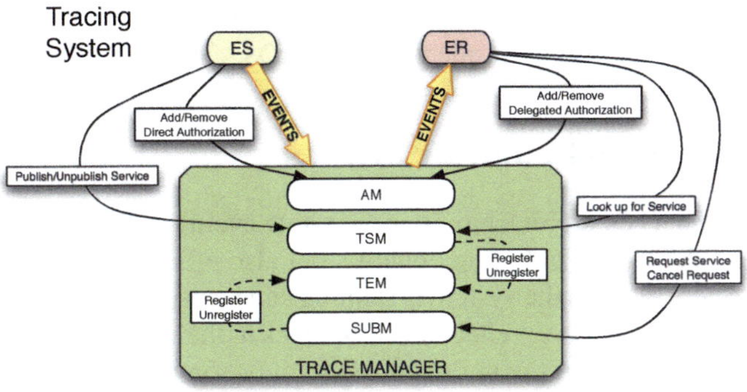

Fig. 1 Architecture model of the tracing system and interactions among tracing entities depending on their tracing roles and the Trace Manager's internal modules

Figure 1 shows how tracing entities interact with the Trace Manager depending on the tracing role that they are playing. These interactions are detailed below:

- **Publish/Unpublish Service:** ES entities have to publish the tracing services they offer so that ER entities can request them. Published tracing services are stored in the TSM. When the ES does not want to offer a tracing service anymore, it has to remove the publication. If the tracing service is the first one offered by the ES entity, then this ES is internally registered in the TEM. In the same way, when an ES entity unpublishes all of its tracing services, it is internally removed from the TEM.
- **Add/Remove Direct Authorization:** ES entities can specify which roles have to be assumed by ER entities in order to request their tracing services. ES entities add and remove direct authorizations for each of their tracing services which they offer. The corresponding authorization graph is stored in the AM.
- **Add/Remove Delegated Authorization:** ER entities which have assumed a role which authorizes them to request a tracing service can also authorize other roles to request that tracing service. In the same way, ER entities can remove those delegated authorizations which they previously added. Modifications in the corresponding authorization graph are registered in the AM.
- **Look up for Service:** ER entities can look up in the TSM to know which tracing services are available and which ES entities offer them.

- **Request Service / Cancel Request:** ER entities which want to receive certain trace events from an ES have to request the corresponding tracing service to the Trace Manager. The Trace Manager verifies against the AM that the ER entity has authorization for that tracing service before adding the subscription to the SUBM. When an ER entity does not want to receive events corresponding to a specific tracing service, it has to cancel the request of that service and the corresponding subscription is also deleted in the SUBM. If the ER entity which requests the tracing service was not subscribed to any other tracing service, then this entity is internally registered and listed in the TEM. In the same way, when an ER entity cancels all of its requests, it is internally removed from the TEM. The trace manager only records and delivers those trace events for which there is at least one tracing service request in the SUBM.

4 Conclusions and Future Work

This paper presents an architecture to incorporate event trace support to multiagent systems, considering not only functionality, but also efficiency and security issues. Trace events generated as the system is running can be used by entities in the system to communicate as well as to reflect those changes in their environment which could require a reorganisation of the multiagent system or some of its components.

References

1. Búrdalo, L., Terrasa, A., García-Fornes, A.: Towards providing social knowledge by event tracing in multiagent systems. In: Corchado, E., Wu, X., Oja, E., Herrero, Á., Baruque, B. (eds.) HAIS 2009. LNCS, vol. 5572, pp. 484–491. Springer, Heidelberg (2009)
2. Dignum, V., Dignum, F., Sonenberg, L.: Towards dynamic reorganization of agent societies. In: Proceedings of Workshop on Coordination in Emergent Agent Societies, pp. 22–27 (January 2004)
3. Omicini, A., Ricci, A., Viroli, M.: Artifacts in the a&a meta-model for multi-agent systems. In: Autonomous Agents and Multi-Agent Systems (January 2008)
4. Valetto, G., Kaiser, G., Kc, G.: A mobile agent approach to process-based dynamic adaptation of complex software systems. In: Ambriola, V. (ed.) EWSPT 2001. LNCS, vol. 2077, pp. 102–116. Springer, Heidelberg (2001)

An UCT Approach for Anytime Agent-Based Planning

Damien Pellier, Bruno Bouzy, and Marc Métivier

Abstract. In this paper, we introduce a new heuristic search algorithm based on mean values for anytime planning, called MHSP. It consists in associating the principles of UCT, a bandit-based algorithm which gave very good results in computer games, and especially in Computer Go, with heuristic search in order to obtain an anytime planner that provides partial plans before finding a solution plan, and furthermore finding an optimal plan. The algorithm is evaluated in different classical planning problems and compared to some major planning algorithms. Finally, our results highlight the capacity of MHSP to return partial plans which tend to an optimal plan over the time.

1 Introduction

The starting point of this work was to apply Upper Confidence bounds for Trees (UCT) [13], an efficient algorithm well-known in the machine learning and computer games communities, and originally designed for planning, on planning problems. A weakness of classical planners is the all-or-nothing property. First, when the problem complexity is low enough, classical planners find the best plan very quickly. Second, when the problem complexity is medium, planners first try to find a solution plan (not the optimal one) and then pursue their search to extract a better solution [6, 4]. This technique is called anytime planning. Finally, when the problem complexity is too high, planners are not able to find any solution plan. In order to answer in part to this weakness, we introduce a new approach based on heuristic search and mean values for anytime planning able to provide partial plans before finding a first plan, and furthermore finding an optimal plan.

Anytime planning can be understood in two meanings. In the planning domain, anytime planning means finding a solution plan, and then refining it to find an optimal plan. There is a good chance that if you stop the planner before finding an

Damien Pellier · Bruno Bouzy · Marc Métivier
Laboratoire d'Informatique de Paris Descartes 45, rue des Saints Pères, 75006 France

Y. Demazeau et al. (Eds.): Advances in PAAMS, AISC 70, pp. 211–220.
springerlink.com

optimal plan, the planner has already a solution plan to provide, and it looks like anytime. However, if you stop the planner before having a first solution plan, the planner is not anytime in a strict meaning. When stopped before having a first solution plan, an anytime planner should be able to give some relevant information, for example the beginning of a plan, a partial plan, or the first action. Until a solution plan is found, the longer the time the longer the partial plan is. In this work, the term anytime refers to the strict meaning. We are interested in partial plans.

Originally, UCT is a bandit-based planning algorithm designed for Markov Decision Process (MDP). UCT builds a tree whose root is the current state on which a decision must be taken. The principal variation of the tree is the current solution, and when a plan is found, the principal variation of the tree is the sequence of actions to perform to reach the goal. As time is going on, UCT builds up its tree adding nodes at each iteration. At any time, UCT has a principal variation which can be considered as a partial plan. However, [13] did not give known successful applications in the planning domain yet. Instead, UCT gave tremendous results in computer games, and specifically in computer go with the Go playing program Mogo [5]. In computer go, UCT is efficient for several reasons. The first reason is that the Go complexity is high, and games are played in limited time. Consequently Go playing programs must find moves that does not need to be optimal, but that need to be the less bad as possible given the limited time. The anytime property is crucial in computer games, and UCT has it. Consequently, studying UCT, anytime algorithm originally designed for planning problems successful in two-player games, was a good starting point to attempt removing the all-or-nothing weakness observed on classical planners. In this attempt, we reached an interesting point to contribute to the planning community. This paper presents the work consisting in associating UCT ideas with heuristics in state space search in order to obtain an anytime planner which provides partial plans before finding a first plan, and furthermore finding the best plan. The paper shows a new heuristic search algorithm based on mean values for anytime planning, called Mean-based Heuristic Search for anytime Planning (MHSP).

The outline of the paper is the following. Section 2 describes previous works. Section 3 presents MHSP. Section 4 shows experimental results. Finally, section 5 discusses this approach and concludes.

2 Previous Works

UCT and Computer Go. UCT worked well in Go playing programs, and it was used under many versions leading to the Monte-Carlo Tree Search (MCTS) framework [3]. A MCTS algorithm starts with the root node as a tree, and while time remains, it iteratively grows up a tree in the computer memory by following the steps below: (a) starting from the root, browse the tree until reaching a leaf by using (1), (b) expand the leaf with its child nodes, (c) choose one child node, (d) perform a random simulation starting from this child node until the end of the game, and get the return, i.e. the game's outcome, and (e) update the mean value of the browsed nodes with this return. With infinite time, the root value converges to the minimax value of the game tree [13]. The Upper Confidence Bound (UCB) selection rule (1) answers the requirement of being optimistic when a decision must be made facing uncertainty [1].

$$N_{select} = \arg\max_{n \in N}\{m + C\sqrt{\frac{\log p}{s}}\} \tag{1}$$

N_{select} is the selected node, N is the set of children, m is the mean value of node n, s is the number of iterations going through n, p is the number of iterations going through the parent of n, and C is a constant value setup experimentally. (1) uses the sum of two terms: the mean value m, and the UCB bias value which guarantees exploration.

Planning under time constraints. Besides, planning under time constraints is an active research domain that results in adaptive architectures [9], real-time control architectures [16], and real-time heuristic search [15].

3 MHSP

This section defines our algorithm MHSP. We made two important choices in designing MHSP after which we give the pseudo-code of MHSP.

Heuristic values replace simulation returns. On planning problems, random simulations are not appropriate. Browsing randomly the state space does not enable the algorithm to reach goal states sufficiently often. Many runs complete without reaching goal states. Replacing the simulations by a call to the heuristic is far better. Not only the algorithm finds the goal, but it may reach it very quickly: on certain classes of problems, MHSP is as fast as a classical planner to find the best solution. In Computer Go, the random simulations were adequate mainly because they always completed after a limited number of moves, and the return values (won or lost) were roughly equally distributed on most positions of a game. Furthermore, the two return values correspond to actual values of a completed game. In planning, one return means that a solution has been found (episode completed), and the other return means that the episode has not been completed. This simulation difference is fundamental between the planning problem, and the game playing problem. Furthermore, heuristic values bring domain-dependent knowledge into the returns. In Computer Go, replacing the simulations by evaluation function calls is forbidden by fifty years of computer Go history. However, in Computer Go, and in other domains, adding proper domain-dependent knowledge into the simulations improves the significance of the returns, henceforth the level of the playing program. Consequently, using heuristic values in our work should be positive bound to the condition that the heuristic value generator is good, which is the case in planning. In MHSP, we replace stage (d) of MCTS above by a call to a heuristic function.

Optimistic initial mean values. Computer games practice shows that the UCB bias of (1) can merely be removed provided the mean values of nodes are initialized with sufficiently optimistic values. This simplification removes the problem of

Algorithm 1. MHSP(O, s_0, g)

1 $C[s_0] \leftarrow \emptyset$; $R[s_0] \leftarrow \Delta(s_0, g)$; $V[s_0] \leftarrow 1$; $\pi \leftarrow nil$
2 **while** *has_time* **do**
3 $\quad s \leftarrow s_0$
4 $\quad$ **while** $g \not\subseteq s$ *and* $V[s] \neq 1$ **do** $s \leftarrow argmax_{s' \in C[s]}(R[s']/V[s'])$
5 $\quad$ $reward \leftarrow (R[s_0]/V[s_0]) + 1$
6 $\quad$ **if** $g \subseteq s$ **then** $reward \leftarrow 0$
7 $\quad$ **else if** $V[s] = 1$ **then**
8 $\quad\quad$ $A \leftarrow \{a \mid a$ ground instance of an operator in O and $precond(a) \subseteq s\}$
9 $\quad\quad$ **foreach** $a \in A$ **do**
10 $\quad\quad\quad$ $s' \leftarrow (s \cup effects^{+}(a)) - effects^{-}(a)$
11 $\quad\quad\quad$ $C[s'] \leftarrow C[s] \cup \{s'\}$; $R[s'] \leftarrow \Delta(s', g)$; $P[s'] \leftarrow s$; $V[s'] \leftarrow 1$
12 $\quad\quad$ **if** $C[s] \neq \emptyset$ **then** $s \leftarrow argmax_{s' \in C[s]}(R[s'])$; $reward \leftarrow R[s]$
13 $\quad$ $i \leftarrow 0$
14 $\quad$ **while** $s \neq s_0$ **do** $s \leftarrow P[s]$; $R[s] \leftarrow R[s] + (reward - i)$; $V[s] \leftarrow V[s] + 1$; $i \leftarrow i + 1$
15 $\quad$ **if** $g \subseteq s$ **then**
16 $\quad\quad$ $\pi' \leftarrow reconstruct\_solution\_plan()$
17 $\quad\quad$ **if** $length(\pi) > length(\pi')$ **then** $\pi \leftarrow \pi'$
18 **if** $\pi = nil$ **then return** *reconstruct_best_plan*() ; **else return** π

tuning C, while respecting the optimism principle. Generally, to estimate a given node, the planning heuristics give a path length estimation. Convergence to the best plan is provided by admissible heuristics, i.e. heuristics ensuring the heuristic value is inferior to the actual distance from the node to the goal, i.e. optimistic heuristics. Consequently, the value returned by planning heuristics on a node can be used to initialize the mean value of this node. In MHSP, the returns are negative or zero, and they must be in the opposite of the distance from s to g. Thus, we initialize the mean value of a node with $\Delta(s, g)$ which is minus the distance estimation to reach g from s. With this initialization policy, the best node according to the heuristic value will be explored first. Its value will be lowered after some iterations whatever its goodness, and then the other nodes will be explored in the order given by the heuristic.

The algorithm. MHSP algorithm is shown in algo. 1 : O is the set of operators, s_0 the initial state, g the goal, $C[s]$ the set of children of state s, $R[s]$ the cumulative return of state s, $V[s]$ the number of visits of state s, and $P[s]$ the parent of s. The outer *while* (line 2) ensures the anytime property. The first inner *while* (line 4) corresponds to stage (a) in UCT. The default reward is pessimistic: $(R[s_0]/V[s_0]) + 1$ is the current pessimism threshold. The first two *if* test whether the inner *while* has ended up with a goal achieved (line 6) or with a leaf (line 7). If the goal is not reached, the leaf is expanded, stage (b) in MCTS. The second *if* corresponds to stage (c). Stage (d) is performed by writing $\Delta(s', g)$ into the return. The second inner *while* (line 14) corresponds to stage (e). Function *reconstruct_solution_plan*() browses the tree by selecting the child node with the best mean, which produces the solution plan. Function *reconstruct_best_plan*() browses the tree by selecting the child node with the best number of visits. The best plan reconstruction happens when the time is over before a solution plan has been found. In this case, it is important to reconstruct a robust plan, may be not the best one in terms of mean value. With the child with the best mean, a plan with newly created nodes could be selected, and the plan would not be robust. Conversely selecting the child with the best number of visits ensures that the plan has been tried many times, and should be robust to this extent.

4 Experimental Results

In this section, we present experimental results in two steps: a first experiment aiming at showing that MHSP can be compared to state-of-the-art planners, and a second experiment aiming at underlining the anytime feature of MHSP (anytime meaning building good partial plans when the running time is shorter than the time necessary to build the first solution plan). We present experimental results obtained in test domains and problems from International Planning Competition, which illustrates the effectiveness of our techniques implemented in MHSP. All the tests were conducted on an Intel Core 2 Quad 6600 (2.4Ghz) with 2 Gbytes of RAM. The implementation of MHSP used for experiments is written in Java based on the PDDL4J library.

First experiment. The experiments were designed in order to show that MHSP: (1) performs almost as well as classical planners on classical planning problems, and (2) returns, given a fixed amount of time, the beginning of the optimal plan that classical planners cannot solve with the same amount of time.

Figure 1(a) shows (on log scale) performance of MHSP-speed on blocksworld domain according to the problem size. The planners used to the comparison were chosen for their planning techniques: IPP for planning graph [14], Satplan 2006 for planning by satisfiability [12], SGPlan-5 for subgoal decomposition planning [11] and FDP for constraint satisfaction techniques [7]. For both domains, MHSP was tested with three heuristics: Hs^{+}, Hs^{max} used by HSP planner [2] and FF-heuristic used by [10]. The CPU-time limit for each run was 10000 seconds, after which termination was forced. The results show that MHSP performs almost the most quickly (except SGPlan which is based on hill climbing and heuristic search and goal agenda techniques). However, the three heuristics do not perform as well. As expected, Hs^{max} is less informative than Hs^{+} and FF-heuristic. Thus, MHSP with Hs^{max} performs more slowly than with the other heuristics. Moreover, Hs^{+} is more efficient than FF-heuristic as displayed by Table 1. This difference can be explained by the fact that Hs^{+} returns values more dispatched than FF-heuristic which is more informative for MHSP. Finally, if we look at the number of actions of the first solution plan found by the different planners, we observe that MHSP finds solution plan of good quality. To conclude this first experimentation, let's consider the figure 1(b) that displays the behavior of MHSP on a specific blocksworld problem

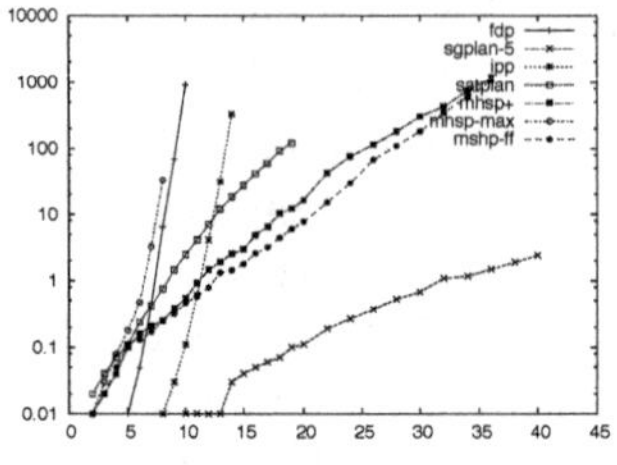

(a) Planning times (sec.) blocksworld domain

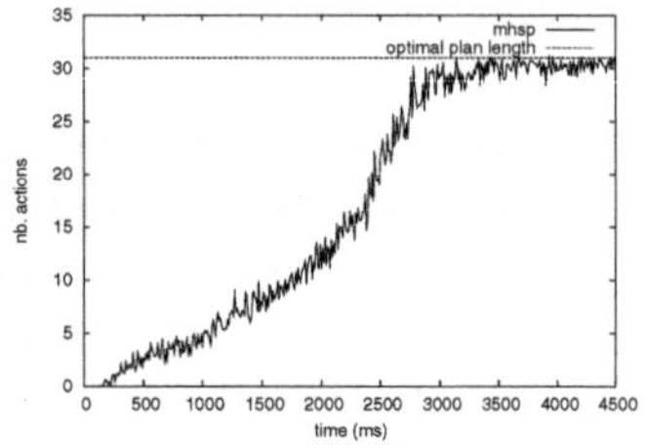

– (b) The number of actions belongs to the optimal solution plan found by MHSP given a fixed amount of time on a specific blocksworld problem containing 17 blocks

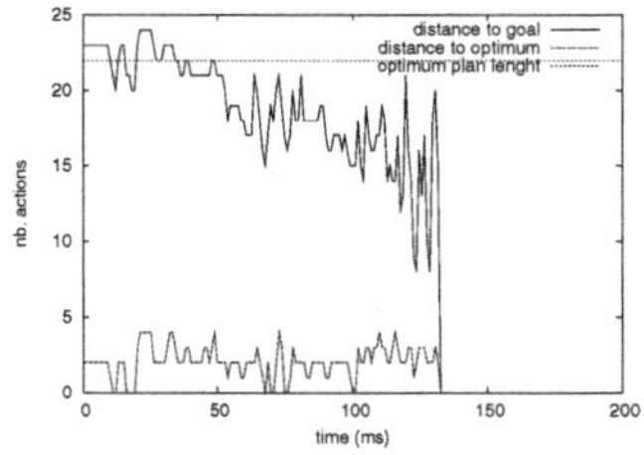

(c) A* Blocksworld problem 12

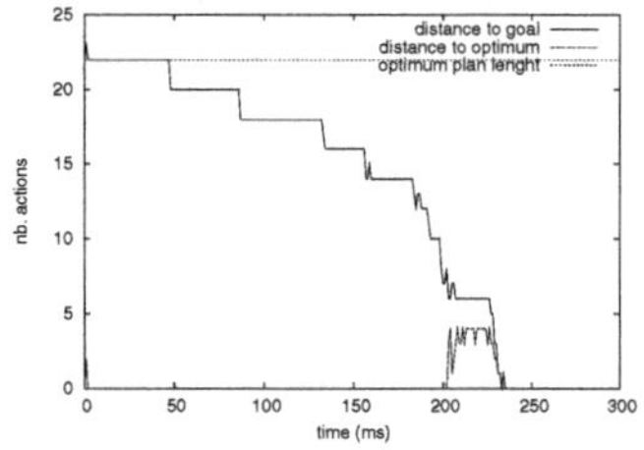

(d) MHSP Blocksworld problem 12

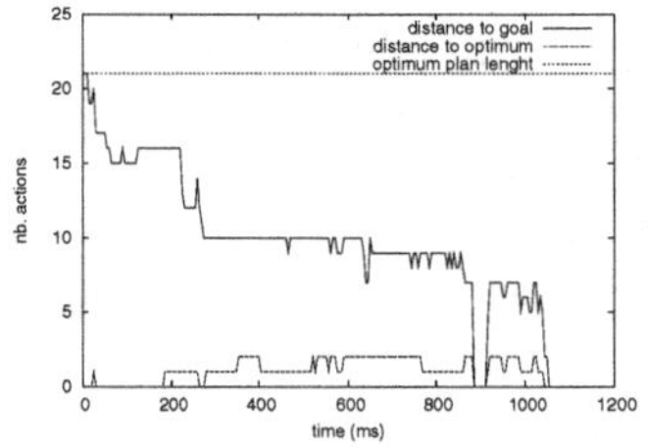

(e) MHSP Ferry problem L6 C9

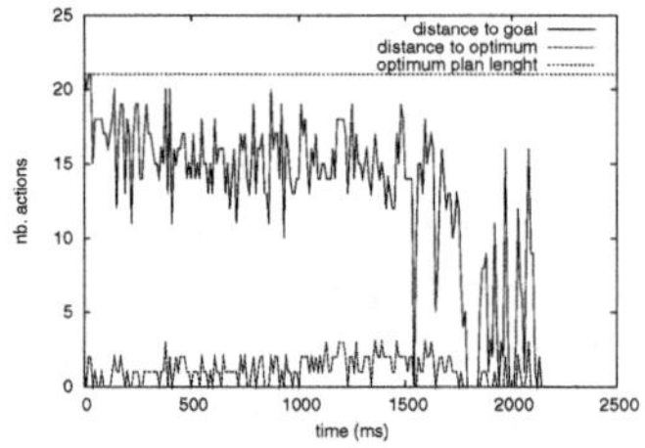

(f) A* Ferry problem L6 C9

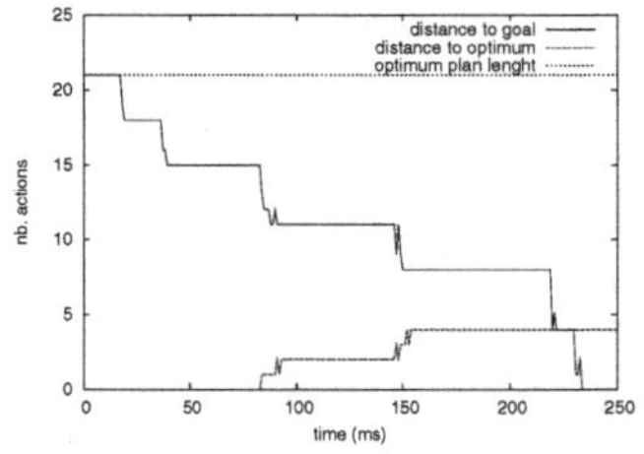

(g) EHC Ferry problem L6 C9

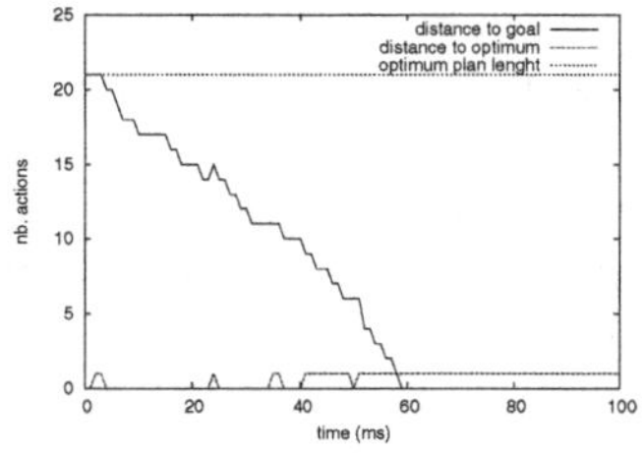

(h) Greedy Search Ferry problem L6 C9

Fig. 1 Experimental results

Table 1 Comparison of the time (sec.) and cost (number of actions) of the plan found by MHSP

domains	best	mhsp+		mhsp-ff		ipp		satplan		sgplan5		fdp	
	plan	time	cost	time	cost	time	cost	time	cost	time	cost	time	cost
gripper-7	21	4.55	23	6.66	21	0.16	21	20.57	21	0.00	21	1.42	15
gripper-8	21	11.91	27	22.51	23	0.43	23	25.38	23	0.01	23	4.10	23
gripper-9	27	28.88	29	85.57	27	1.61	27	> 120	na	0.01	27	15.39	27
satellite-2-4	20	0.71	20	27.87	20	41.36	23	0.45	25	0.00	24	> 120	na
satellite-2-5	29	7.76	29	> 120	na	> 120	na	11.40	43	0.04	35	> 120	na
satellite-2-6	43	28.79	43	> 120	na	> 120	na	> 120	na	0.11	71	> 120	na
zeno-2-6	15	9.71	16	118.79	15	0.03	17	0.18	19	0.01	24	51.61	15
zeno-3-6	11	24.48	11	> 120	na	0.04	18	0.39	18	0.01	15	> 120	na
zeno-3-8	23	24.28	23	> 120	na	77.23	29	0.92	27	0.01	29	> 120	na

containing 17 blocks. This figure shows that the number of actions belongs to the optimal solution plan found by MHSP given a fixed amount of time. Notice that the results are statistically meaningful (MHSP was run 20 times each 10 ms time step using FF-heuristic). We observe that MHSP finds very quickly the first actions of the optimal solution plan. It meaningfully needs only 1500ms to find the first 10 actions of the optimal solution plan that has a length of 31 actions. Of course, MHSP performs only if the heuristic is informative and a complete study of the behavior of MHSP with all the heuristics available in the literature would be necessary.

Partial plan experiment. We present the results obtained by A*, Greedy Search (GS), MHSP-ff, and Enforced Hill-Climbing (EHC) [10] on two problems: Blocksworld problem 12 and Ferry problem L6 C9. The aim of this second experiment is to see whether the partial plans built by the four algorithms are good or not when the running time is shorter that the time to build a first solution plan. To evaluate a partial plan, we define two distances: the distance to the goal and the distance to the optimum.

- *Distance to the goal.* The distance to the goal of a partial plan is the length of the optimal plan linking the end state of this partial plan to the goal state. When the distance to the goal diminishes, the partial plan has been built in the appropriate direction. When the distance to the goal is zero, the partial plan is a solution plan.
- *Distance to the optimum.* The distance to the optimum of a partial plan is the length of the partial plan, plus the distance to the goal of the partial plan, minus the length of the optimal plan. When the distance to the optimum of a partial plan is zero, the partial plan is the beginning of an optimal plan. The distance to the optimum of a solution plan is the diffence between its length and the optimal length. The distance to the optimum of the void plan is zero.

The distance to the goal and the optimal distance of an optimal plan is zero. Conversely, when the distance to the goal and the distance to the optimum of a partial plan are zero, the partial plan is an optimal plan. For each problem, the results are shown with figures giving the distance to the goal and the distance to the optimum of the partial plan in the running time. These distances are computed every ms by calling an optimal planner (i.e. A*).

Partial plans of the four algorithms. The partial plan given by A* at a given time is the path linking the root node to the current expanded leaf node. Given A* manages a list of whole leaf nodes, the partial plan provided by A* varies a lot from a given time to another. To get the partial plan provided by MHSP at a given time, we browse the MHSP tree from the root using the number of visits, and we stop when this number is below a threshold that equals the branching factor of the tree. This way, the partial plan is more stable, but shorter. GS browses and expands the tree starting from the root by selecting the child whose heuristic value is minimal. The weakness of GS is its non-optimality. EHC chooses the first node whose heuristic value is strictly inferior to the heuristic value of the current node, insuring a progress toward the goal is made when selecting this node. The weakness of EHC is its inability to go out of a deadend or a plateau. In such cases, our EHC returns a failure. With the state-of-the-art heuristics, when they find a solution, GS and EHC are actually very fast.

Blocksworld problem 12. In this problem, A* finds the optimal solution in 130ms (see figure 1(c)). When the running time is inferior to 50 ms, the distance to the goal remains at its initial value. Between 50 ms and 130 ms, the distance to the goal decreases but remains high before the optimal plan is found. Along the running time, the distance to the optimum is low but strictly greater than zero. Experimental results of GS and EHC are not shown. Both algorithms go to deep in the search space. Consequently the algorithm used to compute the distance to the optimum (in our experiments A*) falls to find the optimum plan in a reasonable time frame. These results highlight the weakness of both algorithms in that problem. MHSP optimally solves this problem in 230 ms (see figure 1(d)). Like A* does, when the running time is inferior to 50 ms, the distance to the goal remains at its initial value, and between 50 ms and 230 ms, the distance to the goal decreases but remains high before the optimal plan is found. MHSP explores along optimal partial plans for running times inferior to 200 ms. When looking at the distance to the goal, the relative comparison between A* and MSHP on this problem is in favour of A*, but the distance to the goal of MHSP decreases almost monotically while the distance to the goal of A* is decreasing with large oscillations. When looking at the distance to the optimum, the relative comparison between A* and MSHP on this problem is in favour of MHSP (except after 130 ms).

Ferry problem L6 C9. On the Ferry problem L6 C9, MHSP finds the optimal solution in 1050 ms (see figure 1(e)), A* finds the optimal solution in 2100 ms (see figure 1(f)). MHSP is twice faster than A* on this problem. The distance to the goal of MHSP decreases more quickly than it does for A*. MHSP shows a better anytime ability than A* on this problem. However, EHC finds a solution in 230 ms, four times faster than MHSP (see figure 1(g)), but this solution is not optimal. Besides,

GS finds a solution in 60 ms, four times faster than EHC (see figure 1(h)), but this solution is not optimal. EHC and GS are one order of magnitude faster than MHSP and A* and they find solutions not far from optimal. MHSP is the fastest algorithm to find an optimal plan on this problem.

5 Conclusion

Anytime heuristic search has been studied already [8]. However, this work focused on finding a first plan, and refining it to find the best plan. Such a method cannot give any information before a first plan is found, especially a partial plan.

In this paper, we presented MHSP a new anytime planning algorithm which provides partial plans before finding a solution plan. This algorithm combines an heuristic search and the learning principles of UCT algorithm, i.e. states' values based on mean returns, and optimism in front of uncertainty. Of course, when given insufficient time, the partial plan is not garanteed to be a prefix of an optimal plan or of a solution plan. However, on average over our benchmark, the partial plans returned by MHSP are prefix of either solution plans or optimal plans. Evaluated in several classical problems, our first results showed that MHSP performs almost as well as classical planners on classical planning problems. However, MHSP is surpassed by SGPlan that uses a goal agenda (MHSP does not). We defined two distances to evaluate partial plans provided by a planner at a given time: the distance to the goal and the distance to the optimum. With such measures, we experimentally compared MHSP, A*, EHC and GS. Given a fixed amount of time, MHSP provides partial plans which tend to be the beginning of solution plans and then optimal plans when the running time increases. However, given the speed of EHC and GS when they find solution plans, anytime conclusions can hardly be drawn when considering absolute running times.

Averaging in MHSP may be discussed. A possibility is to replace averaging by backing-up the best child value. Then MHSP would look like Greedy Search which may fall into deadends, even with admissible heuristics. Therefore, a first reason for averaging is to avoid deadends. With heuristics admissible or not, MHSP expands a different leaf than A* when, in the upper part of the tree, a mean value is high and leads to a bottom part of the tree where the heuristics values are high on average but inferior to the heuristic value of the node selected by A*. Our partial plan experiment shows that MHSP have both their pros and cons since MHSP is better than A* on Ferry and Gripper, but worse on Blocksworld.

In the future, we want to apply MHSP on problems with non deterministic environments to take avantage of the averaging of MSHP. Furthermore, since MHSP is successfully validated on four benchmarks of classical planning, integrating MHSP into practical applications remains an enlightening future work. In the current study, we used three heuristic functions: seeing which function is best-suited to each problem is another interesting research direction.

References

1. Auer, P., Cesa-Bianchi, N., Fisher, P.: Finite-time Analysis of the Multiarmed Bandit Problem. Machine Learning 47(2–3), 235–256 (2002)
2. Bonet, B., Geffner, H.: Planning as Heuristic Search. Artificial Intelligence 129(1–2), 5–33 (2001)
3. Chaslot, G., Winands, M., van den Herik, H., Uiterwijk, J., Bouzy, B.: Progressive Strategies for Monte-Carlo Tree Search. New Mathematics and Natural Computation 4(3), 343–357 (2008)
4. Chen, Y., Huang, R., Zhang, W.: Fast Planning by Search in Domain Transition Graphs. In: Proc. AAAI, pp. 886–891 (2008)
5. Gelly, S., Wang, Y., Munos, R., Teytaud, O.: Modification of UCT with Patterns in Monte-Carlo Go. Tech. Rep. RR-6062, INRIA (2006)
6. Gerevini, A., Serina, I.: LPG: A Planner Based on Local Search for Planning Graphs with Action Costs. In: Proc. ICAPS, pp. 13–22 (2002)
7. Grandcolas, S., Pain-Barre, C.: Filtering, Decomposition and Search Space Reduction for Optimal Sequential Planning. In: Proc. AAAI (2007)
8. Hansen, E.A., Zhou, R.: Anytime Heuristic Search. JAIR 28(1), 267–297 (2007)
9. Hayes-Roth, B.: An architecture for adaptive intelligent systems. Artificial Intelligence 72, 329–365 (1995)
10. Hoffmann, J., Nebel, B.: The FF Planning System: Fast Plan Generation Through Heuristic Search. JAIR 14(1), 253–302 (2001)
11. Hsu, C.W., Wah, B., Huang, R., Chen, Y.: Handling Soft Constraints and Goals Preferences in SGPlan. In: Proc. of the ICAPS Workshop on Preferences and Soft Constraints in Planning (2006)
12. Kautz, H.A., Selman, B.: Unifying SAT-based and Graph-based Planning. In: Proc. IJCAI, pp. 318–325 (1999)
13. Kocsis, L., Szepesvari, C.: Bandit-based Monte-Carlo Planning. In: Fürnkranz, J., Scheffer, T., Spiliopoulou, M. (eds.) ECML 2006. LNCS (LNAI), vol. 4212, pp. 282–293. Springer, Heidelberg (2006)
14. Koehler, J., Nebel, B., Hoffmann, J., Dimopoulos, Y.: Extending planning graphs to an ADL subset. In: Steel, S. (ed.) ECP 1997. LNCS, vol. 1348, pp. 273–285. Springer, Heidelberg (1997)
15. Korf, R.: Real-Time Heuristic Search. Artificial Intelligence 42(2-3), 189–211 (1990)
16. Musliner, D., Goldman, R., Krebsbach, K.: Deliberation scheduling strategies for adaptive mission planning in real-time environments. In: Proceedings of the Third International Workshop on Self Adaptive Software (2003)

Elitist Ants Applied to the Undirected Rural Postman Problem

María-Luisa Pérez-Delgado

Abstract. Ant Colony Optimization is a metaheuristic inspired by the behaviour of natural ants. In a recent work, the algorithm called Ant Colony System was applied to solve the Undirected Rural Postman Problem. In this paper another ant-based algorithm, called Elitist Ant System, is applied. It generates better results than the previous one, and better than other approximate solutions applied to the same problem.

1 Introduction

Ant Colony Optimization is a metaheuristic that arose from the doctoral thesis of Marco Dorigo, [7]. This approximate solution imitates the behaviour of natural ants when looking for food. It has been proved that biological ants can find the shortest path between their nest and a food source, [6]. They do this by communicating among themselves through a substance called pheromone, which they put on the ground as they walk. When an ant must choose from among several paths, it prefers the paths with high amounts of pheromone. The paths with less pheromone are selected by fewer ants. Pheromone evaporates over time, making the paths selected by a small number of ants less and less desirable.

The first problem considered to prove ant-based algorithms was the well-known Traveling Salesman Problem (TSP). Let $G = (V, E)$ be a weighted graph, where V is the set of n points or cities in the problem, and $E = \{(i, j)/i, j \in V\}$ is the set of connections among cities. Each element (i, j) of E has a nonnegative cost, d_{ij}, associated with it. The aim of the TSP is to find a closed path of minimum cost passing through all the cities once and only once.

María-Luisa Pérez-Delgado
Escuela Politécnica Superior de Zamora. Universidad de Salamanca,
Av. Requejo, 33, C.P. 49022, Zamora, Spain
e-mail: mlperez@usal.es

Y. Demazeau et al. (Eds.): Advances in PAAMS, AISC 70, pp. 221–230.
springerlink.com

Among the ant-based algorithms, one of the best performing when applied to the TSP is the so called Ant Colony System (ACS), [8]. In [24], the ACS algorithm was applied to solve the Undirected Rural Postman Problem (URPP). This NP-hard problem will be stated in section 2. In short, the aim of the URPP is to find a closed path of minimum cost that traverses a set of connections of a graph at least once. To solve the URPP by artificial ants, it was transformed into a TSP, and this new problem was solved by applying ACS. When the TSP was solved, the TSP solution was transformed into a URPP solution. The results obtained were encouraging, but not better than the ones reached by other existing approximate methods.

In this paper another ant-based algorithm is applied: the Elitist Ant System (EAS) algorithm. The solution scheme is similar to the one proposed in [24], transforming the URPP into a TSP. The EAS algorithm generates shorter solutions than ACS. It also generates better solutions, on average, than many of the other approximate methods proposed for the problem.

2 The Undirected Rural Postman Problem

Let $G = (V, E)$ be an undirected weighted graph, where V is the set of points in the graph and $E = \{(i, j)/i, j \in V\}$ is the set of edges, which have a nonnegative cost associated with them. Let $F \subset E$, $F \neq \emptyset$, be a set of required edges. The aim of the URPP is to find a closed tour of minimum length of G containing at least once each edge in F, [22]. Lenstra proved that the problem is NP-hard, [20].

This problem arises in some practical situations, such as: the reading of meters, waste collection, street sweeping, mail delivery, school bus routing, snow plowing, or the optimization of the movements of a plotter. For a review of more applications [11] or [13] can be consulted.

There are several exact methods for solving the URPP. The most important are the solutions proposed by Christofides et al, [4]; Corberán and Sanchís, [27, 5]; Letchford, [21]; and Ghiani and Laporte, [15].

The complexity of the problem makes it necessary to apply approximate methods to solve large instances. One of the best known heuristics for the URPP is the one proposed by Frederickson, [13], which is similar to the heuristic proposed in [3] to solve the TSP. Hertz proposed a family of local search heuristics to improve the solutions obtained by Frederickson, [17, 18]. Groves and van Vuuren applied the 2-opt and 3-opt heuristics to improve the results obtained by Frederickson, [16]. In [23], Pearn and Wu proposed another two heuristics, making two transformations to the solution proposed in [4]. Fernández de Córdoba et al. employed a heuristic based on Monte Carlo methods, [12]. Ghiani and Laporte applied a constructive heuristic which, at each iteration, inserts a connected component of the required edges and performs a local post-optimization, [14]. Several metaheuristics have also been applied to try to solve the problem: genetic algorithms, [19]; memetic algorithms, [26]; the GRASP metaheuristic combined with genetic algorithms, [2]; simulated annealing combined with GRASP and genetic algorithms, [1]; neural networks, [25]; and artificial ants, [24].

3 The Elitist Ant System Algorithm

The EAS algorithm was proposed by Dorigo, [7], [9]. Let us consider a TSP graph with n cities. To solve the TSP by EAS, a set of m ants is considered, and a pheromone trail, τ_{ij}, is associated with every edge (i, j) of the graph. At the beginning of the algorithm, all the pheromone trails are set to the initial value τ_0. The pheromone will be updated as the algorithm proceeds, thus making communication among the ants possible.

Each ant must define a closed tour, visiting each city in the problem once. To define its solution, every ant starts from a randomly selected city. The ant determines the remaining stops in its tour, taking into account both the cost and the pheromone associated with the connections in the graph. Ants prefer to select connections with a low cost and a high amount of pheromone. When ant k is placed in city i, it selects city j as the next stop on its path with probability p_{ij}^k, according to Eq. (1), called the random proportional rule.

$$p_{ij}^k = \frac{\tau_{ij}^\alpha \cdot \eta_{ij}^\beta}{\sum_{l \in N_i^k} \left[\tau_{il}^\alpha \cdot \eta_{il}^\beta\right]} \text{ if } j \in N_i^k \tag{1}$$

$\eta_{ij} = 1/d_{ij}$ is the so-called visibility of the connection (i, j). N_i^k is the feasible neighborhood for ant k when it is placed in city i: the set of cities accessible from city i and not yet visited by the ant. α and β are two parameters that determine the relative influence of the pheromone trail and the visibility associated with the connections of the graph.

When ant k has visited the n cities in the problem, it goes back to the first city in its tour, thus defining a closed tour, S^k. When all the ants have defined a tour, the shortest one is selected as the best solution of the present iteration of the algorithm. If the iteration-best solution is shorter than the one obtained in the previous iterations, it is selected as the best solution so far, S^b.

When all the ants have found a solution, Eq. (2) is applied to update the pheromone associated with the graph. First, the pheromone of all the connections is reduced in a constant factor. Next, each ant k settles pheromone on the connections of the tour it has defined at the present iteration of the algorithm, S^k. Moreover, an amount of pheromone is settled on the connections belonging to the best solution found since the beginning of the algorithm, S^b. This amount is proportional to the number of elitist ants, e, considered by the algorithm.

$$\tau_{ij} = (1-\rho)\tau_{ij} + \sum_{k=1}^{m} \Delta\tau_{ij}^k + e\Delta\tau_{ij}^b \tag{2}$$

$\rho \in (0, 1)$ is the evaporation rate of the pheromone. The increase $\Delta\tau_{ij}^r$ equals $\frac{1}{C^r}$ for $(i, j) \in S^r$, and is 0 otherwise; with $r = b$ or $1 \leq r \leq m$, and C^r being the cost of the tour S^r.

Table 1 Characteristics of the test problems

PR.	$\|V\|$	$\|E\|$	$\|F\|$	*Best*	$\|V_{TSP}\|$	PR.	$\|V\|$	$\|E\|$	$\|F\|$	*Best*	$\|V_{TSP}\|$
p1	11	13	7	76	14	p14	28	79	31	202	62
p2	14	33	12	152	24	p15	26	37	19	441	38
p3	28	58	26	102	52	p16	31	94	34	203	68
p4	17	35	22	84	44	p17	19	44	17	112	34
p5	20	35	16	124	32	p18	23	37	16	146	32
p6	24	46	20	102	40	p19	33	55	29	257	58
p7	23	47	24	130	48	p20	50	98	63	398	126
p8	17	40	24	124	48	p21	49	110	67	366	134
p9	14	26	14	83	28	p22	50	184	74	621	148
p10	12	20	10	80	20	p23	50	158	78	475	156
p11	9	14	7	23	14	p24	41	125	55	405	110
p12	7	18	5	19	10	AlA	102	160	99	10599	198
p13	7	10	4	35	8	AlB	90	144	88	8629	176

Table 2 Result obtained when $\beta = 2$

	$e = 3$			$e = 4$			$e = 5$		
PR.	AV	SD	T(s)	AV	SD	T	AV	SD	T(s)
p1	76.00	0.00	0.05	76.00	0.00	0.05	76.00	0.00	0.00
p2	152.00	0.00	0.10	152.00	0.00	0.10	152.00	0.00	0.15
p3	102.15	0.67	0.75	102.00	0.00	1.00	102.00	0.00	0.90
p4	85.50	0.89	0.45	85.60	0.82	0.60	85.60	0.75	0.60
p5	124.00	0.00	0.15	124.00	0.00	0.15	124.00	0.00	0.15
p6	102.00	0.00	0.30	102.00	0.00	0.45	102.00	0.00	0.45
p7	130.00	0.00	0.50	130.30	0.73	0.75	130.10	0.45	0.60
p8	122.05	0.22	0.55	122.00	0.00	0.75	122.00	0.00	0.55
p9	83.00	0.00	0.10	83.00	0.00	0.10	83.00	0.00	0.10
p10	81.35	1.73	0.10	81.30	1.66	0.05	82.80	1.70	0.05
p11	23.00	0.00	0.00	23.00	0.00	0.05	23.00	0.00	0.05
p12	19.00	0.00	0.00	19.00	0.00	0.00	19.00	0.00	0.00
p13	35.00	0.00	0.00	35.00	0.00	0.00	35.00	0.00	0.00
p14	202.90	1.41	0.95	203.05	1.47	1.25	202.75	1.33	1.00
p15	441.00	0.00	0.25	441.00	0.00	0.25	441.00	0.00	0.25
p16	203.00	0.00	1.15	203.15	0.67	1.45	203.10	0.45	1.15
p17	112.00	0.00	0.20	112.00	0.00	0.20	112.00	0.00	0.20
p18	146.00	0.00	0.15	146.00	0.00	0.20	146.00	0.00	0.15
p19	258.35	1.50	0.90	258.10	1.62	1.15	258.35	1.69	0.80
p20	400.25	2.51	5.55	399.50	1.82	5.95	399.50	2.01	5.55
p21	371.50	2.67	11.30	372.25	4.69	11.10	370.15	3.48	9.90
p22	624.10	3.13	12.40	623.20	2.65	12.15	622.40	1.27	12.40
p23	477.05	1.99	14.15	477.00	1.84	12.25	477.00	1.92	11.90
p24	406.55	2.16	5.85	406.50	2.04	5.25	405.40	0.68	5.35
AlA	11118.35	176.26	40.50	11122.25	146.39	28.85	11125.55	180.44	32.10
AlB	8898.00	112.35	19.80	8906.95	121.16	21.65	8892.60	124.28	21.85

The process is repeated until the solution converges or the prefixed maximum number of iterations has been performed.

Ant-based algorithms usually apply 2-opt exchange to improve the path found by the ants, [10]. We will apply this exchange to the solution reached by every ant, before the pheromone update takes place.

4 Computational Results

The algorithm has been coded in C language. The tests have been performed on a personal computer with Intel Centrino Core 2 Duo processor, 2.2 GHz, with 2G RAM memory and working on Linux Operating System.

Table 1 shows information about the test problems considered, proposed in [4] and [5]: the number of nodes ($|V|$) and edges ($|E|$) in the graph; the number of required connections ($|F|$); the cost of the best solution known for the problem (*Best*),

Table 3 Result obtained when $\beta = 3$

	$e = 3$			$e = 4$			$e = 5$		
PR.	AV	SD	T(s)	AV	SD	T(s)	AV	SD	T(s)
p1	76.00	0.00	0.00	76.00	0.00	0.00	76.00	0.00	0.05
p2	152.00	0.00	0.15	152.00	0.00	0.15	152.05	0.22	0.15
p3	102.00	0.00	0.80	102.00	0.00	0.95	102.00	0.00	0.90
p4	85.70	0.73	0.65	85.60	0.82	0.65	85.15	0.99	0.65
p5	124.00	0.00	0.20	124.00	0.00	0.15	124.00	0.00	0.20
p6	102.00	0.00	0.45	102.00	0.00	0.40	102.00	0.00	0.40
p7	130.10	0.45	0.75	130.00	0.00	0.75	130.00	0.00	0.75
p8	122.00	0.00	0.80	122.00	0.00	0.75	122.05	0.22	0.70
p9	83.00	0.00	0.10	83.00	0.00	0.10	83.00	0.00	0.10
p10	81.25	1.59	0.10	82.10	1.62	0.05	82.40	1.67	0.05
p11	23.00	0.00	0.00	23.00	0.00	0.05	23.00	0.00	0.05
p12	19.00	0.00	0.05	19.00	0.00	0.00	19.00	0.00	0.00
p13	35.00	0.00	0.00	35.00	0.00	0.00	35.00	0.00	0.00
p14	202.90	1.41	1.20	202.75	1.33	1.30	202.75	1.33	1.15
p15	441.00	0.00	0.25	441.00	0.00	0.25	441.00	0.00	0.25
p16	203.00	0.00	1.45	203.00	0.00	1.50	203.00	0.00	1.35
p17	112.00	0.00	0.15	112.00	0.00	0.20	112.00	0.00	0.20
p18	146.00	0.00	0.20	146.00	0.00	0.15	146.00	0.00	0.15
p19	258.00	1.38	1.10	257.70	1.42	1.15	258.10	1.59	1.00
p20	399.30	1.49	6.60	399.40	2.16	6.40	399.40	2.26	6.25
p21	373.60	4.33	9.20	373.50	2.78	10.35	372.60	3.55	9.50
p22	622.95	1.79	11.40	622.30	2.77	10.50	623.55	2.21	11.25
p23	477.50	1.93	13.40	477.65	2.74	14.25	476.90	2.61	14.15
p24	406.10	0.91	5.20	406.80	2.69	6.30	405.75	1.71	5.15
AlA	11221.35	131.50	24.85	11113.90	173.87	33.60	11098.80	189.99	30.60
AlB	8926.25	131.64	18.65	8946.70	102.16	19.30	8849.25	102.92	23.50

and the number of nodes of the TSP associated with each URPP when we apply artificial ants to solve it ($|V_{TSP}|$).

Tables 2 to 6 show the results obtained when twenty independent runs were performed for each problem, and the following values were considered for the parameters: $\alpha = 1$, $\beta = \{2,3,4,5\}$, $\rho = 0.5$, and $\tau_0 = (e+m)/(\rho L_{NN})$, as proposed in [10]; $m = 40$ and $e = \{3,4,5\}$. L_{NN} is the length of a nearest-neighbor solution.

Table 4 Result obtained when $\beta = 4$

	$e = 3$			$e = 4$			$e = 5$		
PR.	AV	SD	T(s)	AV	SD	T(s)	AV	SD	T(s)
p1	76.00	0.00	0.05	76.00	0.00	0.05	76.00	0.00	0.05
p2	152.00	0.00	0.15	152.00	0.00	0.10	152.00	0.00	0.10
p3	102.25	0.79	0.95	102.00	0.00	0.90	102.25	0.79	0.85
p4	85.55	0.83	0.70	86.00	0.00	0.60	85.90	0.45	0.40
p5	124.00	0.00	0.15	124.00	0.00	0.20	124.00	0.00	0.20
p6	102.00	0.00	0.50	102.00	0.00	0.40	102.00	0.00	0.25
p7	130.00	0.00	0.75	130.00	0.00	0.70	130.00	0.00	0.45
p8	122.00	0.00	0.70	122.05	0.22	0.70	122.00	0.00	0.50
p9	83.00	0.00	0.10	83.00	0.00	0.15	83.00	0.00	0.10
p10	81.35	1.73	0.10	81.90	1.80	0.05	81.70	1.78	0.05
p11	23.00	0.00	0.05	23.00	0.00	0.05	23.00	0.00	0.05
p12	19.00	0.00	0.00	19.00	0.00	0.00	19.00	0.00	0.00
p13	35.00	0.00	0.00	35.00	0.00	0.00	35.00	0.00	0.00
p14	202.90	1.41	1.20	203.20	1.51	1.20	203.20	1.51	0.90
p15	441.00	0.00	0.30	441.00	0.00	0.25	441.00	0.00	0.25
p16	203.00	0.00	1.50	203.15	0.49	1.45	203.00	0.00	1.00
p17	112.00	0.00	0.20	112.00	0.00	0.15	112.00	0.00	0.15
p18	146.00	0.00	0.20	146.00	0.00	0.20	146.00	0.00	0.15
p19	257.85	1.60	1.00	258.40	1.70	1.00	257.60	1.19	0.85
p20	399.80	1.79	6.55	399.75	1.21	6.55	399.80	1.94	6.65
p21	372.70	2.49	10.55	372.75	3.52	11.30	371.70	2.05	10.55
p22	622.65	1.42	12.80	623.25	3.11	9.40	622.50	1.54	11.10
p23	476.75	1.89	13.25	527.75	224.62	12.40	477.45	1.64	11.10
p24	405.40	0.75	5.35	405.65	1.42	4.95	406.20	1.88	5.65
AlA	11147.65	129.31	33.65	11109.05	134.03	29.95	11084.80	133.21	32.10
AlB	8950.50	128.29	21.50	8928.05	102.59	23.55	8848.75	136.70	23.45

Each table from 2 to 5 shows the results for a value of β and the three values considered for e. Each row of the tables shows the results for a problem. The columns labeled as AV and SD show the average and the standard deviation, respectively, of the costs of the tours obtained for each problem. The column labeled T shows the average time in second to reach a solution.

All the problems proposed in [4] are solved to optimality for all the combinations of the parameters. Table 6 shows the best solution reached for problems proposed in [5]. Problem AlbaidaB is solved to optimality for two combinations of values of

Table 5 Result obtained when $\beta = 5$

	$e = 3$			$e = 4$			$e = 5$		
PR.	AV	SD	T(s)	AV	SD	T(s)	AV	SD	T(s)
p1	76.00	0.00	0.00	76.00	0.00	0.00	76.00	0.00	0.05
p2	152.00	0.00	0.10	152.00	0.00	0.15	152.00	0.00	0.15
p3	102.10	0.45	0.90	102.00	0.00	0.85	102.10	0.45	0.85
p4	85.50	0.83	0.70	85.05	1.00	0.70	85.65	0.75	0.65
p5	124.00	0.00	0.20	124.00	0.00	0.20	124.00	0.00	0.20
p6	102.00	0.00	0.45	102.00	0.00	0.45	102.00	0.00	0.40
p7	130.10	0.45	0.75	130.00	0.00	0.75	130.10	0.45	0.70
p8	122.00	0.00	0.70	122.00	0.00	0.75	122.00	0.00	0.65
p9	83.00	0.00	0.15	83.00	0.00	0.10	83.00	0.00	0.10
p10	81.45	1.85	0.05	81.85	1.76	0.10	81.40	1.79	0.10
p11	23.00	0.00	0.05	23.00	0.00	0.00	23.00	0.00	0.00
p12	19.00	0.00	0.00	19.00	0.00	0.05	19.00	0.00	0.00
p13	35.00	0.00	0.00	35.00	0.00	0.00	35.00	0.00	0.00
p14	202.60	1.23	1.30	202.75	1.33	1.15	202.60	1.23	1.15
p15	441.00	0.00	0.30	441.00	0.00	0.25	441.00	0.00	0.25
p16	203.00	0.00	1.40	203.00	0.00	1.40	203.00	0.00	1.20
p17	112.00	0.00	0.20	112.00	0.00	0.20	112.00	0.00	0.20
p18	146.00	0.00	0.15	146.00	0.00	0.20	146.00	0.00	0.15
p19	257.80	1.54	1.15	258.20	1.58	1.00	258.60	1.79	1.05
p20	400.35	2.13	5.75	400.20	1.74	5.45	399.50	1.57	4.80
p21	372.75	3.49	10.10	373.30	3.69	11.20	372.55	3.02	11.25
p22	622.75	1.52	9.25	622.20	1.74	11.40	622.60	1.98	11.75
p23	527.45	224.69	12.35	477.05	1.64	13.85	477.20	2.40	12.45
p24	405.75	1.45	4.75	405.55	1.00	5.60	405.55	0.89	4.80
AlA	11104.05	165.79	36.40	11079.50	187.76	29.70	11143.35	170.64	29.95
AlB	8911.55	121.07	22.35	8883.25	149.51	19.30	8845.15	113.67	22.10

Table 6 Best solution reached for problems proposed by Corberán et al.

	β	$e = 3$	$e = 4$	$e = 5$		β	$e = 3$	$e = 4$	$e = 5$
AlA	2	10699	10847	10799	AlB	2	8657	8657	8629
	3	11007	10727	10731		3	8657	8733	8677
	4	10879	10903	10807		4	8737	8797	8629
	5	10671	10775	10867		5	8733	8657	8657

β and e. For the problem AlbaidaA, never solved to optimality, the best solution reached has a percentage over the optimum equal to 0.68, and it is obtained when $\beta = 5$ and $e = 3$.

Table 7 compares the best solution reached by applying EAS, reported in the last column, with the solutions reported by several authors who have applied approximate methods to the same test set: [4] in the column labeled S1; [12] in column S2; [13] in column S3; [16] in columns S4 and S5; [17] in columns S6 and S7; [2] in

Table 7 Percentage over the optimum for the set of test problems by applying several approximate solutions

	S1	S2	S3	S4	S5	S6	S7	S8	S9	ACS	EAS
p01	0	0	0	0	0	0	0	0	0	0	0
p02	7.895	7.24	1.97	0.66	0	0	0	7.24	7.24	0	0
p03	0	0	2.94	0.98	0.98	0	0	0	0	0	0
p04	0	2.38	0	0	0	0	0	0	0	0	0
p05	8.87	4.03	4.84	0	0	0	0	4.03	4.03	0	0
p06	4.90	0	4.90	4.90	0	0	0	0	0	0	0
p07	0	0	0	0	0	0	0	0	0	0	0
p08	0	0	0	0	0	0	0	0	0	0	0
p09	1.21	0	0	0	0	0	0	0	0	0	0
p10	0	5.00	0	0	0	0	0	0	0	0	0
p11	0	0	13.04	0	0	0	0	0	0	0	0
p12	15.79	10.53	15.79	0	0	0	0	10.53	10.53	0	0
p13	8.57	8.57	0	0	0	0	0	8.57	8.57	0	0
p14	4.95	3.47	2.48	0.99	0	0	0	3.47	3.47	0	0
p15	0.91	0.91	0.91	0	0	0	0	0.91	0.91	0	0
p16	0	0	5.91	0.99	0	0	0	0	0	0	0
p17	3.57	0	3.57	0	0	0	0	0	0	0	0
p18	1.37	1.37	0.69	-	-	0.69	0.69	1.37	1.37	0	0
p19	8.95	2.34	6.62	5.45	3.50	0	0	2.34	2.34	1.56	0
p20	0.50	0.25	1.01	0.50	0.50	0.50	0	0	0	6.53	0
p21	1.64	0.55	1.64	1.64	1.64	0	0	1.64	0	7.92	0
p22	1.77	0	1.93	0.16	0.16	0.16	0	2.42	0	5.96	0
p23	1.05	2.95	0.84	0.42	0.42	0	0	2.53	1.05	7.58	0
p24	1.48	0	1.48	0	0	0	0	0	0	2.47	0
AlA	-	1.75	0	0	0	0	0	3.74	0.12	-	0.64
AlB	-	1.07	0	0	0	0	0	2.94	0	-	0

column S8; and [1] in column S9. The column labeled ACS shows the results obtained when Ant Colony System was applied, [24]. This table shows the percentage over the optimum of the best solutions reached. Cells in the table without a value represent values not reported by the author. Only the second solution obtained by Hertz can solve as many problems as EAS. We must point out that Hertz proposed two improvement heuristics applicable to the solution reached by Frederickson. Therefore, his starting point is a good solution to the problem, unlike what happens with EAS.

When comparing the two ant-based algorithms applied to the problem, we observe that EAS can solve to optimality more problems than ACS: EAS solved 25 out of 26 problems, whereas ACS solved only 18. Problems AlbaidaA and AlbaidaB, not considered in [24], are not solved to optimality when ACS is applied. The average time required to reach a solutions is shorter when EAS is applied (Figure 1).

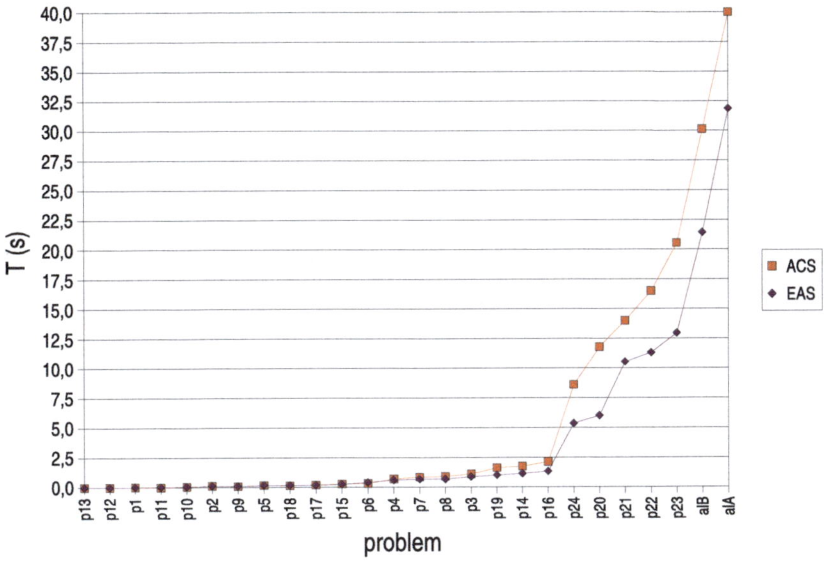

Fig. 1 Average time to reach a solution when ACS and EAS are applied

5 Conclusions

This paper shows the application of the EAS algorithm to the URPP. This algorithm generates good solutions for the set of test problems. It improves the results obtained when ACS is applied: more problems are solved and less time is required. Moreover, these results improve the ones obtained by many of the other approximate solutions proposed for the same problem.

To apply EAS, the input information required is defined by the cost matrix, the set of connections and the set of required connections of the URPP. Therefore, a special formulation of the problem is not necessary to solve it. Moreover, EAS can also be applied to the directed version of the problem, which is also a NP-hard problem.

Acknowledgements. This work has been partially financed by the Samuel Solórzano Barruso Memorial Foundation, of the University of Salamanca.

References

1. Baldoquín, M.G.: Heuristics and metaheuristics approaches used to solve the rural postman problem: a comparative case study. In: Proc. of the Fourth International ICSC Symposium on Engineering of Intelligent Systems (EIS 2004). Maderia, Portugal (2004)
2. Baldoquín, M.G., Ryan, G., Rodriguez, R., Castellini, A.: Un enfoque hibrido basado en metaheurísticas para el problema del cartero rural. In: Proc. of XI CLAIO, Concepción de Chile, Chile (2002)
3. Christofides, N.: Worst case analysis of a new heuristic for the traveling salesman problem. Technical Report 388, Graduate School of Industrial Administration, Carnegie Mellon University, Pittsburgh, PA (1976)

4. Christofides, N., Campos, V., Corberán, A., Mota, E.: An algorithm for the rural postman problem. Technical Report IC-OP-81-5 (1981)
5. Corberán, A., Sanchís, J.M.: A polyhedral approach to the rural postman problem. Eur. J. Oper. Res. 79, 95–114 (1994)
6. Deneubourg, J.-L., Aron, S., Goss, S., Pasteels, J.-M.: The self-organizing exploratory pattern of the argentine ant. J. Insect Behav. 3, 159–168 (1990)
7. Dorigo, M.: Optimization, learning and natural algorithms. Ph.D. Thesis, Dip. Elettronica, Politecnico di Milano (1992)
8. Dorigo, M., Gambardella, L.M.: Ant colony system: A cooperative learning approach to the traveling salesman problem. IEEE Trans. Evol. Comput. 1(1), 53–66 (1997)
9. Dorigo, M., Maniezzo, V., Colorni, A.: Ant system: Optimization by a colony of cooperating agents. IEEE Trans. Systems, Man, Cybern.- B 26(1), 29–41 (1996)
10. Dorigo, M., Stützle, T.: Ant Colony Optimization. The MIT Press, Cambridge (2004)
11. Eiselt, H.A., Gendreau, M., Laporte, G.: Arc routing problems, Part II: The rural postman problem. Oper. Res. 43(3), 399–414 (1995)
12. Fernández, P., García, L.M., Sanchis, J.M.: A heuristic algorithm based on monte carlo methods for the rural postman problem. Comput. Oper. Res. 25(12), 1097–1106 (1998)
13. Frederickson, G.N.: Approximation algorithms for some postman problems. J. ACM 26(3), 538–554 (1979)
14. Ghiani, G., Laganà, D., Musmanno, R.: A constructive heuristic for the undirected rural postman Problem. Comput. Oper. Res. 33(12), 3450–3457 (2006)
15. Ghiani, G., Laporte, G.: A branch-and-cut algorithm for the undirected rural postman Problem. Math. Program 87(3), 467–481 (2000)
16. Groves, G.W., van Vuuren, J.H.: Efficient heuristics for the rural postman problem. Orion 21(1), 33–51 (2005)
17. Hertz, A., Laporte, G., Nanchen-Hugo, P.: Improvement procedures for the undirected rural postman problem. INFORMS J. Comput. 11(1), 53–62 (1999)
18. Hertz, A., Mittaz, M.: Heuristic Algorithms. In: Arc Routing: Theory, Solutions and Applications, pp. 327–386. Kluwer Academic Publishers, Dordrecht (2000)
19. Kang, M.-J., Han,C.-G.: Solving the rural postman problem using a genetic algorithm with a graph transformation. Technical Report, Dept. of Computer Engineering, Kyung Hee University (1998)
20. Lenstra, J.K., Rinnooy-Kan, A.H.G.: On the general routing problem. Networks 6(3), 273–280 (1976)
21. Letchford, A.N.: Polyhedral results for some constrained arc routing problems. Ph.D. Thesis, Lancaster University, Lancaster (1996)
22. Orloff, C.S.: A fundamental problem in vehicle routing. Networks 4, 35–64 (1974)
23. Pearn, W.L., Wu, C.M.: Algorithms for the rural postman problem. Comput. Oper. Res. 22, 815–828 (1995)
24. Pérez-Delgado, M.L.: A solution to the rural postman problem based on artificial ant colonies. In: Borrajo, D., Castillo, L., Corchado, J.M. (eds.) CAEPIA 2007. LNCS (LNAI), vol. 4788, pp. 220–228. Springer, Heidelberg (2007)
25. Pérez-Delgado, M.L., Matos-Franco, J.C.: Self-organizing feature maps to solve the undirected rural postman problem. In: Moreno Díaz, R., Pichler, F., Quesada Arencibia, A. (eds.) EUROCAST 2007. LNCS, vol. 4739, pp. 804–811. Springer, Heidelberg (2007)
26. Rodrigues, A.M., Ferreira, J.S.: Solving the rural postman problem by memetic algorithms. In: MIC 2001 - 4TH Metaheuristics International Conference, Porto, Portugal (2001)
27. Sanchis, J.M.: El poliedro del problema del cartero cural. Ph.D. Thesis, Universidad de Valencia, Spain (1990)

Distributed Bayesian Diagnosis for Telecommunication Networks

Andrés Sedano-Frade, Javier González-Ordás, Pablo Arozarena-Llopis, Sergio García-Gómez, and Alvaro Carrera-Barroso

Abstract. In this paper an innovative approach to telecommunication network and service management is presented. It tries to solve some of the main challenges in this area by combining distributed intelligent agents with probabilistic techniques. The solution focuses on fault diagnosis but could potentially be extended to other management areas. Moreover, a prototype has been implemented and applied to a specific networking scenario. Finally, an evaluation of the reduction of fault diagnosis time by using this prototype has been conducted, showing significant advantages when compared to traditional manual processes.

1 Introduction

Managing telecommunication networks and services is an essential task for a telco operator. It involves many different processes, including diagnosing and troubleshooting problems whenever they are detected or reported. Doing this in an efficient manner is very important for the competitiveness of such companies and for the quality of the service provided to their subscribers. The systems developed in the last decades to support network technicians while performing management tasks are called Operating Support Systems (OSS). In order to reduce operating expenses, telecommunication operators try to minimise the need of human intervention in the management processes.

One of the most important management tasks is the diagnosis of telecommunications networks and services. Whenever a failure is detected, it is reported to a trouble ticketing system, which serves as a task queue for the technical departments in charge of solving it. In order to increase the automation levels of current systems, different Artificial Intelligence techniques

Andrés Sedano Frade · Javier González Ordás · Pablo Arozarena Llopis ·
Sergio García Gómez · Alvaro Carrera Barroso
Telefónica I+D, Spain
e-mail: {andress,javiord,pabloa,sergg,alvaroc}@tid.es

Y. Demazeau et al. (Eds.): Advances in PAAMS, AISC 70, pp. 231–240.
springerlink.com

have been applied to develop OSS's. In the case of diagnosis and troubleshooting tasks, achieving real automation means using systems able to detect and diagnose failures by themselves, and to trigger corresponding repair actions. To achieve this objective, two of the main drawbacks of current diagnosis systems are scalability and management of incomplete or inaccurate information [12]. In order to overcome these challenges, KOWGAR is based on an overlay network of intelligent agents which use bayesian inference as a reasoning mechanism.

The multi-agent approach to build telecommunications diagnosis systems is very common in academic proposals [9], even for Internet based diagnosis [11, 8]. In the autonomic communications [4] and the cognitive networks [5] fields, multi-agent architectures are also considered. However, the practical dificulties of deploying them in real systems hamper their usage. To overcome this issue, KOWGAR has been designed to deploy agents both in already existing IT infrastructure and network devices.

Although the most common reasoning aproach in real diagnosis systems is based on deterministic expert systems, there are a number of proposals that rely on probabilistic reasoning [3, 1]. Furthermore, some papers have been published in the field of probabilistic reasoning using intelligent agents [12]. Regarding network diagnosis, [8] proposes a way of diagnosing problems in the Internet. Since these proposals do not take into account the need to make partial diagnosis in local domains and merge all the results in wider domains, KOWGAR proposes a way of partitioning bayesian networks and distribute inference intelligence among a set of distributed agents.

The second section of the article analyzes the above mentioned challenges and the third section describes the KOWGAR platform. The fourth section summarizes how the platform was adapted to manage Telefonica's MacroLAN service. Finally, in the fifth section the main conclusions about the application of this approach are remarked, together with a description of future evolution paths.

2 Challenges in Networks Diagnosis

One of the key challenges for the next decades in telco operators is to reduce the manpower required to operate network and services. In fact, in the last years the Autonomic Communications initiave has been launched to coordinate the research efforts to create self-managed networks and communications infrastructure [4].

The KOWGAR approach faces the automation issue with regard to the troubleshooting processes in a practical manner. Tightly integrated with existing infrastructure and systems, the objective is to deploy intelligent agents which are able to automate some of the tasks that are currently performed by expert technicians.

While solving this challenge, two other issues arise. First, the decision systems developed to automate the management tasks must handle uncertainty as human technicians do. Second, in order to foster scalability, problems should be handled locally as far as possible, trying to avoid centralized solutions.

2.1 Uncertainty

There are many reasons why information is uncertain for a diagnosis system [12]. Topology and inventory data is scattered among multiple systems and it is not always complete neither consistent. The complexity of the network environment makes it .very difficult or even impossible to observe certain aspects of the domain, and moreover, the relationships between domain events are not always deterministic, particularly between observable and nonobservable ones. Besides, it is often impractical to model and analyze explicitly all the dependencies.

When there are failures, the behaviour of OSS systems is uncertain since they have been conceived to work deterministically. The observations that can be obtained from the domain, when it is feasible to get them, are sometimes innacurate or vague. In other cases, it is not possible to get observations because there are not enough observation resources or they are not accessible. Therefore, the state of the network or the service must be guessed from partial or innacurate information.

In this context, a deterministic system, such as those based on rules or on diagnosis processes, is not able to get accurate results in all cases [3]. Moreover, uncertainty makes efficient automation of troubleshooting tasks impossible.

2.2 Scalability

The classical diagnosis processes usually entails getting observations from remote resources and systems. This approach is limited by the domains reachability from the OSS and the scalability required to manage a huge number of user network domains and devices. Therefore, a new approach is required [2].

In KOWGAR different specialized agents have been designed to remotely gather evidences from the network and related systems. However, one of the biggest challenges is to take into account the customers's infrastructure in the diagnosis process. KOWGAR allows to deploy agents inside a remote network, with full visibility over local devices and their configuration. Thus they can get evidences from that environment and share them with the rest of the agents. This highly distributed approach enables the system to solve problems locally, without involving resources from central systems and thus providing a solution to the scalability handicap of more centralized designs.

3 Description of the KOWGAR Platform

3.1 KOWGAR Principles

KOWGAR is a distributed agent platform for bayesian fault diagnosis in telecommunication networks. KOWGAR's architecture is based on the following design principles:

- Ability to diagnose problems under uncertainty. This is achieved by using Bayesian inference techniques, in particular Bayesian Networks (BNs) [7].
- Reusability: KOWGAR has been conceived as a generic diagnosis platform, easy to adapt to multiple network and service scenarios.
- Automation of diagnosis, thanks to fault surveillance agents that monitor the status of network components and trigger the diagnosis process every time a fault is detected.

3.2 KOWGAR Architecture

KOWGAR has adopted a distributed agent architecture and an ontology-based agent collaboration model. The main components of this architecture are described in the following paragraphs:

- Ontology: KOWGAR has defined a generic diagnosis ontology, which can be instantiated in different scenarios. This ontology contains two types of information:
  1. BN model: the ontology models all fault hypothesis, available evidences and causal probabilistic dependencies between them. The models are built after capturing knowledge from domain experts through an interview and information exchange process. This knowledge is formalized using a BN design tool and converted into an ontology that is fed to all diagnosis agents.
  2. Agent interactions model: the ontology defines the knowledge (obtained evidences and computed probabilities) shared by the agents during a diagnosis operation, as well as the actions that agents can perform.
- Local inference: When a bayesian inference procedure is triggered, a BN is constructed based on the working ontology of the agent, and the available evidences for the given diagnosis operation. There are two types of nodes in KOWGAR BNs. Hypothesis nodes represent all possible root causes of faults, while evidence nodes represent observations that can be taken either by performing a test or querying an external system. Inference is done by obtaining the value of available observations and deriving from them the probability of each hypothesis (so called beliefs). The beliefs obtained from the probabilistic reasoning process are then incorporated into the ontology, so that they can be shared with other agents or stored into the database. Inference is done in a iterative way: whenever a new

evidence is available[1], Bayesian inference is performed. This is repeated until all evidences have been collected or a given confidence threshold has been reached.

- Distributed inference: As the complexity of the diagnosis scenarios grows, scalability may become a problem. To address it, KOWGAR supports the distribution of the inference process by partitioning a complex BN into smaller BNs [12]. Partitioning is done with the Virtual Evidence Method (VEM)[10] algorithm. Each agent is only aware of a fragment of the entire BN, and is able to perform local inference and cooperate with neighbouring agents to reach a valid conclusion. This approach allows each agent to focus on specific areas. For instance, some agents may only diagnose network problems while others service problems. It also facilitates reusing knowledge in other diagnosis processes that share common parts of the overall BN.
- Self-Learning: KOWGAR implements the Expectation Maximization (EM) learning algorithm [6] to enable a dynamic and automatic update of BNs based on the feedback provided by its users ('Good' or 'Bad'), or by external post-mortem analysis systems. EM does not change the structure of a BN; instead, it improves diagnosis quality by changing the conditional probabilities between nodes. EM uses an iterative algorithm that estimates the missing values in the input data representing previous diagnoses. This is relevant since for some diagnoses there may be only a subset of the possible evidences available. Once this is done, statistical methods are used to recalculate the weights in the BN based on the set of previous diagnoses.
- Agents: The following types of agents can be identified:
  - Yellow Pages agent enables agents to discover each other, providing information about agents instantiated in the platform, and their capabilities.
  - Observation agents provide specific information (evidences) about the status of network components. Evidences are obtained by either conducting a test or querying a system. Observation agents publish their capabilities in the Yellow Pages so that other agents know which evidences are available, and their cost (used by diagnosis agents to select which observation to get next).
  - Diagnosis agents are in charge of orchestrating the diagnosis process and gathering evidences from the observation agents in order to infer the root cause. This process is driven by the BN instantiated for a particular diagnosis scenario.
  - Knowledge agents update the system knowledge base, perform self-learning and notify all subscribed diagnosis agents when a BN has been modified.
  - Fault surveillance agents detect faults in the network and trigger the diagnosis process.

[1] The ordering of observations depends on the value and cost of each one.

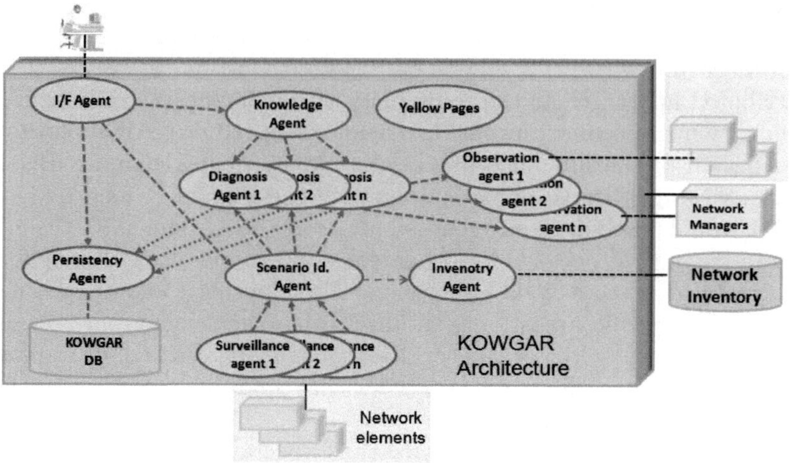

Fig. 1 KOWGAR architecture

- Auxiliary agents have a supporting role in KOWGAR, performing specific tasks.

Figure 1 shows KOWGAR agents and how they interact.

3.3 System Development

KOWGAR development relies to a great extent on the following Open Source tools:

- Web Server: Apache was selected because it is one of the most used and reliable web servers.
- Database: KOWGAR stores a large amount of diagnosis results in a MySQL database.
- MultiAgent Platform: the basis for the multiagent architecture is WADE/JADE[2].
- SamIam[3] is a bayesian inference library that was chosen because it is a Java native implementation so its integration with JADE agents was rather simple and Genie (SMILE) for the edition of BNs.
- Ontology BeanGenerator[4], a Protégé plugin for JADE has been used to implement the agents' ontology.

[2] http://jade.tilab.com/
[3] http://reasoning.cs.ucla.edu/samiam/
[4] http://protege.cim3.net/cgi-bin/wiki.pl?OntologyBeanGenerator

4 KOWLAN Scenario

MacroLAN is Telefonica's solution to build Virtual Private Networks (VPN) that connect multiple enterprise sites over Ethernet based accesses. It supports service speeds from 2 Mbit/s to 1 Gbit/s. By using standard L2 and L3 VPN technologies, MacroLAN enables geographically distant customer sites to communicate as if they belonged to the same LAN, in terms of speed, reliability and transparency. The MacroLAN service is built on top of a diverse set of Telefonica networks. In the local loop from the customer site to the Central Office, it can use either fiber or copper. Ethernet converters or Synchronous Digital Hierarchy (SDH) circuits allow to extend the distance of this access segment. Copper based accesses may also rely on Telefonica's leased line network. MacroLAN traffic is then first aggregated into province-wide Metropolitan Ethernet networks (MANs) and into an IP backbone network that provides national coverage.

Traditionally, MacroLAN diagnosis has followed a centralized approach and is based on remote access to information and test capabilities to come up with a conclusion. KOWLAN is the result of applying KOWGAR concepts and platform to this scenario.

4.1 Architecture

The KOWLAN physical architecture consists of three main blocks:

- Web user interface: it receives diagnosis requests from users and presents the diagnosis results. Users can provide feedback by approving or rejecting the results.
- Multiagent platform: it is in charge of receiving diagnosis requests from the Web interface or Trouble ticketing systems. It perfoms the diagnosis process and stores diagnosis results into the database.
- Database: it mantains a log of diagnosis results. This results can be shown to users by means of the web user interface. They are also used for self-learning purposes.

The KOWLAN logical architecture uses two of the main principles of KOWGAR: Bayesian networks and Distributed agents. In the multiagent platform several types of agents have been created. Two of them are of special relevance:

- Diagnosis agents: in KOWLAN there are six Diagnosis Agents, one per scenario: Fiber, Sdh, and so on.
- Observation agents: there is a set of 18 Observation Agents defined in KOWLAN. Each one of them can obtain one or more evidences.

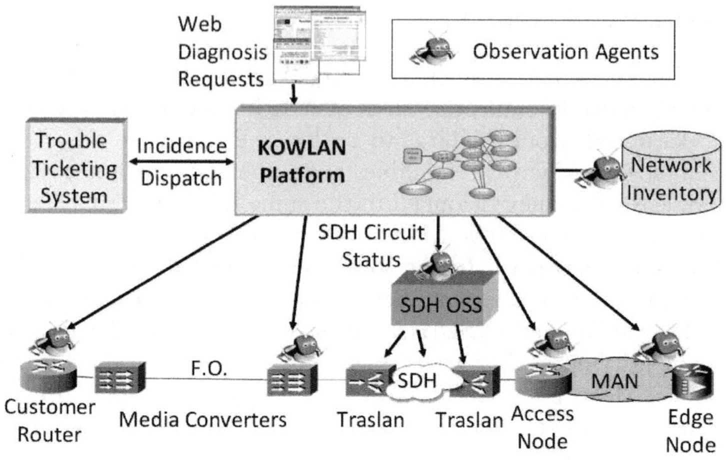

Fig. 2 Top level architecture for a Kowlan SDH scenario

There are other types of auxiliary agents created to resolve the specific requirements of KOWLAN:

- Interface agent: to receive requests from the Web interface and send them to the Switch agent.
- Switch agent: to receive requests from the interface and trouble ticketing agents, identify the topology scenario and route requests to specialized diagnosis agents.
- Inventory agent: to obtain topology scenario parameters from external data systems. These parameters will be used by observation agents to execute tests.
- Trouble ticketing agent: to collect problems from external trouble ticketing systems, and send diagnosis requests to the Switch agent.
- Persistency agent: to store diagnosis results and parameters in the database.

4.2 Contributions

The implementation of KOWLAN was developed in just four months, thanks to the high adaptability of the KOWGAR architecture and to the simplicity of Bayesian network modelling (when compared with rule of flow based reasoning systems). Since KOWLAN started to be used in a real environment, over two thousand diagnosis requests have been fulfilled. Most of them are caused by communication failures with the customer's premises. The second most frequent cause is the decrease in communications speed. KOWLAN takes between one and four minutes to perform an automatic diagnosis, depending on the complexity of the MacroLan scenario and on the number of tests to perform. This is a significant improvement over the performance of manual diagnosis. Since KOWLAN was installed, diagnosis time has dropped an average of 23 minutes, and KOWLAN has contributed to a reduction of 54 minutes in service recovery time. In addition, KOWLAN has reached a high accuracy in diagnosis, with over 38% of very reliable diagnoses, as shown in table 1.

Table 1 Diagnostic Ranking According to its Reliability

Conclusion	Reliability	Frequency
Reliable	$\geq$ 95%	38%
Likely	40% - 90%	18%
Uncertain	$\leq$40%	44%

5 Conclusions and Future Work

In this paper we have demonstrated how probabilistic reasoning in combination with multi-agent systems can be successfully applied to a relevant networking scenario. The evaluation of this prototype shows how such an approach is able to reach valid conclusions without compromising the realibility of the diagnosis process. Besides, deploying a distributed management architecture is facilitated by the use of a multi-agent framework where agents share knowledge and diagnosis capabilities across the network.

Still there are a number of areas of improvement that need to be addressed in the future:

Firstly, in other to further improve the automation of network management tasks, the results of the fault diagnosis process should serve as the basis for triggering self-healing actions. This will encompass designing a new type of agents with the capability of reconfiguring network parameters. In addition, monitoring how well these actions improve network status may be further used to enable self learning features.

Secondly, different management areas could be addressed by applying similar concepts. So far KOWGAR has focused on Fault Management, but we plan to explore its application to performance and configuration management as well.

Thirdly, we should not restrict our work to just Bayesian reasoning as inference mechanism. On the contrary, other Artificial Intelligence technologies that enable reasoning under uncertainty coming from different fields like, for example, biologically inspired algorithms, should also be explored.

Finally, the application of KOWGAR's principles to different network technologies is of utmost importance to demonstrate the generality of the solution. Especially LTE networks represent a significant oportunity, since they are being designed with embedded self-organisation features. Another challenging networking scenario where uncertainty and distribution of management intelligence plays a significant role is the emerging field of Home Area Networks.

References

1. Barco, R., Lazaro, P., Wille, V., Diez, L.: Knowledge acquisition for diagnosis in cellular networks based on bayesian networks. In: Lang, J., Lin, F., Wang, J. (eds.) KSEM 2006. LNCS, vol. 4092, p. 55. Springer, Heidelberg (2006)
2. Bellifemine, F., Caire, G., Greenwood, D., Corporation, E.: Developing multi-agent systems with JADE. Springer, Heidelberg (2007)
3. Ding, J., Krämer, B., Bouvry, P., Guan, H., Liang, A., Davoli, F.: Probabilistic fault management. In: Dargie, W. (ed.) Context-Aware Computing and Self-Managing Systems, pp. 309–347. Chapman & Hall/CRC Press, Boca Raton (2009)
4. Dobson, S., Denazis, S., Fernández, A., Gaïti, D., Gelenbe, E., Massacci, F., Nixon, P., Saffre, F., Schmidt, N., Zambonelli, F.: A survey of autonomic communications. ACM Transactions on Autonomous and Adaptive Systems (TAAS) 1(2), 223–259 (2006)
5. Fortuna, C., Mohorcic, M.: Trends in the development of communication networks: Cognitive networks. Computer Networks 53(9), 1354–1376 (2009)
6. Hastie, T., Tibshirani, R., Friedman, J.: The elements of statistical learning. In: The EM Algorithm, ch. 8.5, pp. 236–243. Springer, Heidelberg (2001)
7. Kjærulff, U., Madsen, A.: Bayesian networks and influence diagrams: a guide to construction and analysis. Springer, Heidelberg (2007)
8. Lee, G.: CAPRI: a common architecture for distributed probabilistic Internet fault diagnosis. Ph.D. thesis, Dept. of Electrical Engineering and Computer Science (Massachusetts Institute of Technology) (2007)
9. Lesser, V.R. (ed.): Encyclopedia of Computer Science, chap. Multi-agent systems, pp. 1194–1196. John Wiley and Sons Ltd., Chichester (2003)
10. Pan, R., Peng, Y., Ding, Z.: Belief update in Bayesian networks using uncertain evidence. In: Proceedings of the 18th IEEE International Conference on Tools with Artificial Intelligence, Citeseer, pp. 441–444 (2006)
11. Sugawara, T., Murakami, K., Goto, S.: A multi-agent monitoring and diagnostic system for TCP/IP-based network and its coordination. Knowledge-Based Systems 14(7), 367–383 (2001)
12. Xiang, Y.: Probabilistic Reasoning in Multiagent Systems: A graphical models approach. Cambridge University Press, Cambridge (2002)

Solving an Arc-Routing Problem Using Artificial Ants with a Graph Transformation

María-Luisa Pérez-Delgado

Abstract. In a recent paper artificial ants were applied to solve the Undirected Rural Postman Problem. The problem was transformed into a Traveling Salesman Problem and artificial ants were applied to this new problem. This paper applies another transformation, proposed by Pearn et al, which also transforms the arc-routing problem into a node routing problem.

1 Introduction

The Undirected Rural Postman Problem (URPP) appears in a variety of practical situations, such as mail delivery, street patrolling, school bus routing or electrical lines inspection [6]. Several exact solutions have been proposed to solve the problem [2], [3], [7], [12]. Since the problem is NP-hard [13], diverse approximate algorithms have been applied to it [1], [2], [5], [6], [9], [10].

In a recent work the application of artificial ants was proposed to solve the URPP [14]. The ant-based algorithms have been successfully applied to solve the Traveling Salesman Problem (TSP) [16], a well-known NP-hard combinatorial optimization problem. Therefore, to apply ants to the URPP, it was transformed into a TSP.

Arc-routing and node-routing problems have a lot of practical applications. Every problem variant in one class has a mirror image in the other one, thus making it possible to translate research results between the two. There exist problems for which research results are much more impressive for node routing than for their arc routing counterparts. The most notable of these cases is probably the pair Capacitated Vehicle Routing Problem (CVRP) [11] / Capacitated Arc Routing Problem (CARP) [8].

María-Luisa Pérez-Delgado
Escuela Politécnica Superior de Zamora, Universidad de Salamanca, Av. Requejo, 33, C.P. 49022, Zamora, Spain
e-mail: `mlperez@usal.es`

Y. Demazeau et al. (Eds.): Advances in PAAMS, AISC 70, pp. 241–246.
springerlink.com

Let $G = (V,E)$ be a graph, where V is the set of nodes and E is the set of connections. c is a function which associates a cost to each element of E. There is a distinguished depot node and a fleet of K vehicles, each having a capacity Q. In the CARP, each connection of E has a demand associated with it, and there is a set of required connections, $F \subset E$, with positive demand. The problem consists of determining a set of K feasible vehicle trips that minimizes the total cost of traversed connections. Each trip starts and ends at the depot, each required connection is serviced by one single vehicle and the total demand serviced by any trip must not exceed vehicle capacity. In the CVRP, all nodes but the depot are the customers, and a demand is associated to each customer. The objective of the problem is to design a least cost set of K routes, all starting and ending at the depot, where each customer is visited exactly once. The total demand of all customers on a route must not exceed vehicle capacity. The CARP and the CVRP are NP-hard problems.

The existing exact methods to solve the CARP are limited to very small problems. This is why larger instances must be tackled in practice using heuristics or metaheuristics. Pearn et al proposed a transformation to reduce a CARP into an equivalent CVRP [15], that allows to solve the CARP via CVRP algorithms.

The URPP and the undirected TSP are special cases of the CARP and the CVRP, respectively, where one vehicle of unlimited capacity is considered and the connections of the graph are undirected. Therefore, the transformation proposed by Pearn et al can be applied to the URPP. In this paper we present the results obtained when this transformation is applied to the URPP and the resulting problem is solved by the ant-algorithm called Ant Colony System (ACS).

2 Transformation of the URPP into an Undirected TSP

When the transformation proposed by Pearn et al is applied to the URPP, an undirected TSP is obtained. A new graph $G' = (V',E')$ is constructed. Each edge $(i,j) \in F$ is replaced with three nodes: s_{ij}, s_{ji} (called side nodes) and m_{ij} (called middle node). One can think of the new nodes as being placed along the original edge at equal spacings (Fig. 1). The new problem is defined on the set of nodes:

$$V' = \bigcup_{(i,j)\in F} \{s_{ij}, s_{ji}, m_{ij}\} \tag{1}$$

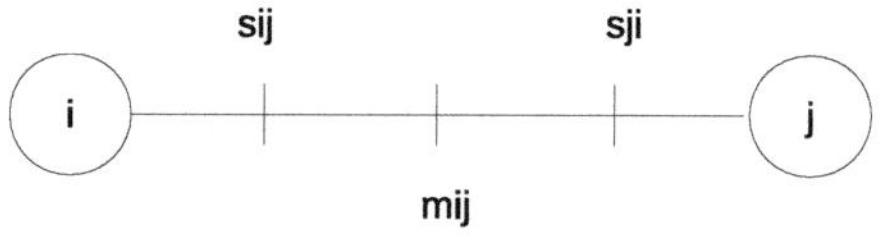

Fig. 1 Edge $(i,j) \in F$ and the new nodes associated with it

E' includes the connections between each pair of nodes in V'. The cost of these connections is calculated according to expressions (2) and (3), where $d(i,k)$ is the cost of a shortest path from node i to node j in G, calculated using the cost c. The expression (2) determines the distance between two side nodes. The expression (3) determines the distance from a middle node to another node; this value is only finite between a middle node an the side nodes associated with the same edge of G. This expression ensures that a middle node m_{ij} is only accessible from nodes s_{ij} and s_{ji}.

$$c'(s_{ij},s_{kl}) = \begin{cases} \frac{1}{4}(c_{ij}+c_{kl})+d(i,k) & \text{if } (i,j) \neq (k,l) \\ 0 & \text{if } (i,j) = (k,l) \end{cases} \tag{2}$$

$$c'(m_{ij},v) = \begin{cases} \frac{1}{4}c_{ij} & \text{if } v = s_{ij} \text{ or } v = s_{ji} \\ \infty & \text{otherwise} \end{cases} \tag{3}$$

When the TSP is solved, a sequence of nodes is obtained. In a feasible solution any middle node, m_{ij}, must always be visited in the sequence s_{ij}, m_{ij}, s_{ji} or s_{ji}, m_{ij}, s_{ij}. Therefore, the solution includes node triplets corresponding to the required connections of the original graph. To transform the TSP solution into a URPP solution, every node triplet is replaced by the associated edge of G. The edge is traversed in the same order as the node triplet is visited: if the first node of the triplet is s_{ij}, the edge is traversed from i to j; otherwise, it is traversed from j to i. Once all the triplets have been replaced by edges of G, the shortest path between each pair of edges must be included in the URPP solution.

3 The Ant Colony System Algorithm

The ant algorithms are based on the behavior of natural ants. Ants communicate among themselves by a substance called pheromone, which they deposit on the ground when they walk. Ants prefer moving by the paths having more pheromone, contributing in this way to accumulate more pheromone. The pheromone evaporates over time, making less desirable the paths selected by a small number of ants.

Several ant algorithms have been proposed and initially tested on the TSP. One of the best performing is the ACS algorithm [4]. We consider a set of m ants that cooperate in the search of a solution to the TSP (a tour), and a pheromone trail, τ_{ij}, associated to each connection (i,j) of the TSP graph. Each ant generates a complete tour starting from a city selected randomly and selecting the next city by means of a probabilistic state transition rule. Let k be an ant placed on node i, $q_0 \in [0,1]$ a parameter and q a random value uniformly distributed in the interval $[0,1]$. The next stop of the path, j, is selected randomly by means of the following probability distribution:

If $q \leq q_0$:

$$p_{ij}^{k} = \begin{cases} 1 & \text{if } j = arg\ \max_{l \in N_k^i} \left\{ \tau_{il}^{\alpha} \cdot \eta_{il}^{\beta} \right\} \\ 0 & \text{otherwise} \end{cases} \tag{4}$$

If $q > q_0$:

$$p_{ij}^k = \begin{cases} \frac{\tau_{ij}^{\alpha} \cdot \eta_{ij}^{\beta}}{\sum_{l \in N_i^k} \left[\tau_{il}^{\alpha} \cdot \eta_{il}^{\beta} \right]} & \text{if } j \in N_i^k \\ 0 & \text{otherwise} \end{cases} \tag{5}$$

η_{ij} is the inverse of the distance associated to the connection (i, j). N_i^k, is the set of cities accessible from city i and not yet visited by the ant k. The parameters α and β determine the relative influence of the pheromone and the distance, respectively.

The pheromone is updated locally when each ant h builds its solution. The ant deposits pheromone on the connections of the tour it has defined, S_h, by applying the expression (6), where τ_0 is the inverse of the length of a tour calculated by applying the nearest-neighbor heuristic, and $0 < \rho_L < 1$ is a value of local persistence.

$$\tau_{ij} = \rho_L \tau_0 + (1 - \rho_L)\tau_{ij} \ \forall (i, j) \in S_h \tag{6}$$

The pheromone is updated globally when each iteration ends. The trail is updated on the connections of the graph belonging to the best global tour (the best tour from the beginning of the algorithm), S_{h_g}, by applying the expression (7), where L_{h_g} is the length of the best global tour and ρ is the global persistence.

$$\tau_{ij} = (1 - \rho)\tau_{ij} + \frac{\rho}{L_{h_g}} \ \forall (i, j) \in S_{h_b} \tag{7}$$

The process is repeated until the solution converges or the prefixed maximum number of iterations has been performed.

4 Computational Results

The proposed solution has been coded using C language. The tests have been performed on a personal computer with Intel Centrino Core 2 Duo processor, 2.2 GHz, with 2G RAM memory and working on Linux Operating System. The solution has been applied to the benchmark problems proposed in [2] and [3]. Fifty independent tests have been performed for each problem, considering the next values for the parameters: $\alpha = 1$, $\beta = 2$, $\rho = \rho_L = 0.1$, $q_0 = 0.9$. The initial pheromone trail takes random values in the interval $(0, 1]$. To obtain a feasible solution, the TSP tour must include a sequence of triplets, each one representing a required connection of the URPP (that is, a middle node must always appear between the two side nodes associated to the same URPP edge). Therefore, the first city selected by an ant must be always a side node; this allows the ant to select next the middle node associated to that side node.

Table 1 summarizes the results. It shows the name of the RPP problem, the number of nodes, $|V|$, the number of required, $|E_r|$, and not required, $|E_{nr}|$, edges, as well as the best known solution for each problem, OPT. The last two columns show the best solution obtained when the ACS algorithm is applied: column labeled ACS_1 shows the results obtained when the transformation proposed in [14] is applied, and

Table 1 Best solution for the set of sample problems by applying different methods ('-' represents values not reported by the authors)

PR.	$\|V\|$	$\|E_r\|$	$\|E_{nr}\|$	OPT	[2]	[5]	[6]	[9]	[10]	[1]-a	[1]-b	ACS_1	ACS_2
p01	11	7	6	76	76	76	76	76	76	76	76	76	76
p02	14	12	21	152	164	163	155	153	152	163	163	152	152
p03	28	26	32	102	102	102	105	103	103	102	102	102	102
p04	17	22	13	84	84	86	84	84	84	84	84	84	84
p05	20	16	19	124	135	129	130	124	124	129	129	124	124
p06	24	20	26	102	107	102	107	107	102	102	102	102	102
p07	23	24	23	130	130	130	130	130	130	130	130	130	130
p08	17	24	16	122	122	122	122	122	122	122	122	122	122
p09	14	14	12	83	84	83	83	83	83	83	83	83	83
p10	12	10	10	80	80	84	80	80	80	80	80	80	80
p11	9	7	7	23	23	23	26	23	23	23	23	23	23
p12	7	5	13	19	22	21	22	19	19	21	21	19	19
p13	7	4	6	35	38	38	35	35	35	38	38	35	35
p14	28	31	48	202	212	209	207	204	202	209	209	202	202
p15	26	19	18	441	445	445	445	441	441	445	445	441	441
p16	31	34	60	203	203	203	215	205	203	203	203	203	205
p17	19	17	27	112	116	112	116	112	112	112	112	112	112
p18	23	16	21	146	148	148	147	-	147	148	148	146	146
p19	33	29	26	257	280	263	274	271	266	263	263	257	257
p20	50	63	35	398	400	399	402	400	400	398	398	403	439
p21	49	67	43	366	372	368	372	372	372	372	366	373	388
p22	50	74	110	621	632	621	633	622	622	636	621	643	637
p23	50	78	80	475	480	489	479	477	477	487	480	477	477
p24	41	55	70	405	411	405	411	405	405	405	405	406	450
alA	102	99	61	10599	-	10784	10599	10599	10599	10995	10612	11695	11585
alB	90	88	56	8629	-	8721	8629	8629	8629	8883	8629	9469	9398

column labeled ACS_2 shows the best solution reached when the transformation described in this paper is used. The optimum is reached for 19 problems in the first case and for 18 in the second case. The computational time needed reduces to a few seconds. The number of nodes of the TSP problem is greater when the new transformation is applied. Therefore, the average time to reach a solution is greater in the second case. Columns from 6 to 12 show the best solution reported for the problems in several papers. The paper of Baldoquín et al [1], reports results for two methods; therefore, two columns are included in the table for this paper. Compared with the other approximate solutions, the ant-algorithm solves more problems.

5 Conclusion

In the last years artificial ants have been applied to solve several NP-hard problems. This heuristic generates aproximate results with a small time consumption. These algorithms can be directly applied to new problems if there exists a transformation from the new problem to the one which can be solved by the ants. In this paper the ACS algorithm has been applied to solve an arc-routing problem. To apply the algorithm, the problem has been transformed into an undirected TSP. The solution reached is better than the one obtained by other methods proposed for the problem. One advantage of the method is that it does not require a complex mathematical representation of the problem to be solved, we simply use the graph that represents the problem. Moreover, it always generates a feasible solution for the problem. Compared with a previous transformation applied to solve the same problem, the time required to reach a solution inscreases, since the TSP obtained with the transformation has more nodes. The solution obtained will be able to get improved by applying some modifications to the basic algorithm, such as the use of candidate lists.

References

1. Baldoquín, M.G., Ryan, G., Rodriguez, R., Castellini, A.: Un enfoque hibrido basado en metaheurísticas para el problema del cartero rural. In: Proc. of XI CLAIO, Concepción de Chile, Chile (2002)
2. Christofides, N., Campos, V., Corberán, A., Mota, E.: An algorithm for the rural postman problem. Imperial College Report, London (1981)
3. Corberán, A., Sanchis, J.M.: A polyhedral approach to the rural postman problem. Eur. J. Oper. Res. 79, 95–114 (1994)
4. Dorigo, M., Gambardella, L.: Ant colony system: a cooperative learning approach to the traveling salesman problem. IEEE Trans. Evol. Comput. 1(1), 53–66 (1997)
5. Fernández, P., García Raffi, L.M., Sanchis, J.M.: A heuristic algorithm based on Monte Carlo methods for the rural postman problem. Comput. Oper. Res. 25(12), 1097–1106 (1998)
6. Frederickson, G.: Approximation algorithms for some postman problems. J. Assoc. Comput. Mach. 26, 538–554 (1979)
7. Ghiani, G., Laporte, G.: A branch and cut algorithm for the undirected rural postman problem. Math. Program 87, 467–481 (2000)
8. Golden, B.L., Wong, R.T.: Capacitated Arc Routing Problems. Networks 11, 305–315 (1981)
9. Groves, G.W., van Vuuren, J.H.: Efficiente heuristics for the rural postman problem. Orion 21(1), 33–51 (2005)
10. Hertz, A., Laporte, G., Nanchen, P.: Improvement procedures for the undirected rural postman problem. INFORMS J. Comput. 1, 53–62 (1999)
11. LenstrA, J.K., Rinnooy Kan, A.H.G.: Complexity of vehicle routing and scheduling problems. Networks 11, 221–227 (1981)
12. Letchford, A.N.: Polyhedral results for some constrained arc routing problemas. PhD Dissertation. Lancaster University, Lancaster (1996)
13. Orloff, C.S.: A fundamental problem in vehicle routing. Networks 4, 35–64 (1974)
14. Pérez-Delgado, M.L.: A solution to the rural postman problem based on artificial ant colonies. In: Borrajo, D., Castillo, L., Corchado, J.M. (eds.) CAEPIA 2007. LNCS (LNAI), vol. 4788, pp. 220–228. Springer, Heidelberg (2007)
15. Pearn, W.L., Assard, A., Golden, B.L.: Transforming arc routing into node routing problems. Comput. Oper. Res. 14(4), 285–288 (1987)
16. Reinelt, G.: The traveling salesman problem: computational solutions for TSP applications. In: Reinelt, G. (ed.) The Traveling Salesman. LNCS, vol. 840. Springer, Heidelberg (1994)

Using Situation Calculus for Normative Agents in Urban Wastewater Systems

Juan-Carlos Nieves, Dario Garcia, Montse Aulinas, and Ulises Cortés

Abstract. Water quality management policies on a river basin scale are of special importance in order to prevent and/or reduce pollution of several human sources into the environment. Industrial effluents represent a priority issue particularly in Urban Wastewater Systems (UWS) that receive mixed household and industrial wastewaters, apart from rainfall water. In this paper, we present an analysis and an implementation of normative agents which capture concrete regulations of the Catalan pollution-prevention policies. The implementation of the normative agents is based on situation calculus.

Keywords: Rational Agents, Environmental Decision Support Systems, Practical Normative Reasoning, Situation Calculus.

1 Introduction

Environmental decision-making is a complex, multidisciplinary and crucial task. As an example, in the water management field, water managers have to deal with complex problems due to the characteristics of processes that occur within environmental systems. In addition to this, water managers have to deal with normative regulations that have to be considered in any decision.

In particular, at European level, Directive 96/61/EC on Integrated Pollution Prevention and Control (IPPC) [4] was developed to apply an integrated environmental

Juan Carlos Nieves · Dario Garcia · Ulises Cortés
Universitat Politècnica de Catalunya, C/Jordi Girona 1-3, E08034, Barcelona, Spain
e-mail: {jcnieves,ia}@lsi.upc.edu, dariogarcia@gmail.com

Montse Aulinas
Laboratory of Chemical and Environmental Engineering, University of Girona, Spain
e-mail: aulinas@lequia.udg.cat

Y. Demazeau et al. (Eds.): Advances in PAAMS, AISC 70, pp. 247–257.
springerlink.com

approach to the regulation of certain industrial activities. This means that, at least, emissions to air, water (including discharges to sewer) and land must be considered together. It also means that regulators must set permit conditions so as to achieve a high level of protection for the environment as a whole. Several national and regional efforts are being done in order to improve water quality management as well as to comply with European regulations. Concretely, we analyse the Catalan experience as a realistic example of adapting European guidelines to manage water taking into account the local/regional reality.

In order to analyse the context of pollution-prevention policies in Catalonia we take a concrete regulation: the Decree 130/2003 [12]. It is a regional regulation appeared as a consequence of the Catalan sanitation programme. One of the aims of this updated programme is to directly link the urban wastewater treatment program with the industrial wastewater treatment program. It pays special attention to the industrial component of urban WWTPs in order to facilitate the connection to the public system of those industries and/or industrial parks that accomplish the requirements, and that is the reason why Decree 130/2003 was developed.

Following the perspective that a software agent is any active entity whose behavior is described by mental notions such as knowledge, goals, abilities, commitments, *etc.*, we have been exploring the definition of intelligent agents provided by a *normative knowledge* in order to manage concrete normative regulations which are in the context of UWS.

A central issue of a successful normative agent implementation is the selection of a formalism for performing *practical normative reasoning*. Although several powerful formalisms exist, finding the right one is a non-trivial challenge, as it must provide a level of expressiveness that serves the practical problems at hand in a tractable way. In the literature, one can find several approaches for performing normative reasoning such as deontic logic, temporal logic, dynamic logic, *etc.* [11]; however, these approaches have a high computational cost. One can agree that formal methods do help in the long run to develop a clearer understanding of problems and solutions; hence, the definition of computable tractable approaches based on formal methods takes relevance for performing practical normative reasoning.

Situation Calculus has shown to be a practical approach for facing real problems [1, 6, 9]. Moreover, situation calculus allows the specification of any set of complex action expressions. There are some results that have shown that situation calculus is flexible enough for performing normative reasoning [5].

In this paper, we describe the implementation of normative agents based on Situation Calculus for performing practical normative reasoning. The normative knowledge structure follows an approach introduced in [2] and explored in [14, 15]. Unlike the approach presented in [15], which extends the action language $\mathscr{A}$ for capturing norms, in this paper we explore a norm's lifecycle by considering states of the world (situations/sets of fluents). This means that our main concern will be to monitor the states of norms being either active, inactive or violated.

We will present an analysis of an existing specific Catalan regulation for providing normative knowledge to normative agents. Also, we will describe the implementation of these normative agents by considering situation calculus.

The rest of the paper is divided as follows: In §2, a realistic hypothetical scenario is described in order to illustrate the role of some regulations for managing industrial discharges. In §3, it is described, at a high level, *how* to introduce normative knowledge in a Situation Calculus specification. In §4, we describe *how* to implement the approach presented in §3. In the last section, we present our conclusions.

2 Realistic Scenario

In this section, a realistic hypothetical scenario is described in order to illustrate the role of some regulations for managing industrial discharges.

At the municipality of Ecopolis a new industry called MILK XXI pretends to set up. As a result of its production processes the main characteristics of its wastewater will be as follows:

- Flow: 60 l/s (5184 m^3/day)
- Suspended solids: 130 mg/l
- BOD5: 450 mg/l
- COD 800 mg/l
- Oils and greases : 275 mg/l

The Milk factory plans to work 16 h/day (two shifts), 225 days per year. It plans to get connection to the municipal sewer system that collects wastewater from a population of 12000 inhabitants and transports it to the municipal WWTP. WWTP complies strictly with regulations. The owner of the industry submits the request to obtain authorization to discharge into the municipal sewer system, which it is compulsory by law (Decree 130/2003 that settles the public sewer systems regulations). Moreover, Milk industry plans to apply BAT[1] in order to reduce the consumption of water, so it applies as well for the consideration of this fact into the final authorization decision.

Accordingly the industry intends to reduce 30% on water consumption, and consequently the increment of pollutant concentrations, is projected to be as follows:

- Flow (reduction): 42 l/s (3628,8 m^3/day)
- Suspended solids: 200 mg/l
- BOD5: 600 mg/l
- COD 1000 mg/l
- Oils and greases : 357,5 mg/l

Several rules are launched to manage this case, from the regulation analyzed in this work. As follows we describe the immediate agents implied in the case as well as the type of norms involved:

[1] BAT: Best Available Techniques [4].

Action	Agent	Type of Norm
Request authorization	Industry agent	Obligation (7.1)
Given authorization: thresholds average flow conditions exceptions *etc.*	Water Catalan Agency Agent	 Obligation (13.1) Obligation (13.2) Obligation (13.2)
Apply BAT (declare this when requesting authorization)	Industry Agent	Obligation(8.3)
Discharge	Industrial Agent	Obligation (8.2)

Due to lack of space, we omit the detailed description of each agent; however, a version of those can be found in [3]. Observe that the behavior of each agent is fixed by the set of norms that it has to observe. In the following section, we are going to describe an approach for expressing these norms in the mental notions of an agent.

3 Normative Specification Based on Situation Calculus

In here we are going to describe, at a high level, the modeling process of norms. Since environmental domains are dynamic domains, a dynamic domain is a domain where truth values change with time, the described approach deals with the specification of norms in dynamic domains. In particular, we have used Situation Calculus [9, 10], which works with the concepts of *fluents* and *actions*.

To formally specify any set of norms, first it is necessary to analyse the domain these norms will work with. In Situation Calculus, the world in a certain moment (represented by a first order term known as *situation*) is defined by the value of a set of predicates, known as *fluents*, in that moment. The change between situations is triggered by *actions*, which can be parameterized. Actions may change the value of one or more fluents, therefore modifying the current situation.

A constant s_0 is defined as the initial situation, for which fluent values are given as to define that state. The binary function $do(X,Y)$ denotes the situation resultant of executing action X in situation Y. The binary function $holds(Z,W)$ denotes the truth value of a fluent Z in a situation W. With the combination of these two functions we can represent all possible worlds and relationships between them, inherent of dynamic domains.

As any approach for temporal reasoning, Situation Calculus must deal with the *Frame Problem* to make its implementation consistent [10]. Aware of that, we present a specification that fully asserts the effects of all actions on every norm.

Before working on how to specify norms we analyse them, following the works of [2, 14, 15] and keeping situation calculus in mind. To fully specify a norm several aspects must be taken into account:

Type of Norm: Norms that oblige to do something, norms that allow/permit something or norms that forbid something. It is important to take special care with allowing norms, since to fully specify their content it seems to be necessary to split them into two norms, one that allows something and one that forbids something, *e.g.*, "*It is allowed to spill black waters into the river if one has the required authorization*". This allowing norm implicitly includes the following forbidding norm: "*It is forbidden to spill black waters into the river without authorization*".

Conditions and Content: Separate the norm conditions and the norm content in order to study the characteristics of situations, in which the norm is active and in which the norm is violated.

States: The set of variables the norm refers to. For each possible value of those variables the norm has a state, *e.g.*, if a norm is applied to an agent then it would have one variable such as $IdAgent$, and if it is applied to an agent's spill then it would have two variables, $IdAgent$ and $IdSpill$.

Actions: A complete list of the domain's actions that may influence the activation state and the violation state, separately, of each norm.

Preconditions: Define the preconditions for each action, that is, in which situations are they possible and the conditions about their parameters.

Now we can start to specify our norms. We follow Reiter's solution presented in [16] to our normative domain. We propose to split the specification in two parts corresponding to the main properties that all norms have:

- The scenarios in which the norm is active.
- The scenarios in which the norm is violated.

To specify the scenarios where a norm is active or violated in, we will state the value of the fluents that will define unequivocally the set of situations that represent those set of states.

The first part of the specification is meant to contain all the possible states in which the norm must be taken in consideration (is active). The second one comprises all the states in which the norm's content is violated. As follows, we present the first part of the specification.

In our proposal a norm N, after doing an action A in situation S, is *active* if and only if it fits in one of these three cases:

i. N was not active before doing A. There is a set of conditions under which A changes the activation state of N from inactive to active. The conditions needed for A to activate N are fulfiled in S.

   The *Activation Condition*: Given a certain norm in a situation where the norm is inactive, the range of A is the actions that may change the values of the fluents on which depends the activation state of the norm, in a way that the resultant situation (defined by the resultant value of the fluents) could belong to the scenarios in which the norm is active.

ii. N was active before doing A. There is a set of conditions under which A changes the activation state of N from active to inactive. The conditions needed for A to deactivate N are not fulfiled in S.

   The *Maintenance Condition*: Given a certain norm in a situation where the norm is active, the range of A is the actions that may change the values of the fluents on which depends the activation state of the norm, in a way that the resultant situation could belong to the scenarios in which the norm is active.

iii. *N* was active. There is no set of conditions that can make *A* change the activation state of *N* from active to deactivated.

The *Non-Termination Condition*: Given a certain norm and a situation where the norm is active, the domain of *A* is the actions that may not change the values of the fluents in a way that the resultant situation could belong to the scenarios in which the norm is inactive.

If we analyse these three rules we can assure that every state in which the norm is active fits into one and only one of these three rules. By checking a certain situation with a proposed action we can assert the activation state of any norm after that action has been performed in the situation.

The second part of the specification contains the scenarios where a norm is *violated.* In this case, we have decided to make a simpler specification.

In the activation condition specification we had the temporal progression integrated into it by the use of a variable that represented the action just performed (variable *A*). In this case we will omit that variable and see the specification of the *violation state* based solely on *the situation's fluents*. It is possible to do that without losing expressivity since the temporal progression in our domain is represented as well in the fluents definition (which specification's looks very much like the activation condition specification), otherwise we would lose the concept of timeline.

By deleting that action variable, the specification becomes much simpler as only the fluent that define the states where the norm is violated have to be stated.

In our proposal a norm *N* is *violated* in situation *S* if and only if it fits in one of these two cases:

i. *N* is active in *S*, *N obliges* to the value of one or more fluents and *S* does not fulfil all of those obliged fluents.

This rule is intended to cover the violations done upon norms that oblige to something.

ii. *N* is active in *S*, *N forbids* the value of one or more fluents and *S* fulfils one of those forbidden fluent.

This rule is intended to cover the violations done upon norms that forbid something.

With those two rules we cover the possible violations that can come upon a norm, as norms that allow something cannot be violated. Once having the two parts of the specification of each norm, we can implement them to see how they work once applied to a real life domain.

4 Normative Implementation Based on Situation Calculus

In the previous section we have proposed a specification for norms working on dynamic domains under the approach of Situation Calculus and a norm's lifecycle such as active, inactive and violated. Now we will see how this specification can specify real laws on standard Prolog. Specifically, our focus is on the temporal progression aspects involved in the activation state part of the specification. The implementation of the violation state is simpler and we will omit it here; however, the interested reader can find a Prolog prototype of our normative knowledge base in http://www.lsi.upc.edu/~jcnieves/software/NormativeKnowledge-PAAMS-2010.pl

We are going to consider the Decree 130/2003 of the *Catalan Water Agency*. Recall that this decree was motivated in §2. As an example, we will study article 8.2 related to an industrial agent of our multiagent system:

> If industrial spills contain limited substances they must respect the established limitations (thresholds).

Following the analysis explained in the previous section, we know that:

- Norm 8.2 is an obliging norm.
- The situations in which the norm is active are those in which an industrial agent makes a spill containing limited substances. The situations in which the norm is violated are those where the substance limitations established are not respected.
- The variables the norm 8.2 applies to are an agent *IdAgent* and a spill *IdSpill*.
- The actions that may change the activation state of the norm are:
  - `set_agent_type`: change the type (domestic, industrial,...) of an agent.
  - `add_substance`: add a certain amount of a substance to a spill.
  - `del_substance`: delete a certain amount of substance from a spill.
  - `del_total_substance`: delete all the contained substance from a spill.
  - `make_spill`: start the execution of a spill by an agent
  - `cancel_spill`: cancel the execution of a spill by an agent
- The preconditions of each action will be defined in a predicate named $poss(A,S)$, following the situation calculus syntax, where A is an action and S is a situation.

The resultant Prolog implementation of applying that information to the specification schema given before for the activation state of a norm is as following: Norm 8.2 is active for agent IdAgent's spill IdSpill after doing action A in situation S.

Actions that may activate the norm when the conditions are set:

```
holds(norm(82,IdAgent,IdSpill),do(A,S)) :-
A = set_agent_type(IdAgent,domestic),
holds(spill_contains_limited_substance(IdSpill,Substance),S),
not holds(norm(82,IdAgent, IdSpill), S), poss(A, S).

holds(norm(82,IdAgent,IdSpill),do(A,S)) :-
A = add_substance(IdSpill,Substance,Quantity),
holds(agent_type(IdAgent,domestic),S),
holds(agent_spill(IdAgent,IdSpill),S),
holds(substance_limitation(Substance,Limit),S),Limit>0,Quantity>0,
not holds(norm(82,IdAgent,IdSpill),S),poss(A,S).
```

```
holds(norm(82,IdAgent,IdSpill),do(A,S)) :-
A = make_spill(IdSpill,IdAgent), holds(agent_type(IdAgent,domestic),S),
holds(spill_contains_limited_substance(IdSpill,Substance),S),
not holds(norm(82,IdAgent,IdSpill),S),poss(A,S).
```

Actions that may terminate it when the conditions are not set

```
holds(norm(82,IdAgent,IdSpill),do(A,S)) :-
holds(norm(82,IdAgent,IdSpill),S),
A = set_agent_type(IdAgent,domestic), poss(A,S).

holds(norm(82,IdAgent,IdSpill),do(A,S)) :-
holds(norm(82,IdAgent,IdSpill),S),
A = del_substance(IdSpill,Substance,Quantity),
holds(spill_contains_limited_substance(IdSpill,Substance2),S),
Substance!=Substance2,poss(A,S).

holds(norm(82,IdAgent,IdSpill),do(A,S)) :-
holds(norm(82,IdAgent,IdSpill),S),
A = del_substance(IdSpill,Substance,QD),
holds(substance_spill_quantity(IdSpill,Substance,QPre),S),
QPre>QD,poss(A,S).

holds(norm(82,IdAgent,IdSpill),do(A,S)) :-
holds(norm(82,IdAgent,IdSpill),S),
A = del_total_substance(IdSpill,Substance),
holds(spill_contains_limited_substance(IdSpill,Substance2),S),
Substance!=Substance2,poss(A,S).
```

Actions that may not terminate the norm

```
holds(norm(82,IdAgent,IdSpill),do(A,S)) :-
holds(norm(82,IdAgent,IdSpill),S),
not A = set_agent_type(IdAgent,Type),
not A = del_substance(IdSpill,Sub,Q),
not A = del_total_substance(IdSpill,Substance),
not A = delete_agent(IdAgent),
not A = cancel_spill(IdSpill,IdAgent),poss(A,S).
```

This implementation, as above justified, fully represents, in *a computable way*, a norm and all the states it may be into: activated, deactivated, violated and respected. When implementing agents that have to interact with a legal framework, we have now a way to integrate the understanding of laws contents and conditions into them.

Now that we have a working implementation of norms, as well as the domain's actions and fluents, we can test the performance of our code by asking Prolog about the norms following the next syntax: to know which norms are applicable to an agent *idAgent* after performing an action *action1* in a given initial state *ini*:

```
holds(norm(X,idAgent),do(action1,ini)).
```

That query will return the complete list of norms active in the state generated after executing $action_1$ in the scenario *ini*. To know which norms are applicable to an agent's (*idAgent*) spill *idSpill* after performing an action *action2* in a given initial state *ini*:

```
holds(norm(X,idAgent,idSpill),do(action2,ini)).
```

In situation calculus a state is defined as the result of executing an action in another state, therefore we can nest a list of actions to check a norm's state:

```
holds(norm(X,idAgent),do(action2,do(action1,ini))).
```

That query will return the list of norms active for agent *idAgent* after performing *action1* and later *action2*.

It is important to notice that we assume a perfect knowledge of our world, which means that the fluents of the initial state (on which posterior actions will be performed) are fully asserted.

5 Conclusions

Since norms in real world are usually defined in an abstract level [17], the modeling process of real norms is not straightforward. Some authors have already pointed out that the instantiation of norms in a context domain helps in the process of representing norms in a normative knowledge [17].

In order to capture the scope of a norm in a context domain, one can fix the observable items that affect the lifecycle of a norm. In particular, the representation of these items in terms of fluents/predicates can help to infer the state of a norm. Since the state of a norm will be affected by changes in observable items, one can monitor the lifecycle of a norm in parallel to the changes of the observable items. In order to explore this idea, we have considered Situations Calculus. Observe that a context domain can be clearly delimited by a set of fluents (a *situation*); and this is one of our main motivations of using Situations Calculus. As a running example, we have analyzed the Catalan Decree 130/2003 as a realistic example for managing urban wastewater systems.

In order to incorporate normative knowledge in a Situation Calculus specification, we proposed to split the specification of norms in two parts: 1.- the scenarios in which a norm is active and 2 .- the scenarios in which a norm is violated. The first part of the specification is meant to contain all the possible states in which the norm must be taken in consideration (is active). The second one comprises all the states in which the norm's content is violated. Since the norms are represented in terms of the fluents of the given domain, the proposed specification represents a natural extension of a situation calculus specification.

In the literature, we can find different approaches for performing normative monitoring [6, 7, 8, 13, 18]. Possibly the approach presented in [8] is closely related to the approach presented in this paper. In [8], the authors perform normative monitoring by considering Event Calculus; however, their approach does not consider a lifecycle of a norm.

Some issues for our future work are: 1.- the consideration of a lifecycle of actions: by the moment we have assumed actions as an atomic event; hence, this assumption has its limitations for capturing temporal aspects as deadlines. 2.- the consideration of conflicts between norms: for this we are exploring the definition of a partial order between norms.

Acknowledgements

Numerous discussions with Javier Vázquez-Salceda help us to clarify our ideas. We are grateful to anonymous referees for their useful comments. This work has been partially supported by the FP7 European project ALIVE IST-215890. The views expressed in this paper are not necessarily those of the ALIVE consortium.

References

1. Albrecht, C.C., Dean, D.D., Hansen, J.V.: Using situation calculus for e-business agents. Expert Systems with Applications 24, 391–397 (2003)
2. Aldewereld, H.: Autonomy vs. Conformity: an Institutional Perspective on Norms and Protocols. PhD thesis, Utrecht University (2007)
3. Aulinas, M.: Management of industrial wastewater discharges through agents' argumentation. PhD thesis, University of Girona (October 2009)
4. CEC. Directive 96/61/EC of 24 September 1996 concerning integrated pollution prevention and control. Official Journal L 257(10), 10 (1996)
5. Demolombe, R.: From belief change to obligation change in the situation calculus. In: ECAI, pp. 991–992 (2004)
6. Demolombe, R., Parra, P.P.: Integrating state constraints and obligations in situation calculus. In: LA-NMR (2006)
7. Digmun, F., Dignum, V., Padget, J., Vázqez-Salceda, J.: Organizing Web Services to develop Dynamic, Flexible, Distributed Systems. In: proceedings of 11th International Conference on Information Integration and Web-based Applications and Services, iiWAS 2009 (2009)
8. Kaponis, D., Pitt, J.: Dynamic specifications in norm-governed open computational societies. In: O'Hare, G.M.P., Ricci, A., O'Grady, M.J., Dikenelli, O. (eds.) ESAW 2006. LNCS (LNAI), vol. 4457, pp. 265–283. Springer, Heidelberg (2007)
9. Lesperance, L., Levesque, H., Reiter, R.: A Situation Calculus approach to modeling and programming agents. In: Foundations and theories of rational agents (1999)
10. McCarthy, J., Hayes, P.J.: Some philosophical problems from the standpoint of artificial intelligence. In: Meltzer, B., Michie, D. (eds.) Machine Intelligence, vol. 4, pp. 463–502. Edinburgh University Press (1969) reprinted in McC90
11. Meyer, J.-J.C., Wieringa, R.J. (eds.): Deontic Logic in Computer Science: Normative System Specification. Wiley, Chichester (1993)
12. Ministry, E.C.: Decree 130/2003, Reglament dels serveis públics de sanejament. In: DOGC, vol. 3894, pp. 11143–11158 (2003)
13. Modgil, S., Faci, N., Meneguzzi, F.R., Oren, N., Miles, S., Luck, M.: A framework for monitoring agent-based normative systems. In: AAMAS, vol. (1), pp. 153–160 (2009)
14. Oren, N., Panagiotidi, S., Vázquez-Salceda, J., Modgil, S., Luck, M., Miles, S.: Towards a formalisation of electronic contracting environments. In: Coordination, Organization, Institutions and Norms in Agent Systems, the International Workshop at AAAI 2008, Chicago, Illinois, USA, pp. 61–68 (2008)

15. Panagiotidi, S., Nieves, J.C., Vázquez-Salceda, J.: A framework to model norm dynamics in answer set programming. In: Proceedings of FAMAS 2009 (Multi-Agent Logics, Languages, and Organisations Federated Workshops). CEUR WS Proceedings, vol. 494 (2009)
16. Reiter, R.: The frame problem in situation the calculus: a simple solution (sometimes) and a completeness result for goal regression. In: Artificial intelligence and mathematical theory of computation: papers in honor of John McCarthy, pp. 359–380. Academic Press Professional, Inc., London (1991)
17. Vázquez-Salceda, J.: The Role Of Norms And Electronic Institutions In Multi-Agent Systems Applied To Complex Domains The Harmonia Framework. In: Artificial intelligence, LSI, Universitat Politècnica de Catalunya, Barcelona, Spain (2003)
18. Vázquez-Salceda, J., Aldewereld, H., Grossi, D., Dignum, F.: From human regulations to regulated software agents' behavior. Artif. Intell. Law 16(1), 73–87 (2008)

Organization Nesting in a Multi-agent Application for Ambient Intelligence

Matthieu Castebrunet, Olivier Boissier, Sylvain Giroux, and Vincent Rialle

Abstract. In the context of Health Smart Homes for aging and cognitively impaired people, the adaptation of the smart assistance to the set of sensors/effectors and to the people present is a crucial problem. Some solutions have been proposed. They can either assist a lone person for a small set of activities in a precise smart space, either assist a person in mobility. However, to our knowledge none of them is able to manage the simultaneous assistance of multiple persons in different smart spaces. In this paper, we are considering this problem. We are proposing a multi-part solution to manage a truly ambient intelligence system. Based on a scenario of daily living in such a situation where conflicts in assistance may arise, we propose to use nested and explicit multi-agent organizations to help to "solve" this problem. After defining the concept of nested organization, we illustrate how it can be used to manage dynamically every aspect of multi-person assistance in daily-living activities in a Smart Home.

1 Introduction

The increasing population of aging and cognitively impaired people together with the development of ICT installing the ubiquitous computing vision [1] has brought the trend of Health Smart Homes [2,3,4,5] and other smart places [6,7,8].

The needs of this population in assistance has lead to smart technologies, often based on and confined in a secured smart home environment. Systems have most of the time been designed to monitor, to assist or to protect a lone person in one dedicated smart home. However, a meaningful life for anybody consists in relationships [9], multi-person and mobility context. Therefore, such a technological help must be enriched with the ability to be deployed in every smart place visited by the assisted patient, inhabited by other persons.

Matthieu Castebrunet · Sylvain Giroux
Domus Laboratory, Université de Sherbrooke, Sherbrooke (QC), Canada

Matthieu Castebrunet · Vincent Rialle
TIMC-IMAG Laboratory (AFIRM), Grenoble, France

Olivier Boissier
École Nationale Supérieure des Mines de Saint-Etienne, Saint-Etienne, France

Y. Demazeau et al. (Eds.): Advances in PAAMS, AISC 70, pp. 259–268.
springerlink.com

From our point of view most of the existing ambient intelligence systems are *ad-hoc* and fine-tuned solutions because of the complexity of each smart place and the variety of end-users client for whom they are designed. In this light, ambient intelligence needs a local analysis and interaction to perform a global and general purpose: to assist and personalize the environment for each and every known user. Such system requires a way to define, install and adapt the services to the current context and needs, according to the local configuration in terms of material and software resources. As inhabitants of such smart places may be mobile, they can move in and out. The global configuration of the provided services must take into account both the local smart place configuration and the user personal configuration that moves along with the inhabitant. Furthermore, we expect from these services the same correctness as we have when a respectful person visits another's home where their actions are restricted by their understanding of correctness: services and assistance must not overstep their bounds in another smart place. This requirement brings another level of configuration to manage. As we can see from this quick overview, an important number of global configurations have to be considered and combined with each other.

The solution proposed in this paper relies on the use of multi-agent technologies [10] to support the management of adaptive configurations in smart home environment. The system description and problematic is not in the scope of this paper, only the organizational part is detailed. Our approach consists in mapping the patterns of interaction and configuration of users and spaces onto multi-agent organization definitions [11]. Smart space configurations are thus described by *local* multi-agent organizations. Users configurations are described by *personal* multi-agent organizations. Each of the *personal* configurations takes into account the other persons with a relational configuration described by a *relational* organization. The resulting global configuration interweaves different personal configurations situated in various local configurations, providing dynamically the needed structure and control.

The structure of the paper is the following. To explain properly our approach, section 2 draws a global picture of the problem based on a scenario of daily living. Section 3 describes the three organizations that we defined: *personal* organization for a person and his daily assistance, *local* organization for the physical structure and *relational* organization to manage inter-persons conflicts. Section 4 details how we developed and managed the nesting of these different organizations. We then conclude and describe our perspectives (see section 5).

2 Global Vision

This section presents a global view of the envisioned system. While introducing the foundations of the approach, it illustrates the different practical motivations that drive our proposal. A scenario of daily living (§2.1) is given to establish a base of reference. The global architecture of the smart home system (§2.2) is then described, focusing on the global description of the multi-agent based software infrastructure that supports users in smart places. We then illustrate the needs of managing and adapting configurations in such environments (§2.3).

2.1 Assisted Daily Living Scenario

In the sequel we consider the following scenario to illustrate a daily-living case in which our approach can be demonstrated. Let's consider a Health Smart building where Alice and Bob are living together in a flat and Carl, their neighbor, is living alone in a similar habitation. All three need special assistance to balance their cognitive impairment. Each flat is equipped with a different physical living environment that makes it able to sense the modifications in the environment and the capacity to bring information in the environment or even to modify it directly. In each of those flat, the multi-agent organization proposed will establish a connection between human users and physical infrastructure. In our domain of assisted living for cognitively impaired people, each user must be independently recognized to avoid mistakes in daily tasks assistance and personalization.

Let's consider a particular time in this scenario where Carl is watching TV at home in the cold room-temperature that he likes; Alice and Bob are in their warmer home, Bob preparing supper and Alice watching a TV show. Alice calls Carl on his landline to invite him over for supper at their place. Carl accepts, gets out of his flat and into theirs. The system assisting Carl in his home should assist him as well in Alice and Bob's smart home without impeding their assistance.

2.2 Smart Home Infrastructure

In our project, even if every smart home has a different physical living environment, it is possible to map a virtual representation on each of them. The ability to perceive the modifications occurring in the physical environment is given by various sensors placed everywhere in the smart home. As the smart homes are in a humanly respectful environment, the sensors in use do not include video cameras but only anonymous sensors like infrared motion sensors, pressure rugs, electromagnetic contact on doors, drawers and closets [12], etc. On top of these sensors and effectors, letting users interact with the physical environment, a multi-agent infrastructure has been deployed. Different agents provide various kinds of services. *Environment agents* connected to the sensors communicate with each other to ensure the proper treatment of the information. *Actuators agents* interface the physical actuators used to communicate and to interact with the users: tactile screens to display information and interact easily, loud speakers coupled with microphones, lights control, heat control, and generally every microcontroller present on daily-living appliances (fridge, oven, phone…). They are mainly used to manage access conflicts. *Context agents* address the main challenge consisting in the inference of the user's activities from this information. When there is only a lone person in the Smart Home, the inference is easier because most occurring events can be attributed to this person. Considering the hypothesis of more than one person at the same time in the flat, the inference becomes much more complicated and unreliable. Context agents also provide high-level contextual information in the smart home (e.g. number of people in the flat, localization, identification…). In the context of health smart homes, the home needs not only to sense but also to personalize both the environment and the assistance for specific tasks for the people living in it. Various assistive agents are thus deployed to make use of the environment agent to provide this assistance and personalization.

In the literature, there are some approaches using a service-based architecture to manage a Smart Home [15] but they do not consider the life and assistance outside the flat. Similarly, other solutions based on multi-agent systems propose the management of the smart home resources [16], some even considering multiple persons [17] but not in the assistive paradigm, thus bringing less attention to the person identification. In our domain, differentiating each resident is important and sometimes compulsory during the personalization and daily living assistance. To our knowledge, there is no other existing organizational solution for these problems.

2.3 Configurations in Daily Life Assisted Living

Due to the increasing number of agents populating this physical environment it is necessary to structure and govern them properly. Given the variety of smart homes, of people living in them, of health troubles requiring different configurations of effectors and sensors, structure and governance must express different and specialized configurations. Each configuration expresses the system interactions for the people present in the smart spaces in consideration.

The basic configuration that we consider first is the one needed for one person in a smart place. This configuration captures every aspect of the governance of the multi-agent infrastructure pertaining to the management of the *person* (profile, preferences, habits, authorization level, etc.) and to the *smart home* itself (sensors, actuators, data storage, computing power and communications). The aim is for the system to be able to assist and personalize the environment at best for one person.

Another kind of configuration is the one occurring when two or more living beings are present in the same smart place (Ex: Alice and Bob). Each user is expecting as much personalization and assistance as if he was alone. It is not a trivial operation, as a lot of conflicts can be declared between the users opposite preferences or needs, and even sometimes because of the same needs. Indeed, two users needing assistance in the exact same location will compete for the same resources onto which to be assisted (Alice needs the TV to display her agenda and Bob need it to remind him to monitor his current activity: preparing the meal). A person moving from a smart place to another would also expect a continuity of services (Carl needs a reminder to take his medicine, even when he is eating at Alice and Bob's place), migrating toward another configuration and so on. Those successive configurations need to be part of a dynamic and adaptive process.

3 Multi-agent Organizations for Daily-Life Configurations

Considering the requirements presented in the previous section, we propose for the assistance and personalization in a smart place to be supported by a multi-agent system governed by an open organization. This latter defines the governance structure of the agents evolving at that place and models the changing configuration. We use the Moise framework [13] to define and manage this organization. Thanks to a declarative and explicit organization specification written in an organizational modeling language (OML), each agent entering the organization is able, at any time, to consult the structure (roles, groups, missions) and governance rules to know the permitted/obligatory/forbidden missions. The Moise Framework OML decomposes the specification of an organization into structural, functional, and normative dimensions. The structural dimension specifies the roles, groups, and links between roles of the organization. The definition of roles states that when an agent chooses to play some role in a group, it is accepting some behavioral constraints and rights related to this role. For example, an agent playing the role of actuator will have the constraint to obey to the assistance agent, and the right to command the devices. The functional dimension specifies how global collective goals (keeping temperature low) should be achieved, i.e. how these goals are decomposed (within global plans), grouped in coherent missions (shut down heater for example) to be distributed among the agents. The decomposition of global goals results in a goal-tree, called scheme, where the agents can individually achieve the leaf-goals. The normative dimension is added in order to bind the structural dimension with the functional one by means of the specification of the roles' permissions and obligations with respect to missions. The functional and deontic specification, that is the missions description, will not be detailed in this paper to better concentrate on the structural specification on which the nesting is operated.

To be able to provide a structure for every configurations described in 2.3, we define three organizations that we interweave in the global organization of a smart place: personal, local and relational. The *personal* organization expresses the governance structure for the management of the personalization by roles/missions related to preferences, habits, agenda and assistance of a user (see §3.1). The *local* organization governs the agents attached to the various kind of sensors, actuator, and data management structure situated in a smart place (see §3.2). The relational organization is dedicated to the governance and management of conflicts arisen from the fact that a smart place can host more than one user (see §3.3).

3.1 Personal Organization

A personal Organization is composed of a group structure "Guardian Angel" where four roles can be played. To each of those roles are associated missions for example to analyze an event, or aiming to fulfill higher goals, like the localization, the identification, the agenda managing, the personalization based on the habits and preferences, and the assistance for various daily-living activities.

As schematized in Fig. 1, the roles "Context Mgr" (stay aware of the person's evolving context and localization), "Agenda Mgr" (manage the person's agenda), "People Mgr" (manage the person's preferences and identification) and "DL Assistant" (Personalize the environment to match the users' needs in assistance) extends the abstract role "Angel". This abstract role defines the common constraints for the sub-roles (for example abiding to the users' preferences for quiet assistance) and provides them with the communications rights toward other "Angel" roles. This personal organization will provide a more efficient structure to personalize the environment and assist the flat's inhabitant by modeling the entire chain of adaptation from the physical world sensed by anonymous sensors to the ultimate personalization of the daily-living environment.

Mapping the scenario described in 2.1 on this organization, when Carl receives the phone call, the agents decide to assist him by displaying on the most visible screen available the name of the caller, and Carl's agenda to remind him of possible previous engagements. For instance suppose Carl is watching TV at this moment. The most visible screen is thus the TV. As a result, two agents are trying to assist Carl, the one adopting the role of Daily Living (DL) Assistant (cf. Fig 2.) which is displaying Carl's favorite show, and the one adopting the Agenda Manager role, trying to display his agenda. Both agents can use the communication link provided by the Angel role toward itself to solve this conflict.

3.2 *Local Organization*

The local organization is used (Fig. 2) to manage and organize the architecture of a smart home as presented in section 2.2. It consists in a group "Intelligent Home" where three roles can be adopted. A "*Device*" role is used for expressing the governance responsibilities in relation with the management of sensors and the access to the actuators while proposing some low level services to enable the dynamic migration of any agent on a sufficiently powerful device.

A "*Contexter*" role is used to express the governance responsibilities for guarantying a real-time knowledge on the context. If only one agent adopts this role, the context will be defined as the overall context, aware of the number of person in the flat, while it could be very precise if there are more agents (context of each room for example).

An "*AssistTask*" role covers the missions dedicated to provide assistance to a user for a specific task. An agent playing this role will have the mission to play mediator between the users' systems and the devices. This role extends the abstract role of "Actuator" which allows communication with the "Device" roles.

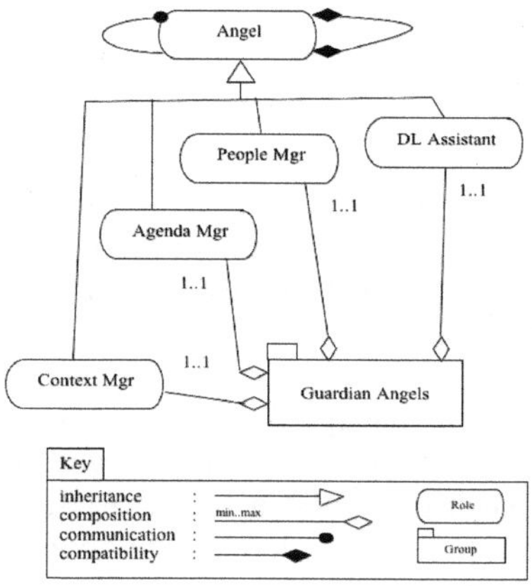

Fig. 1 Graphical representation of the Structural Specification of the Personal organization

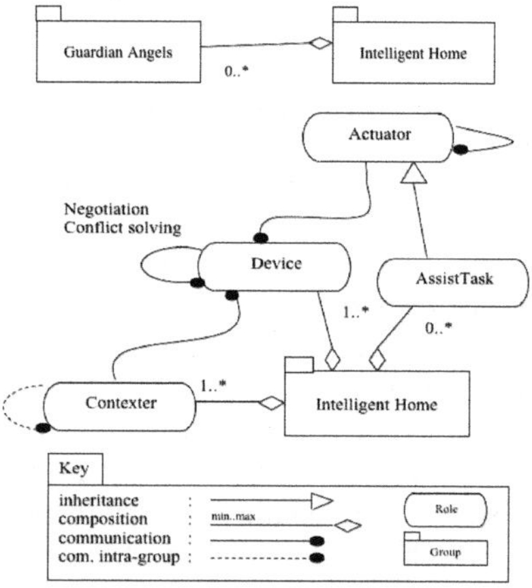

Fig. 2 Graphical representation of the Structural Specification of the Local organization

Agents in the system have to adopt one or more roles (The roles must be compatible) and thus complete the missions associated to these roles. A "Guardian Angel" group needs only one instance of each Angel role.

3.3 *Relational Organization*

To manage multiple persons in a smart home, the global organization needs to specify some channels of communication for each respective roles to negotiate, exchange, and collaborate. A personal organization both has to know if its action will impede another personal organization (by asking the "Actuator" if it is in use and the local "Contexter" if the action would disturb another's context) or the local rules thus take precautions, and has to check if another person is disturbing its assistance.

The inter-group communication link from "Angel" (Fig. 2) to itself allows any "Angel" to communicate with an "Angel" from another group of "Guardian Angels" thus permitting negotiation between two agents assisting two different people. When Alice is watching her show, the sound of the TV can disturb the vocal assistance guiding Bob in the meal preparation. The two agents responsible must negotiate. Furthermore, an "Intelligent Home" group can be composed of any number of "Guardian Angel" group, even none. When Carl arrives in Alice and Bob's smart home, his personal organization tries to integrate itself in the smart home local organization: his "Guardian Angel" group asks for admission in the "Intelligent Home" group.

4 Agents in a Nest of Organizations

The three organizations described previously can be used together in symbiosis. Figure 3 shows the nesting of personal organizations (mobile part of the organization since it follows the person in his/her different moves) in a local organization (configuration of the resources in the smart home).

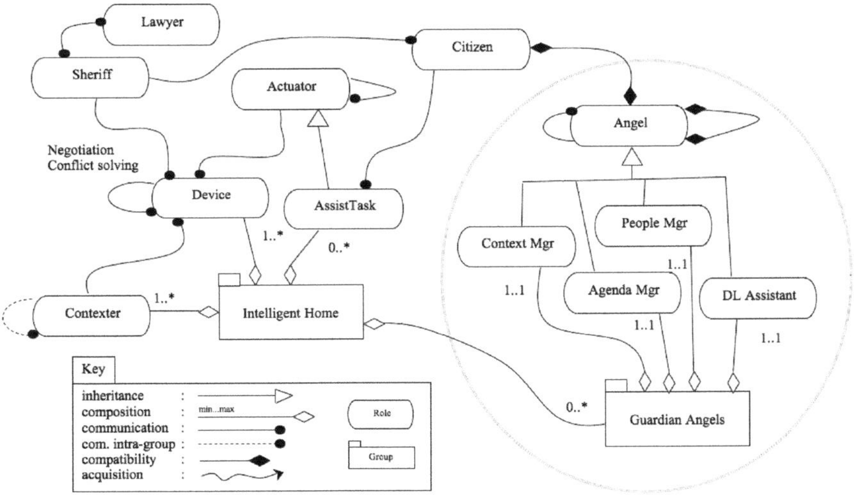

Fig. 3 Graphical representation of the structural specification resulting of the nesting of a personal organization into a local organization

The focus is on the resolution of conflicts issued from the presence of more than one personal organization in more than one smart place : conflicts of rights, access and control. To that end, three roles are introduced for the nesting: Citizen, Sheriff and Lawyer. When a physical person enters for the first time in the smart home, a multi-agent system based on the personal organization is created. This personal organization (Guardian Angel group) is created as part of the local organization (Intelligent Home group) and thus innately obedient to the local rules. When this same person exits the home, the personal multi-agent follows the person (thanks to serialization on his pocket pc for example).

When a person already managed by a multi-agent system enters a smart home, the group “Guardian Angels” ask to be included in the “Intelligent Home” group. To that end, every “Angel” of this group must also adopt a local “Citizen” role (Fig. 3). This “Citizen” role enforces a set of rules: The rules of the home (Ex: No noise after 23h, no access to the bedroom for visitors, etc.).

The nesting of the different organizations can go a step further than this basic vision. To adapt to the need of security and monitoring, it is possible to enrich the local organization with some measures of feedback, control and adaptation. In addition to the role of “Citizen”, this organization adds two roles to control external agents: a “Sheriff” role and a “Lawyer” role (cf. Fig. 4).

The “Sheriff” role has for mission to control the good respect of the local laws (no noise after 23h for example) by every “Citizen”. In case of violation, the sheriff can punish the agent by a loss of privilege (lower priority in negotiations), a probation period (every access to devices is made by a limiting proxy) or an ejection from the local organization. The sheriff can also decide that a rule is obsolete and ask the lawyer to change it. If Carl’s “DL Assistant” (integrated in Alice and Bob’s “Intelligent Home”) command the room temperature’s agent (playing an “AssistTask” role) to lower its level, the “Sheriff” can emit a warning. On relapsing, the “Sheriff” can disable the communication between Carl’s “DL Assistant” and the temperature agent.

The agent adopting the “Lawyer” role has for missions to know the rules, to make them known to the other agents «citizens» and to change them when needed or asked. If Carl becomes a regular presence in Alice and Bob’s home, the lawyer could be told by Alice or Bob to accept lowering a bit the room temperature.

5 Conclusion and Future Works

We described in this paper a solution to manage the various configurations needed to assist and personalize the environment for multiple persons in multiple environments. This application is still an ongoing research and some of the finer points (like some conflicts resolutions for example) aren’t completely rounded out since the testing is actually in motion. This solution proposes the interweaving of personal organization used to manage personal configuration of the environment in local organizations representing each physical environment and its explicit configurations. The resulting nesting provides assurance of a moderated functioning in any space conformingly to the local rules.

Another kind of interlaced configuration that we didn't mention would consider the conflicts between the preferences of a ghost user (not physically present at the moment but for whom the preferences are known) and these of a physically present person or even another ghost. For example, when Bob leaves his home, should the local organization keep a trace of his preferences and force others to negotiate with them? This would be the result of the consideration of multiple persons in multiple places at the same time.

The missions associated to the roles were given as an example in this paper for a Health Smart Home, but the global structure of a local organization stays purposely general to contend with any kind of resources, sensors and actuators that we could find in a smart environment. This research is implemented in the DOMUS laboratory, based on JADE framework deemed the most practical [14] for agents' interactions. Most agents are implemented in Java language and all follow FIPA specifications.

References

1. Weiser, M.: The computer for the 21st century, vol. 265(3), pp. 66–75. Scientific American (International Edition) (1991)
2. Kadouche, R., Mokhtari, M., Giroux, S., Abdulrazak, B.: Personalization in smart homes for disabled people. In: FGCN 2008, Hainan Island, China, December 13 – December 15, pp. 411–415. Inst. of Elec. and Elec. Eng. Computer Society (2008)
3. Kuusik, A.: Engineering approach for realization of small smart home systems. In: First IFAC-Conference on Telematics Applications in Automation and Robotics TA 2001, Kidlington, UK, July 24-26, pp. 403–407. Elsevier Sci., Amsterdam (2001)
4. Rialle, V., Duchene, F., Noury, N., Bajolle, L., Demongeot, J.: Health Smart home: Information technology for patients at home. Telemedicine Journal and e-Health 8(4), 395–409 (2002)
5. Stringer, M., Fitzpatrick, G., Harris, E.: Lessons for the future: Experiences with the installation and use of today's domestic sensors and technologies. In: Fishkin, K.P., Schiele, B., Nixon, P., Quigley, A. (eds.) PERVASIVE 2006. LNCS, vol. 3968, pp. 383–399. Springer, Heidelberg (2006)
6. Aarts, E.: Ambient intelligence: a multimedia perspective. IEEE Multimedia 11(1), 12–19 (2004)
7. Georgalis, Y., Grammenos, D., Stephanidis, C.: Middleware for ambient intelligence environments: Reviewing requirements and communication technologies. In: UAHCI 2009, San Diego, CA, United states, pp. 168–177. Springer, Heidelberg (2009)
8. Martinez, F., Tynan, E., Arregui, M., Obieta, G., Aurrekoetxea, J.: Electroactive pressure sensors for smart structures. In: 3rd International Conference on Smart Materials, Structures and Systems - Emboding Intelligence in Structures and Integrated Systems, CIMTEC 2008, Acireale, Sicily, Italy, June 8 - June 13, pp. 122–126 (2008)
9. Rialle, V., Ollivet, C., Guigui, C., Herve, C.: What do family caregivers of Alzheimer's disease patients desire in smart home technologies? Contrasted results of a wide survey. Methods of information in medicine 47(1), 63–69 (2008)
10. Ferber, J.: Les systèmes multi-agents, vers une intelligence collective. InterEditions édition, InterEditions (1995)

11. Dignum, V.: Multi-Agent Systems - Semantics and Dynamics of Organizational Models. IGI Global (2009)
12. Pigot, H., Mayers, A., Giroux, S.: The intelligent habitat and everyday life activity support. In: Fifth International Conference on Simulations in Biomedicine, Biomedicine 2003, Ljubljana, Slovenia, April 2 - April 4, pp. 507–516 (2003)
13. Hubner, J.F., Boissier, O., Kitio, R., Ricci, A.: Instrumenting multi-agent organisations with organisational artifacts and agents. Giving the organisational power back to the agents (2009)
14. Shakshuki, E., Jun, Y.: Multi-agent development toolkits: an evaluation. In: Orchard, B., Yang, C., Ali, M. (eds.) IEA/AIE 2004. LNCS (LNAI), vol. 3029, pp. 209–218. Springer, Heidelberg (2004)
15. Wu, C.-L., Liao., C.-F., Fu, L.-C.: Service-oriented smart-home architecture based on OSGi and mobile-agent technology. IEEE Transactions on Systems, Man and Cybernetics, Part C (Applications and Reviews) 37(2), 193–205 (2007)
16. Cook, D.J., Youngblood, M., Das, S.K.: A multi-agent approach to controlling a smart environment. In: Augusto, J.C., Nugent, C.D. (eds.) Designing Smart Homes. LNCS (LNAI), vol. 4008, pp. 165–182. Springer, Heidelberg (2006)
17. Hsu, C.-C., Wang, L.-Z.: A smart home resource management system for multiple inhabitants by agent conceding negotiation. In: IEEE International Conference on Systems, Man and Cybernetics, Piscataway, NJ, USA, October 12-15 (2008)

Advantages of MAS for the Resolution of a Power Management Problem in Smart Homes

Shadi Abras*,**, Sylvie Pesty**, Stephane Ploix*, and Mireille Jacomino*

Abstract. This paper contributes to the design of intelligent buildings. A Multi-Agents Home Automation System (MAHAS) is proposed which controls appliances and energy sources in buildings. The objective of this paper is to show that by using intelligent agents related to appliances, it is possible to improve the energy consumption/production in buildings. The proposed MAHAS system is characterised by its openness, its scalability and its capability to manage diversity. In this paper, we show how a multi-agent system, well adapted to solve problems spatially distributed and opened, can dynamically adapt the consumption of energy to various constraints by exploiting the flexibilities of the services provided by domestic devices (services shifting, energy accumulating). Finally, we conclude on the contribution of Multi-Agent approach to the power management problem.

Keywords: Multi-Agent Systems, Home Automation, Power management.

1 Introduction

A home automation system basically consists of household appliances linked via a communication network allowing interactions for control purposes [2]. Thanks to this network, a load management mechanism can be implemented: this is called *distributed control* in [3]. Load management allows inhabitants to adjust power consumption according to expected comfort, energy price variation and CO_2 equivalent emissions. For instance, during consumption peak periods when power plants emitting higher quantities of CO_2 are used and when the energy price is high, some services can be delayed, heater set-points can be reduced or services run according to the weather forecast or inhabitants' requests. Load management offers further advantages when there are local storage and production means. When combined, all

Shadi Abras · Sylvie Pesty · Stephane Ploix · Mireille Jacomino
Laboratoires G-SCOP* & LIG**
46 Avenue Felix Viallet, 38031 Grenoble, France
e-mail: {Shadi.Abras,Sylvie.Pesty}@imag.fr,
{Stephane.Ploix,Mireille.Jacomino}@inpg.fr

Y. Demazeau et al. (Eds.): Advances in PAAMS, AISC 70, pp. 269–278.
springerlink.com

of these possibilities lead to systems offering many degrees of freedom that are very complex for users to manage.

The objective of this study is to design a building electric energy management system able to determine the best energy assignment plan, according to given criteria. A building energy management system can be seen from two points of view: load management and local energy production management. [4] has proposed centralised control strategies for an HVAC (Heating Ventilation and Air Conditioning) system taking into account the natural thermal storage capacity of buildings where the HVAC consumption is shifted from the peak period to the off-peak period.

1.1 Towards a Multi-Agent System for Energy Management?

In the home, the energy system is made up of sources, which produce energy, and loads (household appliances), which consume energy. The energy comes from distant producers (national transmission/distribution networks), but may also come from local energy sources (e.g. solar panels, wind turbines and fuel cells). However, faced with this increasing complexity in buildings, it is necessary to develop control and monitoring systems able to manage multi-source production and understand the dynamic characteristics of the home in order to anticipate energy needs and dialogue with all the components producing and storing energy. Such control and monitoring systems must also be able to plan ahead for the necessarily changing configuration of buildings. Alongside this approach, it is also necessary to find a solution for lowering the energy consumption of existing buildings.

Owing to the number and diversity of devices in the home, leads us to opt for solutions fostering modularity. The need for structural self-adaptation, or more technically the need for plug-and-play appliance, lead us to the Multi-Agent paradigm enabling a minimum amount of information to be shared between modules and asynchronous operation to be possible via message exchanges. Indeed, a home automation system must be open-ended and extendible: it must be possible to add or remove appliances (or new types of appliances) at any time without calling into question the overall operation of the system.

Progress in the field of Multi-Agent Systems, designed to solve spatially distributed and open problems, is likely to lead to a "Multi-Agent Home Automation System" made up of agents embedded in different sources and loads all working together in order to solve the problem of energy consumption control in the home. The agents make joint decisions with respect to the operating requirements of the appliances to which they are connected.

A work closest to our work is that of Davidsson [6], [7]. It presents a MAS for power management in office buildings. The objective of this system is to manage some office services as lighting, heating and ventilation. This work has similar goals as MAHAS system but it is for small office spaces. However, its system does not take into account the multitude energy resources requirements.

The objective of our work is to design an energy management system for the home made up of software agents controlling appliances and sources. Together, all the agents would form a multi-agent home automation system. In other words, they would form an energy management system capable of dynamically setting up a production and consumption policy to suit user criteria and the various load and source requirements.

The following diagram (FIG. 1) presents the overall structure of the MAHAS system.

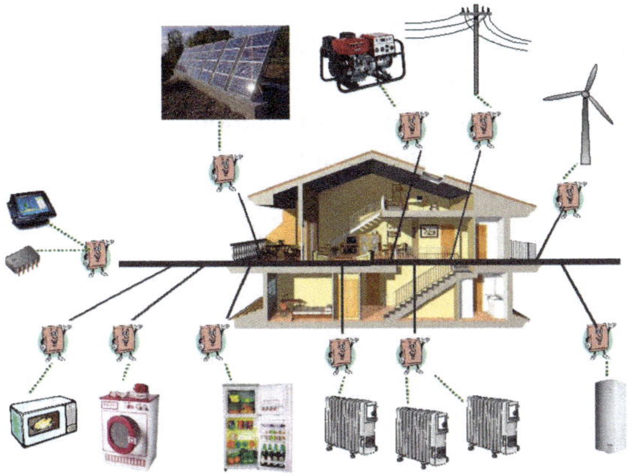

Fig. 1 Overall structure of the MAHAS system for energy management

The paper's objective is to show that the distributed power system based on Multi-Agent System techniques does offer advantages over the centralised approach.

In what follows, in a first part, the paper presents briefly the agents modelling of the proposed system, called MAHAS which is described in details in [1]. In the second part of this paper, we show how a multi-agent system, well adapted to solve problems spatially distributed and opened, can dynamically adapt the consumption of energy to various constraints by exploiting the flexibilities of the services provided by domestic devices (services shifting, energy accumulating). Finally, we conclude on the contribution of Multi-Agent approach to the power management problem.

2 Modelling Principles

In this section, two main notions will be introduced: service and satisfaction.

2.1 Service

Given the broad array of appliances and the fast changes to technology, along with their direct influence on behaviour, it is fairly difficult, if not impossible, to define a model suitable for all the appliances belonging to a specific category. This is why we have chosen to introduce the notion of service as the grouped operation of different appliances. A service (i), denoted as SRV_i, transforms energy in order to meet a user's need via one or several appliances. A service is qualified as *permanent* if its energy consumption / production covers the whole time range of the energy assignment plan such as heating service, otherwise, the service is referred to as a *temporary service* such as cooking or washing service.

A temporary service is characterised by the duration and desired end time of the operation. The flexibility of this service comes from the possibility of shifting its operating time it, i.e. bringing forward or delaying the service.

A permanent service is characterised by a quantity of energy consumed or produced. The flexibility of this service comes from the possibility of modifying the energy quantities consumed/produced throughout all the periods (decrease or increase in energy consumption or production at a given time).

MAHAS system agents are therefore going to be built around this notion of service, i.e. we will refer to "temporary agents" and "permanent agents".

2.1.1 Satisfaction Function

In home automation, "user comfort" is one of the most important aspects to take into consideration. The notion of comfort can be directly linked to the concept of satisfaction function. Since the notion of comfort is not universal, we shall represent different satisfaction functions for each service and these will depend on users' desires. For example, one user will be happy if the temperature in their living room is between 20^oC and 22^oC, while a different user might prefer it to be between 19^oC and 20^oC. In MAHAS system agents, the satisfaction function characterises a user's feelings with respect to a service and is fairly close to the notion of personal satisfaction described by [5]. Satisfaction functions have been defined for each service (energy consumers and energy suppliers).

The satisfaction relative to a service will be expressed by a function defined over an interval $[0, 100\%]$ where zero is "unacceptable" and 100% "perfect". This can be seen as a degree of belonging to the satisfied whole in blurred logic.

The satisfaction function of a temporary agent can be estimated by a piece-wise linear function that depends on the time shift in service offered in relation to the end time desired by the user (FIG. 3). The satisfaction function of a permanent agent can also be estimated by a piece-wise linear function that depends on the characteristic service variable. For example, the satisfaction function for a heating service depends on the temperature variable.

The proposed system is made up of software agents each of which performs a specific task. The components making up the system are energy production sources and household appliances supporting the operation of services. The first idea was to link a software agent to all components. However, we have seen that the notion of service is more generic, allowing us to define two main service categories: temporary services and permanent services. In this system, an agent will thus be able to control one or several appliances and/or sources.

In what follows, we shall present MAHAS system agent modelling.

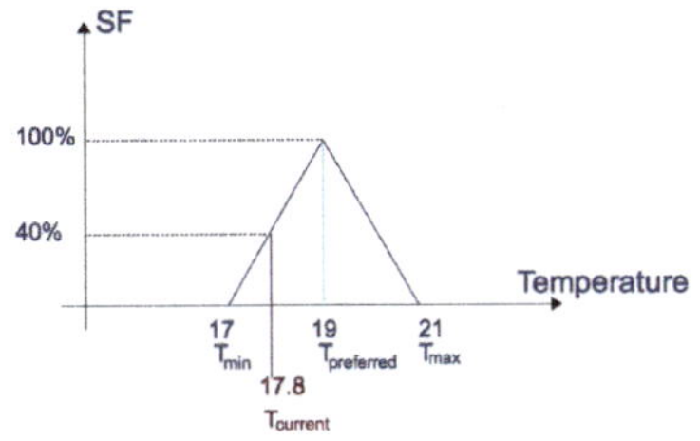

Fig. 2 Example of a heating service temperature satisfaction function

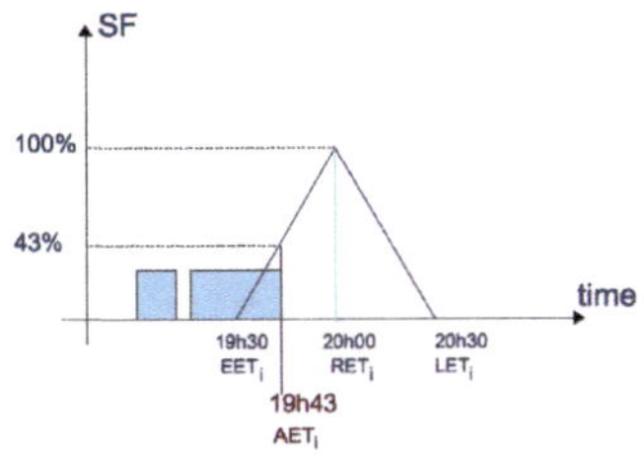

Fig. 3 Temporary service satisfaction function

2.2 *Agent Modelling*

Agent knowledge comprises internal knowledge and some shared knowledge. External (shared) knowledge represents the information exchanged between agents (**temporary and permanent**). This knowledge can be modelled by a realisation:

- a power profile: $\Pi = (P_{i,k}, \dots, P_{i,k+l})$ where $P_{i,m}$ represents the power consumed/produced (negative/positive) by the service over the period stretching from k to $k+l$. l represents the horizon for processing the energy assignment problem. The service duration Δ_i covers the entire length of the profile: $length(\Pi)$.
- a satisfaction function value that depends on the characteristic variable (e.g. the variable of the energy supplied by an energy source).

A realisation therefore corresponds to a power profile denoted as (k, Π, σ) over the time period $[k, k+l]$; σ represents the profile's satisfaction value representing the user's degree of satisfaction with the service accomplished by the agent.

2.2.1 Temporary Agent Modelling

Temporary agent internal (private) knowledge can be modelled by:

- a characteristic variable that depends on the service operation end time AET_i where AET_i is the actual service end time.
- a satisfaction function that depends on the difference between the actual end time AET_i and the required end time RET_i but also on the earliest end time EET_i and the latest end time LET_i (FIG. 3).
- a behaviour model (a finite-state machine) used to define the energy consumption/production for a service rendered. In this part, defines the service operation stages. The overall duration of the temporary service is $\Delta_i = n_i \times \Delta$; $n_i \in \mathbb{N}^*$.

2.2.2 Permanent Agent Modelling

Permanent agent internal (private) knowledge can be modelled as shown below:

- a characteristic service variable (e.g. the temperature for a heating service).

- a behaviour model used to describe the continuous changes in the activity of the service such as heating or air conditioning in relation to the power injected and other contextual magnitudes. As we have already shown, this model can be described as a system of differential equations.
- a satisfaction function that depends on the state variables modelling the service behaviour model.

The agent therefore uses an infinite internal loop, which ensures the agent is persistent (FIG. 4).

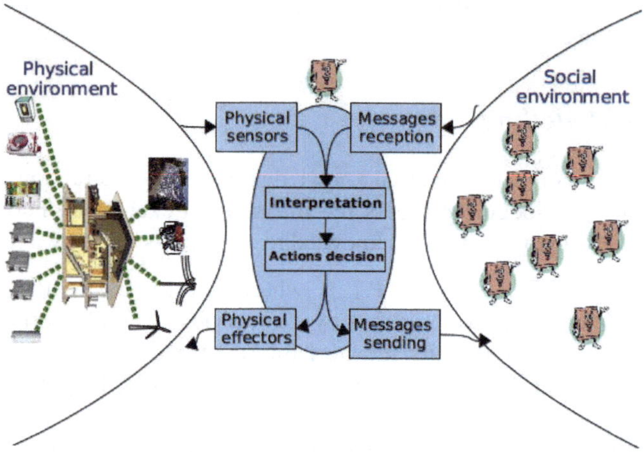

Fig. 4 Agent-environment interaction

3 Multi-level Energy Management System

The biggest difficulty when addressing the problem of assigning energy resources in a home is taking into account very different dynamics. Some phenomena require very short response times, such as violation of the maximum resource requirement calling for swift management of conflicting energy demands. There are also some physical phenomena that are relatively slow such as inertia in the home, cyclical variations such as the price of energy bought or local energy production capacity such as solar energy. Hence, the system's control architecture must be able to manage the periodic cyclical phenomena occurring in the home. The proposed MAHAS system has two main control levels. These levels can be identified according to the different sampling periods and time horizons: the **anticipative mechanism** and the **reactive mechanism**.

3.1 Reactive Mechanism

The reactive mechanism aims to fulfil the anticipative mechanism's assignment plan [1]. It makes it possible to respond to unscheduled events (unexpected lack of power within a time period, unprogrammed consumption, etc.) and to prevent the operation of some predefined appliances from being entirely interrupted (e.g. an oven or a radiator in a room being completely stopped). This makes it possible to deal with urgent situations and maintain a satisfactory level of comfort for the user. This mechanism fulfils what one might call *intelligent load-shedding* since some appliances can be temporarily interrupted. However, unlike conventional load-shedding where the non-priority appliances in an installation are defined, here the agents embedded in the appliances negotiate which ones are to be shed. This mechanism works on real energy values over a relatively short period (of around one minute) because, on the one hand, its objective is to respond to instantaneous unscheduled events and, on the other hand, a closer level to the equipment is needed in order to take into account the actual energy consumption and production values.

Similarly to the reactive Multi-Agent System, the agents of the reactive mechanism behave according to a stimulus-response process (FIG. 4) with communication capacity (dispatch/receipt of messages). The role of an agent is as follows:

- it constantly monitors its level of current satisfaction (e.g. the temperature value for the heating service using a physical sensor).
- when its level of satisfaction falls below a given satisfaction level (critical satisfaction), it warns the other agents (help requested in the messages transmitted).
- when it receives requests from other agents, it analyses these and in turn puts forward proposals.
- when it receives responses to its own requests, it opts for the most interesting proposals.

The agent checks its level of satisfaction during each cycle of its infinite loop. The level of urgency corresponds to the critical satisfaction of the next cycle and determines the reactive period in relation to the satisfaction curve gradient. The agent monitors its level of satisfaction and, if the critical satisfaction level has been reached, it initialises a message exchange process run.

3.2 Anticipative Mechanism

The notion of anticipation has also been used in the field of Artificial Intelligence in different contexts. In the field of Multi-Agent Systems, anticipation is used to build complex adaptive behaviour and planning in dynamic environments.

In the field of energy management, several studies have pointed to the advantage of shifting appliance operation from peak periods to off-peak periods in order to reduce energy costs. [8] suggests a centralised system for managing energy consumption and production in a building. The drawback of this system is the centralisation: the energy assignment plans are set up by collecting all the appliance operation models and all the necessary data. To do this, the models have to be transformed into a set of mixed linear constraints comprising continuous and linear variables. It is difficult to automate such an operation and, because of this, it is difficult to take into account any services that have not been planned in the design of the home automation system. This then limits the possibility of upgrading the system. Furthermore, the centralised approach means that all appliance operation models have to be known, keeping in mind that manufacturers do not always want to supply these for competitive reasons. It is therefore difficult to adapt this approach to real home automation system contexts because it is ill-suited to frequent and varied configurations/reconfigurations. The approach does not allow for an open-ended system where appliances can be added or removed without having to reconfigure the system and without calling into question the overall operation of the optimisation algorithm that should be potentially capable of taking on any type of constraint.

An intelligent power management problem in buildings can be divided into sub-problems involving different agents because, generally, inhabitants do not use their devices most of the time. The basic principle is to divide the whole problem into independent sub-problems then to solve each sub-problem independently in order to find a solution for the whole problem. The advantage of this method is to reduce the complexity of the whole problem which depends on the number of periods for each sub-problem and on the number of devices. When the whole problem is divided into sub-problems, each sub-problem does not involve all of the services (for example: some inhabitants vacuum clean in the morning and not in the evening). Because the number of considered periods in a sub-problem is lower than the number of periods of the whole problem, the complexity of solving the sub-problems is less than the complexity of solving the whole problem at once.

The solving computes a predicted plan for each sub-problem using the **simulated annealing** algorithm. The basic principles of simulated annealing for the MAHAS problem are the following: the search for a solution starts from the initial solution found at the energy distribution step. At each iteration, the solving agent decreases the satisfaction interval (for example 5% for each time) and sends it with the best realisations (according to a combination of cost and comfort criteria) at the previous iteration to the agents; agents computes the neighbourhood of the realisation sent by the solving agent by generating a given number of realisations corresponding to the satisfaction interval. When the solving agent receives the agent realisations, it chooses the realisations that violate the constraints the less as possible. The search is stopped when the collected realisations do not violate the global power constraints and when the global satisfaction has converged. Because the number of realisations corresponding to a satisfaction interval is very high, an agent generates randomly theses realisations corresponding to the satisfaction in performing elementary step from the realisation selected by the solving agent at the previous iteration knowing that an agent realisation will not be generated and sent twice to the solving agent.

The real implementation of the MAHAS system is being constructed in the project PREDIS[1].

4 Results

In this section, we shall compare the anticipative mechanism with the prediction/predicted centralised scheduling system proposed by [8] for the management of energy consumption and production in the home. The criteria used to compare

[1] ANR project: `http://www.tenerrdis.fr/rep-plateformes_competences/rub-predis.html`

these mechanisms are the quality of the solution, linked to the satisfaction function, and the operating time. Our assessment is based on thirty different examples generated at random for both mechanisms.

In figure 5, we can see that the satisfaction of the best solution obtained is 75% for the MAHAS distributed mechanism proposed and 82% for the centralised system. We can also see that the search for a solution takes more time (35% extra) for the mechanism proposed in relation to the centralised system.

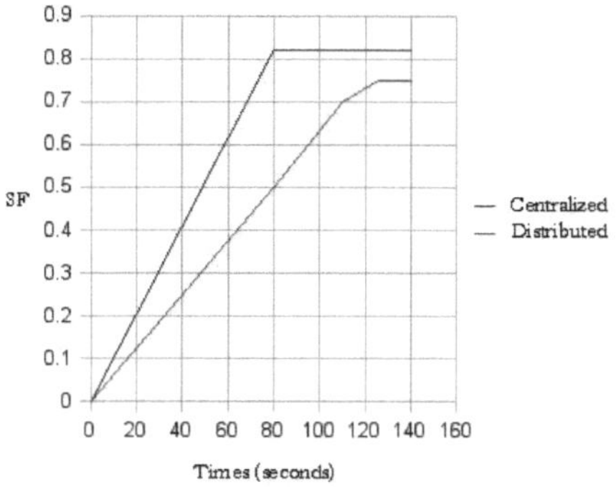

Fig. 5 Comparison of the proposed mechanism and the centralised prediction/predicted scheduling mechanism

Out of the 30 examples, the MAHAS distributed mechanism produces a 40 s to 120 s greater run time than the centralised mechanism. The quality of the solution obtained is lower by 5% to 20% compared with the centralised mechanism. Several factors explain these differences: firstly, satisfaction function: the centralised mechanism looks for the optimum solution unlike the proposed mechanism that aims at an acceptable solution. Secondly, the difference in run time is due to the time needed to send/receive messages so that each agent can analyse and respond to a message.

We can see that the performance of the centralised system is better than that obtained with the proposed distributed system. However, as far as real implementation is concerned, the distributed system based on Multi-Agent System techniques does offer some considerable advantages over the centralised approach:

- it is difficult to adapt the centralised approach to real home automation contexts as this approach is not suitable for frequent and varied configurations / reconfigurations. This means that it is not possible to have an open-ended system unlike with the distributed system where appliances can be added or removed without having to reconfigure the system and without penalising the overall running of the optimisation algorithm that has to be potentially capable of taking on board all kinds of requirements.
- the distributed system based on MAS techniques allows for extendability: new types of appliances can be added to the system without it being necessary to entirely reconfigure it. On the other hand, with the centralised system, the optimisation algorithms must be changed to take into account new appliances:

- the distributed system adapts to the type of home automation system: the physical distribution of different types of energy consumers (such as the oven, the washing machine, the radiator, etc) and the physical distribution of energy sources (such as solar panels, generator sets, fuel cells, etc).

5 Conclusion

In this paper, we have described the architecture of a multi-agent home automation system MAHAS; it adapts to different time scales and is made up of a reactive mechanism and an anticipative mechanism. A hybrid approach was put forward. This combines both methods: a metaheuristic method and an exact method. The aim of the metaheuristic method is to reduce the complexity of a given problem by breaking down the research space into parts. The exact method is then used to identify the best solution for the chosen part. The proposed architecture makes it possible to tackle the phenomena described using different time scales; a solution integrating all the information available at different levels of abstraction can then be built.

We noticed that the performance of the centralised system is better than that obtained with the proposed distributed system. However, as far as real implementation is concerned, the distributed system based on Multi-Agent System techniques does offer advantages over the centralised approach: its openness, its scalability and its capability to manage diversity.

References

1. Abras, S., Ploix, S., Pesty, S., Jacomino, M.: A multi-agent home automation system for power managememnt. In: Third International Conference in Control, Automation, and Robotics, ICINCO 2006, Setubal, Portugal, August 1-5. Springer, Heidelberg (2006)
2. Palensky, P., Posta, R.: Demand side management in private home using lonworks. In: Proceedings.1997 IEEE International Workshop on Factory Communication Systems (1997)
3. Wacks, K.: The impact of home automation on power electronics. In: Applied Power Electronics Conference and Exposition, pp. 3–9 (1993)
4. Zhou, G., Krarti, M.: Parametric analysis of active and passive building thermal storage utilization. Journal of Solar Energy Engineering 127, 37–46 (2005)
5. Lucidarme, P., Simonin, O., Liégeois, A.: Implementation and evaluation of a satisfaction/altruism based architecture for multi-robot systems. In: Proceedings of the 2002 IEEE International Conference on Robotics and Automation, Washington, USA, August 1-5, pp. 1007–1012 (2002)
6. Davidsson, P., Boman, M.: Distributed monitoring and control of office buildings by embedded agents. Inf. Sci. 171(4), 293–307 (2005)
7. Davidsson, P., Boman, M.: A multi-agent system for controlling intelligent buildings. In: ICMAS, pp. 377–378 (2000)
8. Ha, D.L.: Un systme avanc de gestion d'nergie dans le btiment pour coordonner production et consommation. Ph.D. dissertation, Institut Polytechnique de Grenoble (septembre 19, 2007)

A4VANET: Context-Aware JADE-LEAP Agents for VANETS

Mercedes Amor, Inmaculada Ayala, and Lidia Fuentes

Abstract. This paper presents a context-aware agent-based application for Vehicular Ad-hoc Networks. The characteristics of agent technology make them appropriate for developing dynamic and distributed systems for VANETS, and facilitate the addition of context-aware abilities to traditional applications. In this paper we show how a JADE-LEAP agent can process context information and use it to improve the comfort of the vehicle occupants with a minimum user interaction.

Keywords: Agent, VANET, Context-Aware, JADE, AmI.

1 Introduction

Ambient Intelligence (AmI) [1] is a paradigm which proposes the development of intelligent environments, places where users can interact naturally with an environment which is provided with special devices that can assist them. One of the most interesting AmI systems is Vehicular Ad-Hoc Network (VANET), a form of mobile Ad-hoc network that allows information exchanging between users in vehicles and between users in vehicles and service providers, along roads.

Although the reduction of traffic accidents is the focus of many of current research projects in VANETs, but it is also used to provide context-sensitive applications to enhance the driver comfort. These applications use the information from the context to provide the users with services that best fit their interests and profiles. Current context-aware approaches mainly support user location to adapt the provided services, but the information provided by different car sensors and external sources (such as other vehicles and devices of the infrastructure) gives an opportunity to build new context-aware applications able to assist and improve the comfort of the vehicle occupants with a minimum user interaction.

This paper presents the development of an A4VANET [2] Multi-Agent System (MAS) supporting context-aware application for VANETs. A4VANET facilitates the development of agent based applications for mobile devices to be executed in a simulated VANET. This work reports the use of context-aware

Mercedes Amor · Inmaculada Ayala · Lidia Fuentes
Departamento de Lenguajes y Ciencias de la Computación
Universidad de Málaga
Bulevar Louis Pasteur 35, 29071, Málaga, Spain
email: {pinilla,ayala,lff}@lcc.uma.es

Y. Demazeau et al. (Eds.): Advances in PAAMS, AISC 70, pp. 279–284.
springerlink.com

agents implemented in JADE-LEAP [3] that enable the adaptation of the VANET application to changing circumstances and respond according the user's context. In this paper, we propose a context-aware fuel filling up recommender for VANETs.

The paper is structured as follow: First, we provide a brief discussion of related work and we describe a context-aware service used as example. In section 2 we introduce the MAS developed in JADE-LEAP and A4VANET, which is also briefly introduced. In Section 3, we show the execution of the context aware application. The paper concludes with a brief summary and future work.

2 Vehicular Ad-Hoc Networks

A VANET is a special kind of mobile ad-hoc network composed of mobile nodes (mainly vehicles). The specific properties of VANETs allow the development of new services, mainly related to safety and comfort. Safety applications increase the security of passengers by exchanging safety relevant information. Comfort-related applications improve passenger comfort and traffic efficiency. Examples for this category are: traffic information, weather information, gas station or restaurant location, price information, and even Internet access. In a VANET, each vehicle is equipped with a computing device, a short-range wireless interface, a GPS (Global Positioning System) receiver, and different internal sensors, which provide fuel tank level, speed, and information regarding the status of the vehicle.

The development of application for VANETs have to take into account the characteristics of the environment in which such applications are deployed: Devices for VANETS have the same limitations as AmI systems (poor capacity process and memory) and besides, we have to consider problems about the network, for example a low or null connectivity or a limited bandwidth with a low latency. Finally, systems have to be context-aware and intelligent for work with a minimum user interaction. In this section we provide a brief overview of related work and we present the context-sensitive application.

2.1 Related Work

Nowadays, agent technology has been successfully applied to different application domains in which AmI ideas have been exploited such as smarthomes, education, health, emergency services or transport, in which VANETs are included. Existing efforts provide environments and applications that make use of the agent technology to provide context-aware and adaptable services in different application domains: the provision of context awareness services in a airport [4]; the ASK-IT project [5], which provides customized services to physically handicapped person; and an intelligent emotional-aware MAS [6] that is able to support Distributed Decision Making Groups.

In VANETs, the application of agents is usually found as a solution for the collection, management, and communication of routing and traffic information [7, 8]. However, the use of contextual information to enhance VANETs applications has focused in the use of current location of the vehicle and the proximity to other devices of the VANET. Actually, examples of existing context aware applications in VANETs are tourist guides [9], traffic information [10], POI localization and other similar applications. Recent efforts have started to investigate context awareness in personalization and recommender systems in mobile scenarios. Concretely, the proposed approach within the NoW project [11] uses gas station recommendation for cars as an application example, similar to the context-aware application that we propose in next section. Unlike ours, the proposed system has been implemented and simulated in a non-distributed environment, which does not take into account the special system requirements imposed by a VANET.

2.2 Context-Aware Application in VANETS

Context-aware applications are services that are adapted according to the current context. In VANETs, context information is determined with the information provided by internal sensors and information produced by other elements in the VANET. In addition, this contextual information is aggregated with the user preferences, in order to generate a more valuable proposal for the user.

For this paper we propose as an example a context-aware vehicle agent for VANETs which decides when and where the vehicle has to refuel. In a specific scenario, when the agent detects that the level of fuel gas is running low, it decides to refuel. Additional context information, such as vehicle speed, is taken into consideration to make this decision (fuel consume is higher when the speed is low, and it should be acceptable to stop when the tank level is low, and it is not in reserve if the speed is low). Vehicle agents interact with other agent named service agents, which represent road services, to receive information about gas stations in the proximity. When refuel is needed, autonomously, the agent chooses a gas station meeting user preferences (i.e. a specific gas station chain).

3 Context-Awareness in A4VANET

A4VANET is composed of a set of agents, utilities, and components developed to implement a portion of a VANET and simulate MAS for VANETs in an appropriate environment. This section focuses on the description of a context-aware agent in an agent-based VANET application that is included in A4VANET MAS developed in JADE-LEAP. The UML diagram in Fig. 1 shows the architecture of the MAS, providing a general overview of the types of agents of the system and how they interact. In A4VANET, agents are used to represent and simulate meaningful entities in VANETs. According to this, a MAS for VANETs encompasses three different agents: agents inside a vehicle (*VANETAgent* class), and agents representing services (*ServiceAgent* class) and signs of the roads (*SignalAgent* class). Vehicle agents run on board to provide safety and comfort-related services to the vehicle passengers. These agents are context-aware to deal with the reception of events from different information sources (such as the user and internal sensors), and are able to adapt their behavior to current context. Service agents represent the road services (gas stations, cafeterias, and so on) of the road network; while signal agents represent signposting (traffic signs).

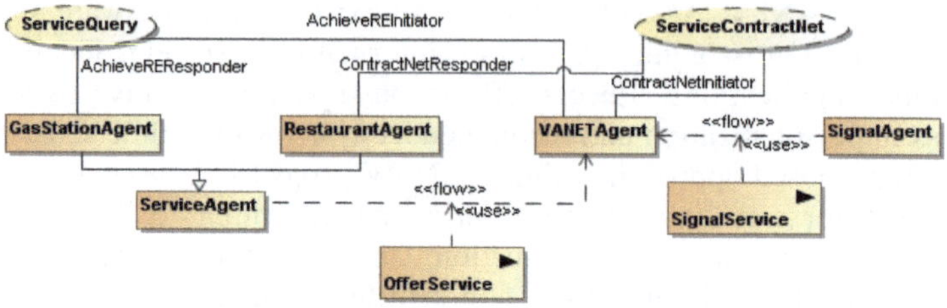

Fig. 1 Architecture of A4VANET MAS

Agent interaction takes place in the following scenarios: Although service and signal agents are registered in the agent platform's directory facilitator (DF), they initiate the interaction with vehicle agents when they are in the proximity (bluetooth-based device discovery and bluetooth communication through the AP). The communication goal is to facilitate information about service offers and signal road information respectively. In addition, vehicle agents interact with service agents to query and negotiate service provision. These interactions follow FIPA interaction protocols (interaction protocols named *ServiceQuery* and *ServiceContractNet*), and exchanges messages are codified using the FIPA ACL supported by JADE-LEAP. The terms (concepts and predicates) used in the content of the exchanged messages are defined in two ontologies (signposting and services).

The UML class diagram in Fig. 2 shows the part of the internal design of the context-aware *VANETAgent* in JADE-LEAP. Context-awareness, which requires functions for gathering, managing, disseminating, and processing context information, is represented by different classes of the agent architecture. Context information (also known as *context model*) is processed and stored by the *RouteControl* and *PetrolControl* classes, which maintain updated information of the current speed and state of the fuel respectively. The current vehicle location is estimated in the class *LocationBehaviour*, which uses Java Location API [13] to update it dynamically. The current location, which varies periodically, is used by the *RouteControl* and *PetrolControl* classes to estimate the current speed and current petrol consume respectively.

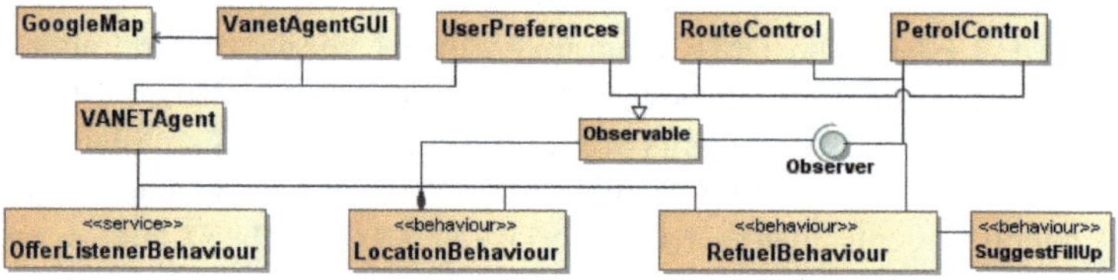

Fig. 2 *VANETAgent* class diagram

An adequate OO approach to cope with context-aware data and behavior is the subject-observer pattern (also known as Observer pattern) [12]. This behavioural pattern defines a one-to-all dependence between objects. When the state of one of the objects (the subject) changes, it notifies the change to all dependant objects (the observers). Regarding context-awareness, the subject is the context and the observers are the behaviors that depend on its current state. The *RouteControl* and *PetrolControl* classes play the role of subjects, while the class *RefuelBehaviour* is the observer. The class *OfferListenerBehaviour* deals with the communication with other agents, and facilitates nearby gas stations when refuelling is required.

The class *RefuelBehaviour* encapsulates the context-aware behavior that decides when it is required to refuel depending on the current fuel level, the speed, and the user preferences. This class listens for events from the *RouteControl* and *PetrolControl* classes, which throw an event the context changes and invokes a specific function to manage the change accordingly. Given a new context, the behaviour decides if it is required to refuel. There are different situations to decide it: (i) if tank level is low and speed is low; (ii) if tank level is in reserve and speed is fast; and (iii) if tank level is in reserve and speed is low. The urgency of the first and second situation is considered minor, and user preference (if it is set up) is considered. When any of these situations occurs the agent recommends the agent to refuel. When refuel is required, current user preferences are consulted to notify the user (left side of Fig. 3) that refuel is required and there is a gas station in the nearby. However, the third situation is considered critical, and the recommendation is made when the next gas station is near (even if it does not meet the user preference). Information about nearby gas station is extracted from the ACL messages received from service agents (handle and processed in the *OfferListener Behaviour* classes).

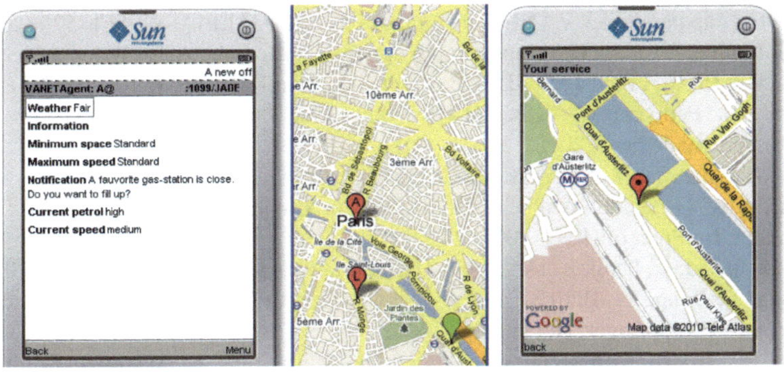

Fig. 3 Vehicle Agent user interface on the system simulation window

The context-aware application implemented in JADE-LEAP has been executed in an A4VANET simulated environment, this provide a map with the position of all the agents of the MAS. Each agent is loaded and executed in a different JAD-LEAP agent platform, which runs for MIDP in split execution mode. Each agent presents a graphical interface to interact (change user preferences) and present information to the user (such as information received from service and signal agents). Snapshots of the user interface of a vehicle are included in Fig. 3.

4 Conclusions

In this work, a context-aware MAS for developing applications for VANETS has been presented. The MAS encompasses different software agents representing the entities of a VANET (vehicles, road signal and services), which provide a distributed environment for the simulation of context-aware agent-based applications for VANETs. The contribution of the paper is twofold. First, we present the distributed and mobile JADE-LEAP implementation of our previously work [2], which fits better to the conditions found in a real VANET. Secondly, we propose context-aware agent-based applications for VANETs that is able to recommend when and where to fill up gasoline.

We are currently improving the current implementation of the context-aware vehicle agent in order to consider more context variables such as weather conditions. In addition, a vehicle simulator, for the practical experimentation of agents for VANETs, is currently under construction. This simulator would provide a more appropriate simulated environment for the vehicle agents.

Acknowledgements. This work has been supported by the Spanish Ministry Project RAP TIN2008-01942, the regional project FamWare P09-TIC-5231.

References

1. Cook, D.J., Augusto, J.C., Jakkula, V.R.: Ambient intelligence: Technologies, applications, and opportunities. Pervasive and Mobile Computing 5, 277–298 (2009)
2. Amor, M., Ayala, I., Fuentes, L.: A4VANET: Una aplicación basada en Agentes JADE-Leap para redes VANET. In: Actas de la XIII CAEPIA, pp. 561–570 (2009)
3. Bellifemine, F., et al.: Developing multi-agent systems with JADE. Wiley Series (2008)
4. Sánchez-Pi, N., et al.: JADE/LEAP Agents in an AmIDomain. In: Corchado, E., Abraham, A., Pedrycz, W. (eds.) HAIS 2008. LNCS (LNAI), vol. 5271, pp. 62–69. Springer, Heidelberg (2008)
5. ASK-IT, http://www.ask-it.org/
6. Marrieros, G., et al.: Ambient Intelligence in Emotion Based Ubiquitous Decision Making. In: Proc. of IJCAI 2007 (2007)
7. Xiaonan Liu, y., Fang, Z.: An Agent-Based Intelligent Transport System. In: Shen, W., Yong, J., Yang, Y., Barthès, J.-P.A., Luo, J. (eds.) CSCWD 2007. LNCS, vol. 5236, pp. 304–315. Springer, Heidelberg (2008)
8. Yin, M., Griss, M.: SCATEAgents: Context-aware Software Agents for Multi-Modal Travel. In: Applications of Agent Technology in Traffic and Transportation (2005)
9. Cheverst, K., et al.: Developing a context-aware electronic tourist guide: some Issues and experiences. In: Proc. of CHI 2000 (2000)
10. Santa, J., Gómez-Skarmeta, A.F.: Sharing Context-Aware Road and Safety Information. Pervasive Computing, 58–65 (2009)
11. Woerndl, W., et al.: Context-aware Recommender Systems in Mobile Scenarios. In: Proc. of IJITWE, pp. 67–85 (2009)
12. Gamma, E., et al.: Design Patterns: Elements of Reusable Object-Oriented Software. Addison-Wesley, Reading (1994)
13. Location Based Services, http://developers.sun.com/mobility/apis/articles/location/

Adaptive Multi-agent System for Multi-sensor Maritime Surveillance

Jean-Pierre Mano, Jean-Pierre Georgé, and Marie-Pierre Gleizes

Abstract. Maritime surveillance is a difficult task as its aim is to detect any threatening event in a dynamic, complex and hugely distributed system. As there are many different types of vessels, behaviours, situations, and because the system is constantly evolving, classical automated surveillance approaches are unrealistic. We propose an adaptive multi-agent system in which each agent is responsible for a vessel. This agent perceives local anomalies and combines them to maintain a criticality value, used to decide when an alert is appropriate. The importance of the anomalies and how they are combined emerges from a cooperative self-adjusting process taking user feedback into account. This software is currently under development in ScanMaris, a project supported by the French National Research Agency.

Keywords: Multi-Agent Systems, Self-Adaptation, Self-Adjustment, Complex Systems, Maritime Surveillance.

1 Maritime Surveillance as a Complex Multi-sensor System

Supported by the French National Research Agency, ScanMaris project 1 aims to increase the efficiency of maritime surveillance and management activities by producing a prototype system assisting human operators, indicating suspicious vessels and quantifying this suspicion.

A Complex and Dynamic Problem is there in concern as there are many different types of vessels, behaviours, situations, and because the system is constantly evolving and hugely distributed. Ultimately, each vessel can be seen as a multi-sensor data source as our prototype collects information from different sensors (AIS or Automatic Identification System, HFSWR or High Frequency Surface Wave Radar and classical radars) and databases (environmental, LLOYD Insurance, TF2000 Vessel DB, Paris MoU, ICCAT) and regroups them for each vessel. By analysing each vessels situation using a rule-based inference engine, anomalies can be detected, like a sudden change of direction, abnormal speed, specific zones

---

Jean-Pierre Mano · Jean-Pierre Georgé · Marie-Pierre Gleizes
IRIT, Institut de Recherche Informatique de Toulouse
Université Paul Sabatier, 118 route de Narbonne
31062 Toulouse cedex 9, France
email: `{mano,george,gleizes}@irit.fr`

Y. Demazeau et al. (Eds.): Advances in PAAMS, AISC 70, pp. 285–290.
springerlink.com

entering... These usually simple anomalies seldom constitute a threat by themselves but can be cumulated for a given vessel and the operator is informed of the state of the vessel. This mechanism of alert raising is based upon numerous parameters which adjustments enable to avoid false positives and be efficient in the detection of threatening situations.

Typical alert raising systems 2, which can be found in the medical, network or manufacturing domains, rely on the definition of "normal" behaviours and scenarios. This can be done using probabilistic and statistic methods, or neural networks. Any behaviour outside these specific boundaries raises an alert. This can not be applied to maritime surveillance where the heterogeneity of vessels and behaviours, and the constant evolution of the behaviours render the definition of behaviours or scenarios too costly. Few multi-agent system approaches exist in the maritime domain 3 and they are currently not able to adapt and learn. More exist in network intrusion detection 4, for meteorological alerts 5 or medicine but there are notable differences between these domains and the maritime domain precluding a simple transposition as they rely heavily on expert knowledge.

For a system to be able to adapt, its functional parameters will need to be adapted or to self-adapt. Manual adjustment by an expert is the most forward but of course unrealistic in a large-scale system. Using mathematical analytical models requires being able to produce an exact model. Genetic Algorithms and Neural Networks as with many bio-inspired approaches are time consuming and delicate to instantiate to a given problem.

Self-adaptive multi-agent systems have already been successfully applied to flood forecast 6 and preliminary aircraft design 7 for instance. A similar approach has been used in the ScanMaris prototype with the design of a self-adaptive multi-agent system (MAS8) able to efficiently raise alarms by self-adjusting the parameters regulating its functioning. We built a first MAS where each agent is responsible for a vessel and defined the agent's behaviour so that it perceives local anomalies and combines them. We then build a second MAS, responsible for adjusting each vessel-agents functional parameters dynamically (during runtime) and autonomously (without supervision) in response to environmental feedbacks. Both MAS will be presented below.

2 A Multi-agent System for Multi-sensor Surveillance

In a MAS, Each agent is an autonomous entity able to observe its environment, acts upon it, communicates directly or indirectly with other agents, and possesses its own representation (or knowledge), skills, state and goals.

Self-adaptation is achieved in our work by relying on the AMAS (Adaptive Multi-Agent Systems) theory 9 which is based on the well-discussed advantages of cooperation 10. The cooperative attitude between agents constitutes the engine of self-organization. Depending on the real-time interactions the multi-agent system has with its environment, the organization between its agents emerges and constitutes an answer to the difficulties of complex systems modelling (indeed, there is no global control of the system) 11.

Sensors, Data and the Architecture of *ScanMaris* are compound to emphasize, among a huge amount of data, elements that characterize abnormal behaviours or situations. Raw data from sensors are combined with data from databases to identify any detected ship. Then the inference rule engine (RE) analyse the ship status in its current environment from the space-time enriched map (EM): location, sea depth, presence or not in regulated areas (fishing zone, traffic separation scheme...), and neighbourhood. Other close ships are screened for collision headings and other abnormal behaviours like chasing, sudden heading or speed alteration. If an event breaks a rule, it becomes an anomaly and is sent to concerned vessel-agents.All those multimodal data are transferred to agent-vessels. Each of them, by weighting new events with its previous status, will compute a criticality value.

The Operative Multi-Agent System (OpMAS) communicates with the JBOSS application server encompassing IS, RE and EM 1 via the AgentComLink. This core class of OpMAS links vessel-agents between them and with their environment, carrying out message transmission, vessel-agent creation and updates according EM data.

The vessel-agent model has been designed using MAY (Make Agents Yourself: www.irit.fr/MAY) and is composed in first hand of 4 groups of communication components (dedicated respectively to IS, RE, EM, other agents), and in a second hand of representations: a module of representations of its properties, a module of representations of its environment, a module of representations of anomalies.

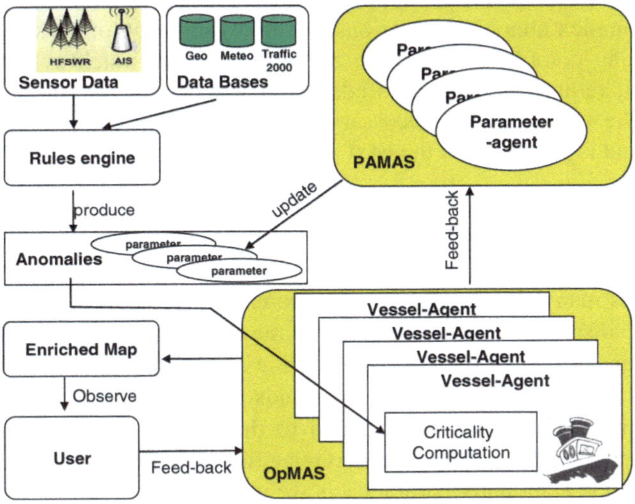

Fig. 1 General architecture of ScanMaris Project

The vessel-agent's life cycle grants its autonomy in an individual thread. The initializing step is associated with a request to RE and then, all threads are synchronized by the push messages from EM. At each cycle, a vessel-agent carries out the data processing of received messages. After analyzing three kinds of messages (events from EM and invariance server including mark detection, anomalies from others vessel-agents and anomalies from the rule engine), a vessel-agent computes its own criticality. This can lead to the triggering of an alert if its criticality rises over the alert threshold.

Detection of Illegal Transhipment is requiered when a reefer ship that does not transmit AIS signals positions itself beyond the bounds of territorial water, which means out of RADAR detection. During several hours, several fishing ship will sail round trips between fishing quota regulated areas and the meeting point defined by the reefer ship. The transhipment occurs when a fishing ship stops close to the unnoticeable reefer ship, moreover; this brief stopping of a fishing ship in sea should not by itself trigger an alert at the RE level, and the period of time between two fishing vessels converging to the meeting place is not easily noticeable by a human operator.

Cooperating vessel-agents of the OpMAS can detect such an abnormal/illegal behaviour by sending marks to the EM. When a vessel-agent slows down and stops, it sends a request to the RE and simultaneously drops a stop mark containing its ID number in the EM. The RE answers with a stop anomaly. When later another vessel-agent slows down and stops at the same place, it will be informed besides its own stop anomaly, of the presence of one or more previous stop marks. So the current "transhipping" vessel-agent drops two marks in the EM, its own stop mark and a collective stop mark. As a result, vessel-agents mark the place of the undetectable reefer vessel with a collective stop and all of themthat have participated this illegal activity, receive a warning flag.

The interest of the ScanMaris approach lies in the distributed and collective detection of complex abnormal situations by the organisation of vessel-agents rather than producing complicated rules in a centralized overwatch system. Complete OpMAS will require several hundreds of parameters and an automated process for tuning the parameters is necessary. A second MAS is in charge with their adaptation and is presented in the next section.

3 Self-adaptive Parameter Adjustment

OpMAS uses many parameters as each anomaly consists of 3 criticality parameters: an "initialisation" parameter related to the occurrence of the anomaly, an "increase" parameter related to anomaly enduring and a "decrease" parameter used to reduce the criticality when the anomaly ends. Marks consist of 2 criticality parameters: initialisation parameter related to the drop, and decreasing parameter used to reduce mark intensity in a pheromone like evaporation.

Moreover, each anomaly parameter is weighted by a vessel-agent parameter in order to discriminate that the same anomaly, for example a stop, has not the same meaning for a fishing vessel, as for a tanker, or another category of vessels. A second parameter combination concerns marks interpretation. In the end, with currently 17 recorded anomalies, 6 categories of marks and 4 categories of vessels, there are 137 interrelated parameters evolving in given intervals, defining a 10242 wide search space that **PAMAS** (Parameter Adjustment MAS) has to resolve.

Parameter-agents of PAMAS have been designed following the Adelfe methodology 12. To produce a valid criticality function for the vessel-agents, a parameter-agent will cooperate with other parameter-agents to adjust its value or propagate requests of adjustment.

The criterion of adjustment/propagation is based on the parameter-agent's own notion of criticality. The more a parameter-agent is used in the vessel-agent criticality function, the higher its own criticality is. By cooperating, they prevent the criticality of the worst parameter-agent to get worse while they are negotiating to absorb a feed-back. As a result, the organisation of parameter-agents continuously acts to ensure the global converging of the parameters represented by PAMAS.

4 Discussion

In the final OpMas, about 10.000 vessel agents are expected. Nevertheless, adaptation will only concern the same 137 interrelated parameters. At the moment, PAMAS is tested with about 60 parameter agents on a period of time of 5 hours (300 step). This choice fulfil both a realism criterium (a maximum of 15 anomalies affecting a single vessel-agent during 5 hours) and an acceptable time of learning. The next step for the ScanMaris prototype will be to test it on real-world large-scale situations. Currently, DCNS is recording all data on several large portions of the French coast during several days. The alert detection system will then be able to "learn" on a selection of situations and its relevance tested on the other situations. A demonstration is planned for January.

This prototype is only a first proof-of-concept and validation before launching two following project which will start shortly. The first, SisMaris, supported by the French Industry Ministry, will lead the prototype towards deployable ready-to-use software. The second, *I2C*, a European FP7 Security Research project starting later aims at enhancing the prototype's functionalities even further to tackle specific needs of other European partner states.

Acknowledgement. The authors would like to thank the other partners of the consortium working on the ScanMaris project: SOFRESUD, ARMINES, ONERA, ECOMER and CDMT. Please refer to the official website1 for further details.

References

1. Jangal, F., Giraud, M.-A., Littaye, A., Morel, M., Mano, J.-P., Napoli, A.: Extraction of Suspicious Behavior of Vessels in the Exclusive Economic Zone. In: International Symposium on Antennas and Propagation, Taipei, Taiwan (2008)
2. Lee, W., Mé, L., Wespi, A.: RAID 2001. LNCS, vol. 2212. Springer, Heidelberg (2001)
3. Tan, O.: A multi-agent system for tracking the intent of surface contacts in ports and waterways. Master's thesis, Naval Postr-graudate School (2005)

[1] `www.irit.fr/scanmaris`

4. Gorodetsky, V., Karsaev, O., Samoilov, V., Ulanov, A.: Asynchro-nous Alert Correlation in Multi-agent Intrusion Detection Systems. In: Gorodetsky, V., Kotenko, I., Skormin, V.A. (eds.) MMM-ACNS 2005. LNCS, vol. 3685, pp. 366–379. Springer, Heidelberg (2005)
5. Mathieson, I., Dance, S., Padgham, L., Gorman, M., Winikoff, M.: An Open Meteorological Alerting System: Issues and Solutions. In: The proceedings of the 27th Australasian Computer Science Conference (2004)
6. Georgé, J.-P., Peyruqueou, S., Régis, C., Glize, P.: Experiencing Self-Adaptive MAS for Real-Time Decision Support Systems. In: International Conference on Practical Applications of Agents and Multiagent Systems, Salamanca. Springer, Heidelberg (2009)
7. Welcomme, J.-B., Gleizes, M.-P., Redon, R.: Self-Regulating Multi-Agent System for Multi-Disciplinary Optimisation Proc-ess. In: CEUR Workshop Proceedings, European Workshop on Multi-Agent Systems (EUMAS), Lisbon (2006)
8. Weiss, G.: Multiagent systems. A modern approach to distributed artificial intelligence. The MIT Press, Cambridge (1999)
9. Gleizes, M.-P., Camps, V., Georgé, J.-P., Capera, D.: Engineering Systems which Generate Emergent Functionalities. In: Weyns, D., Brueckner, S.A., Demazeau, Y. (eds.) EEMMAS 2007. LNCS (LNAI), vol. 5049, pp. 58–75. Springer, Heidelberg (2008)
10. Axelrod, R.: The Evolution of Cooperation. Basic Books, New York (1984)
11. Georgé, J.-P., Edmonds, B., Glize, P.: Making Self-Organising Adaptive Multiagent Systems Work. In: Methodologies and Software Engineering for Agent Systems. Kluwer, Dordrecht (2004)
12. Rougemaille, S., Arcangeli, J.-P., Gleizes, M.-P., Migeon, M.: ADELFE Design, AMASML in Action. In: Artikis, A., Picard, G., Vercouter, L. (eds.) ESAW 2008. LNCS, vol. 5485, pp. 105–120. Springer, Heidelberg (2009)

Author Index

GPSR Compliance

The European Union's (EU) General Product Safety Regulation (GPSR) is a set of rules that requires consumer products to be safe and our obligations to ensure this.

If you have any concerns about our products, you can contact us on ProductSafety@springernature.com

In case Publisher is established outside the EU, the EU authorized representative is:

Springer Nature Customer Service Center GmbH
Europaplatz 3
69115 Heidelberg, Germany

Batch number: 10755023

Printed by Printforce, the Netherlands